VOLUME ONE HUNDRED AND FIFTY TWO

CURRENT TOPICS IN DEVELOPMENTAL BIOLOGY

Gene-Environment Interactions in Birth Defects and Developmental Disorders

CURRENT TOPICS IN DEVELOPMENTAL BIOLOGY

"A meeting-ground for critical review and discussion of developmental processes"
A.A. Moscona and Alberto Monroy (Volume 1, 1966)

VOLUME ONE HUNDRED AND FIFTY TWO

Current Topics in DEVELOPMENTAL BIOLOGY

Gene-Environment Interactions in Birth Defects and Developmental Disorders

Edited by

ROBERT J. LIPINSKI
*Department of Comparative Biosciences,
School of Veterinary Medicine, University of Wisconsin,
Madison, WI, United States*

ROBERT S. KRAUSS
*Department of Cell, Developmental, and Regenerative
Biology, Icahn School of Medicine at Mount Sinai,
New York, NY, United States*

ELSEVIER

ACADEMIC PRESS
An imprint of Elsevier

Academic Press is an imprint of Elsevier
50 Hampshire Street, 5th Floor, Cambridge, MA 02139, United States
525 B Street, Suite 1650, San Diego, CA 92101, United States
The Boulevard, Langford Lane, Kidlington, Oxford OX5 1GB, United Kingdom
125 London Wall, London, EC2Y 5AS, United Kingdom

First edition 2023

ISBN: 978-0-12-820157-2
ISSN: 0070-2153

For information on all Academic Press publications
visit our website at https://www.elsevier.com/books-and-journals

Publisher: Zoe Kruze
Developmental Editor: Naiza Mendoza
Production Project Manager: Vijayaraj Purushothaman
Cover Designer: Greg Harris

Typeset by STRAIVE, India

Contents

Contributors

Patrick Allard
Institute for Society and Genetics; Molecular Biology Institute, University of California, Los Angeles, Los Angeles, CA, United States

Abigail A. Armstrong
Department of Obstetrics/Gynecology and Division of Reproductive Endocrinology and Infertility, University of California, Los Angeles, CA, United States

Nathan G. Burns
Department of Human Genetics, University of Utah, Salt Lake City, UT, United States

Carlo Donato Caiaffa
Center for Precision Environmental Health, Department of Molecular and Cellular Biology, Baylor College of Medicine, Houston, TX, United States

Johann K. Eberhart
Department of Molecular Biosciences, School of Natural Sciences; Waggoner Center for Alcohol and Addiction Research, School of Pharmacy, University of Texas at Austin, Austin, TX, United States

Joshua L. Everson
Department of Molecular Biosciences, School of Natural Sciences; Waggoner Center for Alcohol and Addiction Research, School of Pharmacy, University of Texas at Austin, Austin, TX, United States

Richard H. Finnell
Center for Precision Environmental Health, Department of Molecular and Cellular Biology; Departments of Molecular and Cellular Biology, Molecular and Human Genetics, and Medicine, Baylor College of Medicine, Houston, TX, United States

Sharien Fitriasari
Stowers Institute for Medical Research, Kansas City, MO, United States

Cristiane Sá Roriz Fonteles
Postgraduate Program in Dentistry, Faculty of Pharmacy, Dentistry and Nursing, Federal University of Ceara, Fortaleza, Brazil

Gabrielle Kardon
Department of Human Genetics, University of Utah, Salt Lake City, UT, United States

Kimberly Keil-Stietz
Department of Comparative Biosciences, University of Wisconsin-Madison, School of Veterinary Medicine, Madison, WI, United States

Robert S. Krauss
Department of Cell, Developmental, and Regenerative Biology, Icahn School of Medicine at Mount Sinai, New York, NY, United States

Pamela J. Lein
Department of Molecular Biosciences, University of California, Davis, School of Veterinary Medicine, Davis, CA, United States

Robert J. Lipinski
Department of Comparative Biosciences, School of Veterinary Medicine, University of Wisconsin, Madison, WI, United States

Mary L. Marazita
Center for Craniofacial and Dental Genetics; Oral and Craniofacial Sciences, School of Dental Medicine; Human Genetics, School of Public Health; Clinical and Translational Science Institute, Department of Psychiatry, School of Medicine, University of Pittsburgh, Pittsburgh, PA, United States

Paul A. Trainor
Stowers Institute for Medical Research, Kansas City, MO; Department of Anatomy and Cell Biology, University of Kansas Medical Center, Kansas City, KS, United States

Roxane Verdikt
Institute for Society and Genetics, University of California, Los Angeles, Los Angeles, CA, United States

Lei Yunping
Center for Precision Environmental Health, Department of Molecular and Cellular Biology, Baylor College of Medicine, Houston, TX, United States

Preface

Birth defects and developmental disorders affect millions of newborns around the world each year; they are both a public health concern and a source of suffering to the afflicted and their families. Many types of birth defects are syndromic and can now be attributed to a defined genetic cause. However, the most commonly occurring birth defects (e.g., neural tube defects and orofacial clefts) are nonsyndromic, and it is not possible to identify a clear etiology in the majority of individual cases. It is widely believed that these birth defects, along with common developmental disorders (e.g., autism spectrum disorder and even fetal alcohol spectrum disorder), are caused by a complex interaction between genetic and environmental risk factors. Among the relevant environmental risk factors are maternal health and naturally occurring and synthetic chemical compounds. The ultimate goal of prevention will be served by elucidation of genetic landscapes so as to highlight high-susceptibility populations and individuals, but success will also require a refined understanding of modulable environmental factors that contribute to the overall risk. The challenges and opportunities for understanding gene–environment interactions in the etiology of birth defects and developmental disorders, and related issues such as transgenerational inheritance, are the topics addressed in this volume.

Robert J. Lipinski
Department of Comparative Biosciences
School of Veterinary Medicine
University of Wisconsin
Madison, WI, United States

Robert S. Krauss
Department of Cell, Developmental, and Regenerative Biology
Icahn School of Medicine at Mount Sinai
New York, NY, United States

CHAPTER ONE

Gene-environment interactions in birth defect etiology: Challenges and opportunities

Robert J. Lipinski[a,*] and Robert S. Krauss[b,*]

[a]Department of Comparative Biosciences, School of Veterinary Medicine, University of Wisconsin, Madison, WI, United States

[b]Department of Cell, Developmental, and Regenerative Biology, Icahn School of Medicine at Mount Sinai, New York, NY, United States

*Corresponding authors: e-mail address: robert.lipinski@wisc.edu; robert.krauss@mssm.edu

Contents

Abstract

Birth defects are relatively common congenital outcomes that significantly impact affected individuals, their families, and communities. Effective development and deployment of prevention and therapeutic strategies for these conditions requires sufficient understanding of etiology, including underlying genetic and environmental causes.

Current Topics in Developmental Biology, Volume 152
ISSN 0070-2153
https://doi.org/10.1016/bs.ctdb.2022.10.001

Tremendous progress has been made in defining the genetic basis of familial and syndromic forms of birth defects. However, the majority of birth defect cases are considered nonsyndromic and thought to result from multifactorial gene-environment interactions. While substantial advances have been made in elucidating the genetic landscape of these etiologically complex conditions, significant biological and technical constraints have stymied progress toward a refined knowledge of environmental risk factors. Defining specific gene-environment interactions in birth defect etiology is even more challenging. However, progress has been made, including demonstration of critical proofs of concept and development of new conceptual and technical approaches for resolving complex gene-environment interactions. In this review, we discuss current views of multifactorial birth defect etiology, comparing them with other diseases that also involve gene-environment interactions, including primary immunodeficiency and cancer. We describe how various model systems have illuminated mechanisms of multifactorial etiology and these models' individual strengths and weaknesses. Finally, suggestions for areas of future emphasis are proposed.

Abbreviations

CHD congenital heart defects
CM cardiomyopathy
HPE holoprosencephaly
MSMD Mendelian susceptibility to mycobacterial disease
NGS next generation sequencing
PAE prenatal alcohol exposure
RA retinoic acid
TPS tumor predisposition syndrome

1. Introduction

Birth defects occur in approximately 1 in 33 babies born in the United States and affect around 8 million newborns worldwide each year (Christianson, Howson, & Modell, 2006; Prevention, 2008). Defined broadly as any congenital structural or functional anomaly with a measurable impact on physical, intellectual, or social well-being, birth defects are a major cause of morbidity and mortality (CDC, 1992; Prevention, 2017). Defining the causes of birth defects is a considerable challenge. Despite the notable progress summarized below, up to 80% of birth defect cases still cannot be attributed to a specific cause (Feldkamp, Carey, Byrne, Krikov, & Botto, 2017; Toufaily, Westgate, Lin, & Holmes, 2018). Reducing the incidence and impact of birth defects through development of prevention and therapeutic strategies will require a more complete understanding of these underlying causes.

It has long been recognized that the inborn errors of development that manifest as birth defects have both genetic and environmental origins (Lancaster, 2011). A community of researchers, including pediatricians, geneticists, developmental biologists, and epidemiologists, have made substantial progress in elucidating the etiology of a subset of birth defects that are caused by individual genetic or environmental factors. However, it is now widely recognized that the vast majority of birth defects are multifactorial in origin and likely result from complex gene-environment interactions (Beames & Lipinski, 2020; Krauss & Hong, 2016; Lovely, Rampersad, Fernandes, & Eberhart, 2017; Nishimura & Kurosawa, 2022). Over the last several decades, our understanding of the genetic landscape of birth defect etiology has steadily progressed, with even greater resolution on the immediate horizon as emergent technological advancements are deployed. On the other hand, major gaps in our understanding of the environmental contribution to birth defect etiology remain. Moving forward, there is also increasing need—and opportunity—for integration across disciplines (Leslie, 2020; Sparrow, 2022). While a daunting challenge, defining specific gene-environment interactions and elucidating the nature of their mechanisms of action holds great promise for identifying high-risk individuals and advancing prevention strategies based upon communication of risk.

2. Defining the genetic landscape of birth defect etiology

A significant genetic component to birth defect etiology has long been recognized, but our current knowledge has largely been driven by an ever-expanding and increasingly powerful toolset developed over the last 50 years. Application of these tools has defined the genetic landscape of birth defect etiology to a remarkable degree, including the effective solving of most rare congenital disorders and the identification of dozens of genetic risk factors that contribute to the complex etiology of the most common human birth defects.

2.1 Syndromic birth defects

Syndromic birth defects are typically recognized by multiple anomalies affecting more than one organ system, and often segregate as autosomal dominant, autosomal recessive, or X-linked traits. Advances made in defining the genetic causes of these conditions were recently summarized by Innes and Lynch (Innes & Lynch, 2021), who framed their analyses around the classic textbook, *Smith's Recognizable Patterns of Human Malformation*, now in its 8th edition (Jones, Jones, & del Campo, 2021). The authors note that of the 147 conditions described in the seminal edition of *Smith's* published in 1970, the majority are now "solved." This is a true success story that can largely be credited to the individual and concerted efforts of clinicians and medical geneticists. Solving these conditions typically began with careful physical examination by the astute pediatrician to identify specific major and minor dysmorphologies. Subsequent ascertainment of a cohort of individuals sharing similar patterns of dysmorphology facilitated investigation of genetic underpinnings through cytogenetic assays, linkage analysis, array CGH, and SNP homozygosity mapping (Innes & Lynch, 2021). Over the last decade, next generation sequencing (NGS) approaches have become the main driver of discovery in this arena. With increasing power and decreasing cost, NGS promises to further elucidate the genetic architecture of syndromes, including mosaic disorders and ultra-rare conditions.

2.2 Nonsyndromic birth defects

The remarkable progress that has been made in delineating the etiology of syndromes and rare disorders should be celebrated, but it must also be recognized that the vast majority of human birth defects are considered nonsyndromic and cannot be attributed to specific known causes (Feldkamp et al., 2017). Nonsyndromic or isolated birth defects (i.e. those not associated with multiorgan syndromes) account for up to 75% of all birth defect cases and encompass the majority of the most prevalent malformations, including heart and neural tube defects, orofacial clefts, and limb defects (Feldkamp et al., 2017; Parker et al., 2010). Nonsyndromic birth defects are etiologically complex traits generally thought to result when multiple genetic and environmental influences reach a critical threshold of insult (Dixon, Marazita, Beaty, & Murray, 2011; Krauss & Hong, 2016; Lovely et al., 2017; Murray, 2002). The relatively high prevalence of these conditions has enabled large population-based genetic epidemiology studies comparing cases and controls (Hobbs et al., 2014). These studies have been dominated by candidate gene approaches, like SNP genotyping, that are typically directed at genes linked to particular outcomes through animal model investigation or knowledge of the signaling pathways that regulate specific developmental processes. Genome-wide association studies (GWAS) on the other hand do not require pre-existing knowledge and have been applied to discover new candidate genes and novel risk loci for several nonsyndromic birth defects including congenital heart defects, orofacial clefts, and hypospadias (Lupo, Mitchell, & Jenkins, 2019; Webber et al., 2015). Increasing application of NGS to nonsyndromic birth defects has already proven useful, particularly in identifying the contribution of relatively rare genetic variants.

Collectively, these efforts have been fruitful in discovering susceptibility loci associated with complex traits, thereby informing genetic counseling and family planning strategies; this should be counted among the major successes in birth defects research. At the same time, nonsyndromic birth defects are complex disorders that in most cases will not be accounted for by a single gene or chromosomal abnormality. Moreover, considering the multifactorial nature of nonsyndromic birth defects, it could be argued that even a complete understanding of the genetic architecture of these conditions is unlikely to be sufficient to drive prevention or correction strategies. While the ultimate goal of birth defect prevention is supported by elucidation of their genetic landscape to highlight high-risk populations and individuals, success will also require a refined understanding of modulable environmental factors that contribute to birth defect risk.

3. Elucidating the environmental landscape of birth defect etiology

Formal investigation into how the environment influences development and contributes to birth defect etiology began in earnest in the 1950s, following recognition that the uterus is not impervious to such influences. In the context of birth defect etiology, the "environment" broadly encompasses non-genetic factors, including drugs, medications, contaminants, pollution, and maternal disease, infection, and nutritional status. Relative to the genome, which provides a permanent record that can be retrospectively examined and defined to the individual base pair, our current knowledge of the prenatal environment and ability to assess individual and combinatorial environmental risk factors during critical periods of development is extremely limited.

3.1 Human teratogens

Perhaps the most well-known human teratogen is thalidomide, a drug marketed to pregnant women that caused limb malformations in thousands of children in the 1950s and 1960s. Early concern over thalidomide was raised by clinicians who noticed a dramatic increase in cases of an otherwise rare malformation involving shortened or absent limbs (Lancaster, 2011).

Subsequently, alcohol and several other major human teratogens were identified by the astute clinician approach. As described by Jones and Carey, this approach relies upon observation, documentation, and delineation of cases in which a specific phenotype follows a particularly rare and known exposure (Jones & Carey, 2011). Prospective cohort studies follow a related path, in which tracking of prenatal exposures is paired with careful examination of children by a dysmorphologist to identify major and minor malformations. Facilitated by organizations like the Organization of Teratogen Information Services (OTIS), this approach has been particularly effective in evaluating the potential teratogenicity of drugs that have recently been marketed, as well as those that are commonly used by pregnant women but have not been adequately evaluated. While effective in identifying several major human teratogens, including the Zika virus, these approaches are best suited to examination of clearly defined and relatively rare prenatal exposures (Jones & Carey, 2011). While teratogen exposure can account for a small percentage of birth defects cases, the majority are thought to be multifactorial. For these etiologically complex outcomes, it is important to consider environmental factors that, while not individually causative, contribute to overall risk (Feldkamp, Botto, & Carey, 2015).

3.2 Environmental risk factors

Traditional epidemiological studies are equipped to identify common environmental influences that alter the risk for specific birth defect outcomes (Hernandez-Diaz & Oberg, 2015). Designed to assess the correlation between risk factors and specific outcomes, epidemiological studies have been central in establishing several key environmental risk factors, most notably folic acid. Epidemiological studies have proven particularly useful in examining intentional exposure to general classes of risk factors, like multivitamin use and smoking, as discussed in detail in the chapters on orofacial clefting and craniofacial malformations within this volume. Many epidemiological studies, however, rely on the recall of study participants of events that may have occurred years previously. Most structural birth defects result from insult during the 3rd to the 8th week of embryogenesis and have specific and narrow critical periods of susceptibility that may span only days (Finnell, Waes, Eudy, & Rosenquist, 2002; Shenefelt, 1972). Ascribing exposures to these windows of susceptibility is difficult, particularly for non-persistent chemicals (Kamai, McElrath, & Ferguson, 2019), as many pregnancies are not even recognized at these early stages. Moreover, the capacity of traditional epidemiological approaches is overwhelmed by the sheer number of chemicals used in global commerce, recently estimated in the tens or even hundreds of thousands (Bond & Garny, 2019; Wang, Walker, Muir, & Nagatani-Yoshida, 2020). Efforts to define the prenatal exposome (i.e. the totality of the environment) are in the very early stages and mostly limited to individual studies examining particular classes of chemicals (e.g. flame retardants, pesticides, metals etc.) from placenta or cord blood (Rager et al., 2020). These studies have produced evidence supporting relationships between prenatal exposure to select chemicals and adverse developmental outcomes. However, analyses to-date have generally been limited to a priori hypotheses, account for only a small fraction of the exposome, and were not equipped to examine chemical interactions. Moving forward there is a clear need to more comprehensively assess the prenatal exposome and to develop sophisticated approaches for assessing the risk of environmental factors in isolation and in combination.

3.3 High-throughput predictive approaches

The sheer number of chemicals in the environment outmatches the capacity of animal-based testing, which has traditionally been relied upon for developmental and reproductive toxicity assessment. The limited throughput and resource demands of these approaches, along with the desire to reduce animal usage, has spurred the development of alternative approaches. At the forefront of these efforts is Tox21, a collaboration between the National Toxicology Program, the Environmental Protection Agency National Center for Computational Toxicology, the National Institutes of Health Chemical Genomics Center, the National Center for Advancing Translational Sciences, and the Food and Drug Administration (Attene-Ramos et al., 2013). The Tox21 approach integrates high throughput robotic platforms, chemical profiling strategies, and hundreds of individual *in vitro* biological assays to build a foundation for mechanism-based toxicity prediction (Jeong, Kim, & Choi, 2022). While not specifically targeted toward developmental toxicity, assays examining endocrine disruption and other developmentally important signaling pathways and mechanisms are represented within the Tox21 repertoire. Meanwhile, the same conceptual and experimental approaches are being applied with specific focus on assessing teratogenic potential, via rodent whole embryo culture, zebrafish embryo, and stem cell assays (Zhang, Ball, Panzica-Kelly, & Augustine-Rauch, 2016). While the throughput nature of these assays addresses a central need for toxicity evaluation, key challenges and limitations remain to be fully addressed including ability to account for chemical metabolism, multiple exposures, and gene-environment interactions. These are particularly important challenges to address with respect to assessment of developmental toxicity given the dynamic nature of embryogenesis, multiple molecular and cellular targets of teratogens, and the unique profile of chemical metabolism during pregnancy. As such, high throughput screens are ideally positioned as one approach to developmental toxicity assessment that serves as a complement to traditional animal-based testing and human epidemiological approaches.

4. Gene-environment interactions

As described above, the challenges to understanding gene-environment interactions are many and include the historical siloing of scientific disciplines and the overwhelming number of potential permutations of genetic and environmental factors. But addressing these challenges promises great reward, including potential development of prevention strategies based upon identification of high-risk individuals and communication of culpable environmental risk factors. Toward that goal, it is instructive to consider the progress made in understanding other maladies, such as infectious diseases and cancer, that are also influenced by gene-environment interactions.

4.1 Gene-environment interactions in infectious diseases

Humans display enormous variability in the clinical outcome to primary infections with specific microbes (Casanova & Abel, 2013, 2020). Multiple variables are at play here, including route of infection, size of inoculum, and microbial variability. However, for most microbes, most people survive infections, with lethal disease a rare outcome (Casanova & Abel, 2013, 2020). In many cases, such severe reactions to specific microbial infections arise in individuals with germline mutations that cause specific primary immunodeficiencies (Casanova & Abel, 2013, 2020). Therefore, as is likely to be true for many birth defects, genetic variations predispose individuals to extreme outcomes in response to an environmental exposure (in this case, a specific microbial infection).

Almost 300 inborn errors of immunity have been identified that link monogenic mutations to rare phenotypes arising after specific infections, and these mutations may or may not result in phenotypes in the absence of infection (Casanova, 2015). In the simplest cases, the severity of the genetic defect is directly related to the tendency for clinical manifestation. A classic example is Mendelian susceptibility to mycobacterial disease (MSMD), in which susceptibility to environmental mycobacteria is associated with genetic deficiency in interferon (IFN)-γ production, signaling, and/or responses (Bustamante, 2020; Noma, Mizoguchi, Tsumura, & Okada, 2022). Eighteen different genes have been identified in different forms of MSMD, with various inheritance patterns, including autosomal recessive, autosomal dominant, and X-linked recessive (Bustamante, 2020; Noma et al., 2022). A spectrum of sensitivity exists in MSMD. Patients with autosomal recessive null mutations invariably develop environmental mycobacterial infections by 5 years of age, whereas partial (often autosomal dominant) genetic deficiencies result in a range of responses, from a longer period without symptoms, to mild or no disease at all (Bustamante, 2020; Gruber & Bogunovic, 2020; Noma et al., 2022). Therefore, MSMD shows incomplete penetrance and variable expressivity, dependent on the nature of the genetic alteration. Additionally, MSMD can be divided into "isolated" and "syndromic" forms, the former displaying selective susceptibility to mycobacterial and related infections and the latter also having additional phenotypes (Bustamante, 2020; Noma et al., 2022). Multiple other Mendelian and monogenic susceptibilities to infection display similar patterns (Casanova & Abel, 2020).

Primary immunodeficiencies therefore hold several features in common with birth defects associated with gene-environment interactions, including genetic heterogeneity, a spectrum of phenotypic severity, and isolated vs. syndromic disease. Furthermore, there are additional points of overlap between birth defects and infectious disease in less-well understood aspects of gene-environment interactions. In scenarios with partial penetrance, such as familial cases of the birth defect holoprosencephaly (HPE) and common variable immunodeficiency, respectively, cases may be digenic or polygenic in nature (Gruber & Bogunovic, 2020; Kim et al., 2019). Direct evidence for polygenic disease has been found in only a small number of cases of each, however, and the variants implicated tend to be rare ones (Hong et al., 2017). The role of common variants as modifier genes is still largely unknown for both birth defects and infectious diseases, but they may play significant roles when paired with specific mutations or environmental exposures (Gruber & Bogunovic, 2020; Timberlake et al., 2016). Finally, it has recently been appreciated that somatic mutation, i.e., genetic mosaicism, may underlie cases of both primary immunodeficiency and birth defects, and this phenomenon is a potential explanation for variable penetrance and severity of these defects (Gruber & Bogunovic, 2020; Latorre-Pellicer et al., 2021; Rodin & Walsh, 2018).

Despite these similarities, the study of gene-environment interactions in infectious disease holds certain advantages to such studies in birth defects. Most significantly, it is easier to unambiguously identify a specific microbial infection in a patient than it is to identify a potential teratogenic exposure occurring early in gestation, especially when the clinical outcome—which triggers initial interest in the question—is many months later. Additionally, discovery of such potential teratogens lags far behind our knowledge of the specific common microbes that are risk factors in primary immunodeficiencies. Finally, medicinal intervention may, in some cases, be simpler in primary immunodeficiencies. For example, some forms of MSMD result in suboptimal production of IFN-γ (e.g., those carrying mutations in IL-12Rb2 or IL-23R) and these patients may benefit by IFN-γ therapy (Bustamante, 2020; Noma et al., 2022). Nevertheless, these successes offer potential ways forward for study of developmental disorders.

4.2 Gene-environment interactions in cancer

Birth defects and cancer can both arise from defects in major signaling pathways, including the Hedgehog (HH), WNT, and receptor tyrosine kinase pathways. Frequently, loss of function of a pathway is associated with birth defects, while gain of function of the same pathway is associated with cancer (e.g., HH signaling deficiency results in HPE, while unregulated HH signaling underlies both sporadic and inherited predisposition to basal cell carcinoma and medulloblastoma (Ingham, 2022; Jiang & Hui, 2008)). There are clear exceptions to this generality, however, including activating mutations in the RAS-MAP kinase pathway, which are frequent occurrences in sporadic adult cancers and also responsible for the syndromes collectively designated "RASopathies" (we note that, although gain-of-function in nature, the specific mutations identified in these two scenarios are overlapping but not identical) (Hebron, Hernandez, & Yohe, 2022).

There are more than 100 distinct tumor predisposition syndromes (TPSs) (Carbone et al., 2020). The classical mechanism—seen with such tumor suppressor genes as *RB1*—involves inheritance of one mutant copy of a gene, with somatic mutation of the other copy invariably found in tumor cells (Dimaras et al., 2012). Therefore, inheritance is autosomal dominant, but the tumor cells lack any wild type protein. Most cancers are caused by accumulation of specific somatic mutations, and many known carcinogens are DNA-damaging agents (Totsuka, Watanabe, & Lin, 2021; Wu, Powers, Zhu, & Hannun, 2016; Wu, Zhu, Thompson, & Hannun, 2018). It is not surprising then, that a subset of TPSs are caused by germline mutations in genes involved with DNA repair and maintenance of genomic integrity (Carbone et al., 2020). For example, Lynch syndrome, the most common autosomal dominant TPS, is caused by heterozygous mutations in genes that control DNA mismatch repair (Carbone et al., 2020; Lynch, Snyder, Shaw, Heinen, & Hitchins, 2015). Consequently, Lynch syndrome patients have high mutation rates and high incidences of GI tract and other tumors. Interestingly, the prevalence of tumor types that develop in Lynch syndrome varies with geography; those in Asian countries develop stomach cancer more frequently than those in Western countries, suggesting an environmental impact in tumor tropism (Park et al., 2016; Wei et al., 2010).

TP53 (encoding the tumor suppressor, p53) is the most frequently mutated gene in human cancer (Levine, 2020). Heterozygous germline mutations in *TP53* cause Li-Fraumeni syndrome, a highly penetrant TPS (Guha & Malkin, 2017). Humans and mice carrying *TP53* mutations are highly susceptible to ionizing and UV radiation, and to tobacco-related carcinogens and cigarette smoke (Heymann et al., 2010; Hwang et al., 2003; Jiang, Ananthaswamy, Muller, & Kripke, 1999; Kilroe-Smith, 1976). Furthermore, p53-deficient mice exposed to the tobacco-related carcinogen, benzo[a]pyrene had higher levels of carcinogen-DNA adducts than control mice, demonstrating a gene-environment interaction at a molecular level that is relevant to disease causation (Lorente et al., 2004).

Finally, a striking gene-environment interaction in cancer was observed in villages in Cappadocia, Turkey, where >50% of the population died of mesothelioma (Carbone et al., 2020). Mesothelioma is associated with exposure to asbestos, but only a relatively small fraction of workers with high occupational exposure develop the disease (Miller, 1991). Erionite, a potent fiber similar to asbestos, was used in housing and roads in the Turkish villages, but cases and deaths clustered in families. These families carried heterozygous germline mutations in *BAP1* (encoding a protein which binds to BRCA1, another tumor suppressor); furthermore, tumor cells had somatic inactivation of the wild type copy, consistent with a classical tumor suppressor mechanism (Testa et al., 2011). Additional *BAP1* families in the United States and elsewhere show a similar susceptibility to mesothelioma and additional cancers (Abdel-Rahman et al., 2011). While mesothelioma can occur in *BAP1* mutation carriers not exposed to asbestos, and asbestos can cause mesothelioma in individuals without such mutations, the combination of the two results in much higher incidence than either alone.

Gene-environment interactions are therefore important in both birth defects and cancer. In each case, identifying the most significant genes has been more straightforward than identifying rate-limiting environmental exposures. For both, one complicating issue is that clinical manifestation is observed sometime after exposure. For birth defects and developmental disorders, many key exposures occur early in the first trimester, with anomalies found later in gestation, shortly after birth or, sometimes in the cases of behavioral phenotypes, years later. Latent periods in cancer are longer, and decades may pass before disease is detected. Nevertheless, epidemiological and population studies in cancer have been highly successful (Clapp, Jacobs, & Loechler, 2008; Doll, Peto, Wheatley, Gray, & Sutherland, 1994), and they offer hope that similar studies will illuminate birth defect etiology. Many carcinogens were identified via cohort studies of occupational exposure and lifestyle factors (Irigaray et al., 2007; Ward et al., 2003). As discussed above, this is also true of some known human teratogens, but fewer have been discovered to date.

Taken together, successes in studying gene-environment interactions in infectious diseases and cancer offer some lessons for birth defects. In particular, identification of individuals or populations with severe outcomes to specific environmental exposures can be extremely informative. Additionally, identification of genetic variants that clearly predispose to specific environmental insults will lead to recommendations for behavior modification and risk avoidance.

5. Model systems for the study of gene-environment interactions in birth defects

Combined genetic and epidemiological analyses have broken important ground in methodology and understanding of gene-environment interactions in human birth defects (for example, see (Finnell et al., 2021; Tai et al., 2015)). However, these studies have been limited in what they can conclude about mechanistic interactions between candidate genes and non-genetic risk factors. Consequently, use of model organisms has been important to explore concepts underlying gene-environment interactions, as well as identifying new potential risk factors (Krauss & Hong, 2016). Both invertebrate and vertebrate models have shed light on important questions and have therefore helped in design of human studies.

5.1 Invertebrate models

Invertebrate model organisms such as the roundworm *Caenorhabditis elegans* and the fruit fly *Drosophila melanogaster* offer some major advantages for any studies of developmental biology. Among these are short generation time, sophisticated genetic tools, very well-characterized developmental stages, and the ability to screen large numbers of organisms, thereby permitting genome-wide and forward genetic screens. These species have been used successfully to study a wide-range of issues, such as sensitivity to hypoxia, response to ethanol, and a variety of complex phenomena, including trans-generational inheritance and life span (Bellier, Chen, Kao, Cinar, & Aroian, 2009; Gao et al., 2018; Hong, Choi, & Lee, 2008; Jha et al., 2016; Mabon, Scott, & Crowder, 2009; Morozova, Shankar, MacPherson, Mackay, & Anholt, 2022; Ojelade et al., 2015; Weinhouse, Truong, Meyer, & Allard, 2018). They have also been used to model difficult, multifactorial medical conditions like Parkinson disease (Kim, Perentis, Caldwell, & Caldwell, 2018; Sarkar & Feany, 2021).

Invertebrate body plans are obviously different in important ways from humans. Nevertheless, fundamental signaling pathways that drive embryonic patterning are evolutionarily conserved, and they are often perturbed in human developmental disorders. Among these are the Notch, WNT, HH, BMP, and receptor tyrosine kinase pathways. While it is not possible to use invertebrates to study the specific structural patterning defects seen in common human birth defects, the affected pathways can be genetically perturbed in worms and flies to study mechanisms (Dorsett & Krantz, 2009; Link & Bellen, 2020). Furthermore, genetic variants identified in patients can be verified for loss- or gain-of-function in these organisms, taking advantage of the wealth of knowledge about their development (Humphries, Narang, & Mlodzik, 2020; Marcogliese et al., 2022). Some teratogens directly target components of such pathways, and it may be possible to combine these systems to study interactions between specific gene variants and non-genetic insults within animal models amenable to high throughput analyses.

5.2 Zebrafish

The zebrafish is an attractive model organism for studying birth defects. They are vertebrates, with organ systems closer to those of humans than are those of worms or flies. Furthermore, zebrafish embryos are transparent and develop rapidly and externally, allowing live imaging approaches. They have a long history of forward genetic screens for developmental phenotypes

(Driever et al., 1996; Haffter et al., 1996), and are now also an efficient model organism for chemical screens (Wiley, Redfield, & Zon, 2017). These properties make them accessible to analyses of gene-environment interactions.

Zebrafish have been particularly useful in addressing genetic susceptibility to teratogenesis associated with prenatal alcohol exposure (PAE). This topic is covered by another chapter in this volume, so we will highlight only a few significant studies here. First, Eberhart and colleagues identified an interaction between *pdgfra* mutation and PAE in development of craniofacial defects (McCarthy et al., 2013). A small genome-wide association study provided evidence that this interaction may be of significance to specific individuals with fetal alcohol spectrum disorders (McCarthy et al., 2013). Zebrafish mutants in several other genes associated with craniofacial defects in humans are also sensitized to ethanol-induced phenotypes (McCarthy et al., 2013; Yoon et al., 2022). Most significantly, Swartz et al. took advantage of the tractability of zebrafish to perform the first vertebrate forward genetic screen to identify new mutants with enhanced susceptibility to ethanol teratogenesis (Swartz, Lovely, McCarthy, Kuka, & Eberhart, 2020). This and similar follow-up studies may permit a much more complete understanding of the genetic basis of ethanol's actions as a teratogen, yielding insight into a major clinical and social health issue.

Only a few other gene-environment interaction screens have been performed in zebrafish to date. Ding et al. screened for genetic modifiers of doxorubicin-induced cardiomyopathy (CM) (Ding et al., 2017). Four modifiers were identified, three of which had been linked to CM in humans or mice. The fourth, *dnajb6b(L)*, led to discovery of human CM-associated variants in the orthologous gene, *DNAJB6* (Ding et al., 2017). These results offer proof of principle that such studies are both practicable and likely to illuminate disorders with complex etiology.

Zebrafish have been used more frequently to address specific candidate gene-environment interactions. Long QT syndrome (LQTS, a heart ventricle disorder that can lead to fatal arrhythmias) is associated with heterozygous mutations in *KCNH2* (encoding a potassium channel). Zebrafish with a heterozygous mutation in the orthologous *kcnh2* gene were sensitized to treatment with the potassium channel blocking drug, terfenadine; this mimics observations in humans (Arnaout et al., 2007). Everson et al. demonstrated that piperonyl butoxide (PBO), a component of commonly used pesticides that also inhibits HH pathway signaling, induced craniofacial defects in heterozygous Sonic HH (*shha*) mutants (Everson, Batchu, & Eberhart, 2020). Finally, 2,3,7,8-tetrachlorodibenzo-*p*-dioxin (TCDD) impairs craniofacial development in zebrafish and other species and heterozygosity for the pro-chondrogenic transcription factor, *sox9b*, sensitized to fish to TCDD-induced jaw malformations (Xiong, Peterson, & Heideman, 2008).

5.3 Mice

Mice have been the animal model of choice for most studies on gene-environment interactions in birth defects. They are genetically tractable, including the ability to knock out genes in a cell type-specific and temporal manner. As mammals, they are anatomically closer to humans than are zebrafish. Forward genetic screens are possible, but challenging to the point that we are unaware of any that involved inclusion of a teratogen. Nevertheless, studies with candidate genes and teratogens have been highly illuminating, discovering mechanisms of gene-environment interaction and identifying potential non-genetic risk factors for further study in humans. Here, we briefly focus on two birth defects not covered elsewhere in this volume, HPE and congenital heart defects (CHD).

5.3.1 Gene-environment interactions in mouse models of HPE

HPE has served as a model birth defect for exploring gene-environment interactions in mice (Grinblat & Lipinski, 2019; Hong & Krauss, 2018; Lo, Hong, & Krauss, 2021). It occurs at very high frequency, conservatively estimated at 1 in 250 conceptions, with >97% succumbing in utero (Matsunaga & Shiota, 1977; Shiota, 2021). HPE is caused by a failure to define the midline of the forebrain and/or midface and characterized by a broad spectrum of phenotypic severity (Grinblat & Lipinski, 2019; Lo et al., 2021). HPE is associated with heterozygous mutations in the HH, Nodal, and FGF pathways that display incomplete penetrance and variable expressivity; in the great majority of cases, these mutations are considered necessary but not sufficient to produce the associated clinical phenotypes (Lo et al., 2021). Additional "hits" are therefore invoked, and these may be genetic, environmental, or both. In 1996, *SHH* mutations were first identified in HPE patients, and it is the most frequently mutated gene in HPE (Roessler et al., 1996; Roessler, Hu, & Muenke, 2018). *Shh* knockout mice were also first reported in 1996, and they have severe HPE (Chiang et al., 1996). Subsequently, cyclopamine, an HPE-inducing teratogen, was shown to block HH signaling by directly binding and inhibiting Smoothened (SMO), a key HH signal transducer (Chen, Taipale, Cooper, & Beachy, 2002; Cooper, Porter, Young, & Beachy, 1998). Cyclopamine therefore became a prototype for clinically useful SMO inhibitors, such as Vismodegib, which also induces HPE in mice (Heyne et al., 2015). Taken together, disruption of HH signaling by either genetic or teratogenic mechanisms can produce HPE, and such risk factors may interact, helping explain the complex etiology of human HPE.

This notion has received support in mouse studies of HPE, which offer a set of guideposts on how gene-environment interactions may be systematically pursued. A good initial step in pursuing gene-environment interactions in mice is to assess whether partially penetrant mutations in a given signaling pathway can synergize with subthreshold doses of chemical inhibitors of the same pathway, to produce a specific developmental defect. *Gli2* encodes a HH pathway-responsive transcription factor. Heyne et al. showed that mice heterozygous for *Gli2*, which resemble wild type animals, were sensitized to HPE induced by ineffective, or weakly effective, doses of Vismodegib (Heyne et al., 2016). Therefore, genetic and non-genetic insults each targeting the HH pathway, but insufficient on their own, synergized to produce HPE. These types of observations validate mouse models for gene-environment studies.

A logical next step is to test environmental risk factors implicated in HPE but for which a clear teratogenic mechanism may be lacking (as opposed to Vismodegib, whose teratogenic mechanism is clear). A good example here is PAE, which is associated with HPE in some but not all epidemiological studies (see Lo et al., 2021). *Cdon* encodes a HH coreceptor mutated in some HPE patients (Bae et al., 2011). When placed on a genetically resistant background (129S6 mice), homozygous mutation of *Cdon* has very mild phenotypic effects due to redundancy with other HH coreceptors; PAE is also non-teratogenic to these mice (Hong & Krauss, 2012). A combination of *Cdon* mutation and transient in utero ethanol exposure, however, resulted in a complete spectrum of HPE phenotypes in ~75% of mice (Hong & Krauss, 2012). Similarly, heterozygosity for *Shh* or *Gli2* on a C57BL/6J genetic background also sensitized mice to PAE-induced HPE (Kietzman, Everson, Sulik, & Lipinski, 2014). Therefore, PAE synergizes with subthreshold genetic deficits in HH signaling to induce HPE. It may be that some of the inconsistent results in epidemiological studies arose because PAE requires a predisposing genetic background to exert its ability to induce HPE.

These types of studies with mice also allow for mechanistic insight. Gestational windows of sensitivity to teratogens can be identified, shedding light on key developmental processes occurring during peak sensitivity. PAE is most effective when delivered at E7.0, whereas direct SMO inhibitors are most effective at E7.5 (Heyne et al., 2015; Hong, Christ, Christa, Willnow, & Krauss, 2020). These and additional data argue that PAE exerts its HPE-inducing effects upstream of HH signaling, inhibiting HH as an indirect outcome of Nodal pathway inhibition, which is a proximal, perhaps direct, target of PAE (Hong et al., 2020). PAE inhibits HH signaling indirectly in zebrafish also (Sidik et al., 2021). When PAE in mice occurs later, at E9.0, HH signaling is disrupted as well. This may also occur indirectly by affecting expression of factors that regulate primary cilia, organelles which are the subcellular sites of HH signal transduction (Boschen, Fish, & Parnell, 2021).

Finally, some well-studied compounds with known pharmacological targets outside the HH pathway, were recently discovered to target HH signaling as well. Such compounds are therefore potential risks for HPE teratogenicity. Among these are PBO and Δ9-tetrahydrocannabinol (THC), which were shown to inhibit SMO many years after discovery of their primary targets (Khaliullina, Bilgin, Sampaio, Shevchenko, & Eaton, 2015; Wang et al., 2012). PBO and THC were then each demonstrated to synergize with HH pathway mutations in mice to produce HPE (Everson et al., 2019; Lo et al., 2021). Such studies in mice have potential public health value, as they suggest PBO and THC should be further assessed as potential risk factors for human HPE.

5.3.2 Gene-environment interactions in mouse models of CHD

Like HPE, most cases of CHD (of which there are many subtypes) have a complex etiology thought to be best explained by gene-gene and/or gene-environment interactions (Kalisch-Smith, Ved, & Sparrow, 2020; Kodo, Uchida, & Yamagishi, 2021; Majumdar, Yasuhara, & Garg, 2021). Variants in many genes are associated with CHD, and many non-genetic risk factors have also been implicated. Studies with mice have been instrumental in identifying how these classes of risk factors may interact. As one example, we discuss gestational hypoxia, an environmental stressor that promotes multiple types of birth defects.

Maternal exposure to hypoxia is a potential risk factor for CHD (Hutter, Kingdom, & Jaeggi, 2010; Kalisch-Smith et al., 2020). Mice carrying several mutations found in human CHD, including heterozygosity for *Nkx2-5*, *Tbx1*, *Tbx5*, and *Notch1*, are sensitized to CHD induced by short-term gestational hypoxia (Chapman et al., 2020; Moreau et al., 2019). Percentages of mouse embryos with CHD correlate with degree of hypoxia in these scenarios, suggesting that low oxygen may be one cause of variable penetrance and expressivity in human CHD. Additional work has provided insight into mechanisms underlying mutation + hypoxia-induced phenotypes. Gestational hypoxia led to a prolonged hypoxia inducible factor 1α (HIF1α) response in the developing myocardium, which in turn depressed levels of *Nkx2-5* expression (Moreau et al., 2019). Furthermore, hypoxia rapidly induced an unfolded protein response (UPR) in cardiac progenitors, which reduced levels of FGFR1, a key signaling receptor for cardiac development (Shi et al., 2016). Therefore, gestational hypoxia may exert its effects via multiple mechanisms that are capable of cooperating with predisposing genetic variants. Unsurprisingly, therefore, mechanisms underlying the interaction between even a single mutation and single environmental insult may be multiple and complex.

An important question raised by these mouse studies is the identity of potential clinical conditions or teratogens that may cause gestational hypoxia. Mice are again a useful system for assessing candidate factors. Chapman et al. found that the anti-arrhythmic drug dofetilde, known to cause hypoxia in rodent embryos, induced a higher incidence of CHD in *Notch1* heterozygotes (Chapman et al., 2020). Anti-arrhythmic drugs that cause brachycardia, leading to hypoxia, may therefore be contraindicated during pregnancy. It is critical, of course, to consider the relative doses required for a therapeutic effect vs. a teratogenic effect.

Another possible cause of hypoxia could be maternal anemia, which is a worldwide health problem. To test the possibility that anemia might be a teratogenic cause of CHD, Kalisch-Smith et al. assessed the effects of maternal iron deficiency in mice. Mature female mice maintained on a low iron diet were anemic; when mated, embryos of low-iron diet females developed multiple types of CHD and displayed embryonic lethality (Kalisch-Smith et al., 2021). The effects of an iron-deficient diet were linked to increased retinoic acid (RA) signaling (Kalisch-Smith et al., 2021); RA is a known cardiac teratogen (D'Aniello & Waxman, 2015; Kalisch-Smith et al., 2020). Surprisingly, short-term gestational hypoxia leads to reduced RA signaling (Walton et al., 2018). Therefore, iron deficiency is a condition that may exert teratogenic effects independent of its ability to produce hypoxia. These results show that, even in the absence of a cooperating genetic variant, mechanisms underlying teratogenesis may be complex, requiring rigorous analysis at the level of signaling pathways and transcriptomics.

5.4 Human systems

Animal models have been critical to addressing questions of gene-environment interactions in birth defect etiology, and they will remain so for the foreseeable future. Nevertheless, there are some significant differences in human vs. e.g., mouse development (Rayon et al., 2020; Shahbazi & Zernicka-Goetz, 2018). Furthermore, the high number of combinatory genome-by-exposome permutations makes screening a challenge, even for the most tractable model organisms (Beames & Lipinski, 2020). The ability to generate complex organoids from human pluripotent cells is likely to help bridge this gap. Use of organoid systems is increasingly common, and it is possible to perform genome-wide CRISPR screens with them (Kampmann, 2020). Furthermore, it is possible to generate organoids from patient-derived cells. Nevertheless, it is still early days with this technology, and very few gene-environment interaction studies have been performed. In one interesting example, Modaferri et al. used CRISPR to generate an iPSC line with a heterozygous mutation in *CHD8*, similar to a mutation found in some autism patients (Modafferi et al., 2021). Brain organoids derived from $CHD8^{+/+}$ and $CHD8^{+/-}$ lines were compared for their response to Chlorpyrifos, a pesticide with adverse effects on brain development, including pathways affected by autism (Grandjean & Landrigan, 2014). The $CHD8^{+/-}$ line was sensitized to several negative effects of chlorpyrifos, revealing an interaction between a defined genetic predisposition and a putative environmental risk factor.

It is possible to imagine that organoids of various types (e.g., brain, heart, palate) could be used for automated, high-throughput assays involving genome-wide CRISPR screens and multiple teratogens for a given organ system. iPSC-derived differentiated cells and organoids tend to be "immature" (i.e., at embryonic or fetal stages of development), something that is a hindrance for studies of adult-onset diseases. In contrast, this may be a benefit for studies of birth defects and developmental disorders. An important, but surmountable, caveat with use of organoids for such studies is that protocols are often not fully standardized, and there is much lab-to-lab variability. It is important that attention be paid to this issue in general, but it will be critical for complex studies assessing multiple variables, so as to ensure reproducibility. Once standardized, it will be possible to design studies that take advantage of the enormous amount of information available on genetic variants in birth defects, combined with relevant exposures to environmental conditions.

6. Future directions: Potential mechanisms of gene-environment interaction

Understanding potential mechanisms of gene-environment interaction is important as it may open prevention strategies. Among the likely molecular targets are critical morphogenetic signaling pathways and their associated gene regulatory networks, and cellular structural elements (e.g., components and regulators of cell polarity and the cytoskeleton) (Anvarian, Mykytyn, Mukhopadhyay, Pedersen, & Christensen, 2019; Krauss & Hong, 2016; Nikolopoulou, Galea, Rolo, Greene, & Copp, 2017). Myriad possible mechanisms exist and these have recently been reviewed (Krauss & Hong, 2016) and are presented in subsequent chapters in this volume. One area of likely importance in this realm is epigenetic regulation of gene expression.

6.1 Epigenetics and DNA methylation

Epigenetic mechanisms that interface the environment and the genome have emerged as a relatively underexplored but promising domain of birth defects research. DNA methylation is a particularly intriguing focus for investigation into etiologically complex birth defects as this epigenetic mechanism is inherently sensitive to environmental cues and a practical target for prevention and therapeutic strategies. DNA methylation is influenced by diverse environmental influences including maternal diet and stress, and exposure to drugs, toxins, and environmental pollutants (Breton-Larrivee, Elder, & McGraw, 2019; Lee, 2015; Martin & Fry, 2018) Several lines of investigation suggest that DNA methylation may influence birth defect risk. For example, periconceptional intake of the dietary methyl donor, folic acid, is well established to reduce the risk for neural tube defects and appears to lessen the risk for nonsyndromic orofacial clefts as well (Millacura, Pardo, Cifuentes, & Suazo, 2017; Zhou et al., 2020). Studies have identified differential methylation profiles in tissue from individuals with and without clefts (Gonseth et al., 2019; Howe et al., 2019; Xu, Lie, Wilcox, Saugstad, & Taylor, 2019; Young, Slifer, Hecht, & Blanton, 2021), and differential DNA methylation has been suggested to underlie discordance among monozygotic twins for orofacial clefts and several forms of congenital malformations (Lyu et al., 2018; Weksberg et al., 2002; Young et al., 2021; Zhang et al., 2020). Animal studies have shown that DNA methylation is required for embryonic development and that environmentally mediated DNA methylation changes can influence development and drive phenotypic changes, even among littermates (Dolinoy, Huang, & Jirtle, 2007). Additional animal studies have demonstrated that pharmacologic inhibition of DNA methyltransferase activity can cause structural birth defects, including orofacial clefts (Branch, Chernoff, Brownie, & Francis, 1999; Branch, Francis, Brownie, & Chernoff, 1996; Bulut, Ozdemir, Başimoglu-Koca, Korkmaz, & Atalay, 1999; Rogers et al., 1994).

While these findings provide evidence linking epigenetic modification to birth defect risk, substantial knowledge gaps remain. A fundamental question not yet fully resolved is which developmental processes are most sensitive to epigenetic influences. For individual birth defect outcomes, it will be critical to determine pathogenic epigenetic events, and specific environmental triggers. Most studies have been limited to analysis of blood or other surrogate tissue, and whether DNA methylation changes in these tissues reflect pathogenic changes remains unclear. A foundation for filling these knowledge gaps is being built through initiatives like the Toxicant Exposures and Responses by Genomic and Epigenomic Regulators of Transcription (TaRGET) program funded by the National Institute of Environmental Health Sciences, as well as independent efforts to develop epigenetic-focused cell based assays that could be leveraged for high content chemical screening (Stelzer, Shivalila, Soldner, Markoulaki, & Jaenisch, 2015; Wang et al., 2018, 2021).

7. Concluding remarks

In this article we have described challenges to understanding the complex etiology of commonly occurring, non-syndromic birth defects. While identifying the specific combinations of genetic and environmental insults that underlie individual cases remains difficult, there is reason to be hopeful about opportunities for progress. Lessons can be drawn by comparison with knowledge of other diseases with multifactorial causation. Multiple animal and in vitro models for birth defects have been successfully generated and display ever-increasing accuracy. Together, these approaches offer promise for identification of new risk factors and fundamental mechanisms. Such discoveries are vital to design of epidemiological studies and to development of preventive and risk management strategies.

Acknowledgments

We thank Dr. John Carey and Tyler Beames for thoughtful review and input on this manuscript. RJL is supported by the National Institutes of Environmental Health Sciences and National Institute of Dental and Craniofacial Research under award numbers R01ES026819 and R56DE030917. RSK is supported by the National Institute of Dental and Craniofacial Research under award number R01DE024748.

References

(CDC), C.f.D.C. (1992). Years of potential life lost before ages 65 and 85—United States, 1989–1990. *Morbidity and Mortality Weekly Report*, *41*, 313–315.

Abdel-Rahman, M. H., Pilarski, R., Cebulla, C. M., Massengill, J. B., Christopher, B. N., Boru, G., et al. (2011). Germline BAP1 mutation predisposes to uveal melanoma, lung adenocarcinoma, meningioma, and other cancers. *Journal of Medical Genetics*, *48*, 856–859.

Anvarian, Z., Mykytyn, K., Mukhopadhyay, S., Pedersen, L. B., & Christensen, S. T. (2019). Cellular signalling by primary cilia in development, organ function and disease. *Nature Reviews. Nephrology*, *15*, 199–219.

Arnaout, R., Ferrer, T., Huisken, J., Spitzer, K., Stainier, D. Y., Tristani-Firouzi, M., et al. (2007). Zebrafish model for human long QT syndrome. *Proceedings of the National Academy of Sciences of the United States of America*, *104*, 11316–11321.

Attene-Ramos, M. S., Miller, N., Huang, R., Michael, S., Itkin, M., Kavlock, R. J., et al. (2013). The Tox21 robotic platform for the assessment of environmental chemicals—From vision to reality. *Drug Discovery Today*, *18*, 716–723.

Bae, G. U., Domené, S., Roessler, E., Schachter, K., Kang, J. S., Muenke, M., et al. (2011). Mutations in CDON, encoding a hedgehog receptor, result in holoprosencephaly and defective interactions with other hedgehog receptors. *American Journal of Human Genetics*, *89*, 231–240.

Beames, T. G., & Lipinski, R. J. (2020). Gene-environment interactions: Aligning birth defects research with complex etiology. *Development*, *147*.

Bellier, A., Chen, C. S., Kao, C. Y., Cinar, H. N., & Aroian, R. V. (2009). Hypoxia and the hypoxic response pathway protect against pore-forming toxins in C. elegans. *PLoS Pathogens*, *5*, e1000689.

Bond, G. G., & Garny, V. (2019). Inventory and evaluation of publicly available sources of information on hazards and risks of industrial chemicals. *Toxicology and Industrial Health*, *35*, 738–751.

Boschen, K. E., Fish, E. W., & Parnell, S. E. (2021). Prenatal alcohol exposure disrupts Sonic hedgehog pathway and primary cilia genes in the mouse neural tube. *Reproductive Toxicology*, *105*, 136–147.

Branch, S., Chernoff, N., Brownie, C., & Francis, B. M. (1999). 5-AZA-2'-deoxycytidine--induced dysmorphogenesis in the rat. *Teratogenesis, Carcinogenesis, and Mutagenesis*, *19*, 329–338.

Branch, S., Francis, B. M., Brownie, C. F., & Chernoff, N. (1996). Teratogenic effects of the demethylating agent 5-aza-2'-deoxycytidine in the Swiss Webster mouse. *Toxicology*, *112*, 37–43.

Breton-Larrivee, M., Elder, E., & McGraw, S. (2019). DNA methylation, environmental exposures and early embryo development. *Animal Reproduction*, *16*, 465–474.

Bulut, H. E., Ozdemir, O., Başimoglu-Koca, Y., Korkmaz, M., & Atalay, A. (1999). Effects of a DNA demethylating agent—5-Azacytidine—On testicular morphology during mouse embryo development. *Okajimas Folia Anatomica Japonica*, *76*, 47–53.

Bustamante, J. (2020). Mendelian susceptibility to mycobacterial disease: Recent discoveries. *Human Genetics, 139*, 993–1000.

Carbone, M., Arron, S. T., Beutler, B., Bononi, A., Cavenee, W., Cleaver, J. E., et al. (2020). Tumour predisposition and cancer syndromes as models to study gene-environment interactions. *Nature Reviews. Cancer, 20*, 533–549.

Casanova, J. L. (2015). Human genetic basis of interindividual variability in the course of infection. *Proceedings of the National Academy of Sciences of the United States of America, 112*, E7118–E7127.

Casanova, J. L., & Abel, L. (2013). The genetic theory of infectious diseases: A brief history and selected illustrations. *Annual Review of Genomics and Human Genetics, 14*, 215–243.

Casanova, J. L., & Abel, L. (2020). The human genetic determinism of life-threatening infectious diseases: Genetic heterogeneity and physiological homogeneity? *Human Genetics, 139*, 681–694.

Chapman, G., Moreau, J. L. M., Ip, E., Szot, J. O., Iyer, K. R., Shi, H., et al. (2020). Functional genomics and gene-environment interaction highlight the complexity of congenital heart disease caused by Notch pathway variants. *Human Molecular Genetics, 29*, 566–579.

Chen, J. K., Taipale, J., Cooper, M. K., & Beachy, P. A. (2002). Inhibition of Hedgehog signaling by direct binding of cyclopamine to smoothened. *Genes & Development, 16*, 2743–2748.

Chiang, C., Litingtung, Y., Lee, E., Young, K. E., Corden, J. L., Westphal, H., et al. (1996). Cyclopia and defective axial patterning in mice lacking sonic hedgehog gene function. *Nature, 383*, 407–413.

Christianson, A., Howson, C. P., & Modell, B. (2006). *March of Dimes global report on birth defects: The hidden toll of dying and disabled children*. White Plains, NY: The March of Dimes Birth Defects Foundations.

Clapp, R. W., Jacobs, M. M., & Loechler, E. L. (2008). Environmental and occupational causes of cancer: New evidence 2005–2007. *Reviews on Environmental Health, 23*, 1–37.

Cooper, M. K., Porter, J. A., Young, K. E., & Beachy, P. A. (1998). Teratogen-mediated inhibition of target tissue response to Shh signaling. *Science, 280*, 1603–1607.

D'Aniello, E., & Waxman, J. S. (2015). Input overload: Contributions of retinoic acid signaling feedback mechanisms to heart development and teratogenesis. *Developmental Dynamics, 244*, 513–523.

Dimaras, H., Kimani, K., Dimba, E. A., Gronsdahl, P., White, A., Chan, H. S., et al. (2012). Retinoblastoma. *Lancet, 379*, 1436–1446.

Ding, Y., Long, P. A., Bos, J. M., Shih, Y. H., Ma, X., Sundsbak, R. S., et al. (2017). A modifier screen identifies DNAJB6 as a cardiomyopathy susceptibility gene. *JCI Insight, 2*.

Dixon, M. J., Marazita, M. L., Beaty, T. H., & Murray, J. C. (2011). Cleft lip and palate: Understanding genetic and environmental influences. *Nature Reviews. Genetics, 12*, 167–178.

Dolinoy, D. C., Huang, D., & Jirtle, R. L. (2007). Maternal nutrient supplementation counteracts bisphenol A-induced DNA hypomethylation in early development. *Proceedings of the National Academy of Sciences of the United States of America, 104*, 13056–13061.

Doll, R., Peto, R., Wheatley, K., Gray, R., & Sutherland, I. (1994). Mortality in relation to smoking: 40 years' observations on male British doctors. *BMJ, 309*, 901–911.

Dorsett, D., & Krantz, I. D. (2009). On the molecular etiology of Cornelia de Lange syndrome. *Annals of the New York Academy of Sciences, 1151*, 22–37.

Driever, W., Solnica-Krezel, L., Schier, A. F., Neuhauss, S. C., Malicki, J., Stemple, D. L., et al. (1996). A genetic screen for mutations affecting embryogenesis in zebrafish. *Development, 123*, 37–46.

Everson, J. L., Batchu, R., & Eberhart, J. K. (2020). Multifactorial genetic and environmental hedgehog pathway disruption sensitizes embryos to alcohol-induced craniofacial defects. *Alcoholism, Clinical and Experimental Research*, *44*, 1988–1996.

Everson, J. L., Sun, M. R., Fink, D. M., Heyne, G. W., Melberg, C. G., Nelson, K. F., et al. (2019). Developmental toxicity assessment of piperonyl butoxide exposure targeting sonic hedgehog signaling and forebrain and face morphogenesis in the mouse: An in vitro and in vivo study. *Environmental Health Perspectives*, *127*, 107006.

Feldkamp, M. L., Botto, L. D., & Carey, J. C. (2015). Reflections on the etiology of structural birth defects: Established teratogens and risk factors. *Birth Defects Research. Part A, Clinical and Molecular Teratology*, *103*, 652–655.

Feldkamp, M. L., Carey, J. C., Byrne, J. L. B., Krikov, S., & Botto, L. D. (2017). Etiology and clinical presentation of birth defects: Population based study. *BMJ*, *357*, j2249.

Finnell, R. H., Caiaffa, C. D., Kim, S. E., Lei, Y., Steele, J., Cao, X., et al. (2021). Gene environment interactions in the etiology of neural tube defects. *Frontiers in Genetics*, *12*, 659612.

Finnell, R. H., Waes, J. G., Eudy, J. D., & Rosenquist, T. H. (2002). Molecular basis of environmentally induced birth defects. *Annual Review of Pharmacology and Toxicology*, *42*, 181–208.

Gao, A. W., Uit de Bos, J., Sterken, M. G., Kammenga, J. E., Smith, R. L., & Houtkooper, R. H. (2018). Forward and reverse genetics approaches to uncover metabolic aging pathways in Caenorhabditis elegans. *Biochimica et Biophysica Acta - Molecular Basis of Disease*, *1864*, 2697–2706.

Gonseth, S., Shaw, G. M., Roy, R., Segal, M. R., Asrani, K., Rine, J., et al. (2019). Epigenomic profiling of newborns with isolated orofacial clefts reveals widespread DNA methylation changes and implicates metastable epiallele regions in disease risk. *Epigenetics*, *14*, 198–213.

Grandjean, P., & Landrigan, P. J. (2014). Neurobehavioural effects of developmental toxicity. *Lancet Neurology*, *13*, 330–338.

Grinblat, Y., & Lipinski, R. J. (2019). A forebrain undivided: Unleashing model organisms to solve the mysteries of holoprosencephaly. *Developmental Dynamics*, *248*, 626–633.

Gruber, C., & Bogunovic, D. (2020). Incomplete penetrance in primary immunodeficiency: A skeleton in the closet. *Human Genetics*, *139*, 745–757.

Guha, T., & Malkin, D. (2017). Inherited. *Cold Spring Harbor Perspectives in Medicine*, 7.

Haffter, P., Granato, M., Brand, M., Mullins, M. C., Hammerschmidt, M., Kane, D. A., et al. (1996). The identification of genes with unique and essential functions in the development of the zebrafish, Danio rerio. *Development*, *123*, 1–36.

Hebron, K. E., Hernandez, E. R., & Yohe, M. E. (2022). The RASopathies: From pathogenetics to therapeutics. *Disease Models & Mechanisms*, *15*.

Hernandez-Diaz, S., & Oberg, A. S. (2015). Are epidemiological approaches suitable to study risk/preventive factors for human birth defects? *Current Epidemiology Reports*, *2*, 31–36.

Heymann, S., Delaloge, S., Rahal, A., Caron, O., Frebourg, T., Barreau, L., et al. (2010). Radio-induced malignancies after breast cancer postoperative radiotherapy in patients with Li-Fraumeni syndrome. *Radiation Oncology*, *5*, 104.

Heyne, G. W., Everson, J. L., Ansen-Wilson, L. J., Melberg, C. G., Fink, D. M., Parins, K. F., et al. (2016). Gli2 gene-environment interactions contribute to the etiological complexity of holoprosencephaly: Evidence from a mouse model. *Disease Models & Mechanisms*, *9*, 1307–1315.

Heyne, G. W., Melberg, C. G., Doroodchi, P., Parins, K. F., Kietzman, H. W., Everson, J. L., et al. (2015). Definition of critical periods for Hedgehog pathway antagonist-induced holoprosencephaly, cleft lip, and cleft palate. *PLoS One*, *10*, e0120517.

Hobbs, C. A., Chowdhury, S., Cleves, M. A., Erickson, S., MacLeod, S. L., Shaw, G. M., et al. (2014). Genetic epidemiology and nonsyndromic structural birth defects: From candidate genes to epigenetics. *JAMA Pediatrics, 168*, 371–377.

Hong, M., Choi, M. K., & Lee, J. (2008). The anesthetic action of ethanol analyzed by genetics in Caenorhabditis elegans. *Biochemical and Biophysical Research Communications, 367*, 219–225.

Hong, M., Christ, A., Christa, A., Willnow, T. E., & Krauss, R. S. (2020). mutation and fetal alcohol converge on nodal signaling in a mouse model of holoprosencephaly. *eLife, 9*.

Hong, M., & Krauss, R. S. (2012). Cdon mutation and fetal ethanol exposure synergize to produce midline signaling defects and holoprosencephaly spectrum disorders in mice. *PLoS Genetics, 8*, e1002999.

Hong, M., & Krauss, R. S. (2018). Modeling the complex etiology of holoprosencephaly in mice. *American Journal of Medical Genetics. Part C, Seminars in Medical Genetics, 178*, 140–150.

Hong, M., Srivastava, K., Kim, S., Allen, B. L., Leahy, D. J., Hu, P., et al. (2017). BOC is a modifier gene in holoprosencephaly. *Human Mutation, 38*, 1464–1470.

Howe, L. J., Richardson, T. G., Arathimos, R., Alvizi, L., Passos-Bueno, M. R., Stanier, P., et al. (2019). Evidence for DNA methylation mediating genetic liability to non-syndromic cleft lip/palate. *Epigenomics, 11*, 133–145.

Humphries, A. C., Narang, S., & Mlodzik, M. (2020). Mutations associated with human neural tube defects display disrupted planar cell polarity in. *eLife, 9*.

Hutter, D., Kingdom, J., & Jaeggi, E. (2010). Causes and mechanisms of intrauterine hypoxia and its impact on the fetal cardiovascular system: A review. *International Journal Of Pediatrics, 2010*, 401323.

Hwang, S. J., Cheng, L. S., Lozano, G., Amos, C. I., Gu, X., & Strong, L. C. (2003). Lung cancer risk in germline p53 mutation carriers: Association between an inherited cancer predisposition, cigarette smoking, and cancer risk. *Human Genetics, 113*, 238–243.

Ingham, P. W. (2022). Hedgehog signaling. *Current Topics in Developmental Biology, 149*, 1–58.

Innes, A. M., & Lynch, D. C. (2021). Fifty years of recognizable patterns of human malformation: Insights and opportunities. *American Journal of Medical Genetics. Part A, 185*, 2653–2669.

Irigaray, P., Newby, J. A., Clapp, R., Hardell, L., Howard, V., Montagnier, L., et al. (2007). Lifestyle-related factors and environmental agents causing cancer: An overview. *Biomedicine & Pharmacotherapy, 61*, 640–658.

Jeong, J., Kim, D., & Choi, J. (2022). Application of ToxCast/Tox21 data for toxicity mechanism-based evaluation and prioritization of environmental chemicals: Perspective and limitations. *Toxicology In Vitro, 84*, 105451.

Jha, A. R., Zhou, D., Brown, C. D., Kreitman, M., Haddad, G. G., & White, K. P. (2016). Shared genetic signals of hypoxia adaptation in Drosophila and in high-altitude human populations. *Molecular Biology and Evolution, 33*, 501–517.

Jiang, W., Ananthaswamy, H. N., Muller, H. K., & Kripke, M. L. (1999). p53 Protects against skin cancer induction by UV-B radiation. *Oncogene, 18*, 4247–4253.

Jiang, J., & Hui, C. C. (2008). Hedgehog signaling in development and cancer. *Developmental Cell, 15*, 801–812.

Jones, K. L., & Carey, J. C. (2011). The importance of dysmorphology in the identification of new human teratogens. *American Journal of Medical Genetics. Part C, Seminars in Medical Genetics, 157C*, 188–194.

Jones, K., Jones, M., & del Campo, M. (2021). *Smith's recognizable patterns of human malformation* (8th ed.). Philadelphia: Elsevier, Inc.

Kalisch-Smith, J. I., Ved, N., & Sparrow, D. B. (2020). *Environmental risk factors for congenital heart disease* (p. 12). Cold Spring Harbor Perspectives in Biology.

Kalisch-Smith, J. I., Ved, N., Szumska, D., Munro, J., Troup, M., Harris, S. E., et al. (2021). Maternal iron deficiency perturbs embryonic cardiovascular development in mice. *Nature Communications*, *12*, 3447.

Kamai, E. M., McElrath, T. F., & Ferguson, K. K. (2019). Fetal growth in environmental epidemiology: mechanisms, limitations, and a review of associations with biomarkers of non-persistent chemical exposures during pregnancy. *Environmental Health*, *18*, 43.

Kampmann, M. (2020). CRISPR-based functional genomics for neurological disease. *Nature Reviews. Neurology*, *16*, 465–480.

Khaliullina, H., Bilgin, M., Sampaio, J. L., Shevchenko, A., & Eaton, S. (2015). Endocannabinoids are conserved inhibitors of the Hedgehog pathway. *Proceedings of the National Academy of Sciences of the United States of America*, *112*, 3415.

Kietzman, H. W., Everson, J. L., Sulik, K. K., & Lipinski, R. J. (2014). The teratogenic effects of prenatal ethanol exposure are exacerbated by sonic hedgehog or GLI2 haploinsufficiency in the mouse. *PLoS One*, *9*, e89448.

Kilroe-Smith, T. A. (1976). Linearization of calibration curves with the HGA-72 flameless cuvette for the determination of lead in blood. *Analytica Chimica Acta*, *82*, 421–425.

Kim, H., Perentis, R. J., Caldwell, G. A., & Caldwell, K. A. (2018). Gene-by-environment interactions that disrupt mitochondrial homeostasis cause neurodegeneration in C. elegans Parkinson's models. *Cell Death & Disease*, *9*, 555.

Kim, A., Savary, C., Dubourg, C., Carré, W., Mouden, C., Hamdi-Rozé, H., et al. (2019). Integrated clinical and omics approach to rare diseases: Novel genes and oligogenic inheritance in holoprosencephaly. *Brain*, *142*, 35–49.

Kodo, K., Uchida, K., & Yamagishi, H. (2021). Genetic and cellular interaction during cardiovascular development implicated in congenital heart diseases. *Frontiers in Cardiovascular Medicine*, *8*, 653244.

Krauss, R. S., & Hong, M. (2016). Gene–environment interactions and the etiology of birth defects. *Current Topics in Developmental Biology*, *116*, 569–580.

Lancaster, P. A. (2011). Causes of birth defects: Lessons from history. *Congenital Anomalies (Kyoto)*, *51*, 2–5.

Latorre-Pellicer, A., Gil-Salvador, M., Parenti, I., Lucia-Campos, C., Trujillano, L., Marcos-Alcalde, I., et al. (2021). Clinical relevance of postzygotic mosaicism in Cornelia de Lange syndrome and purifying selection of NIPBL variants in blood. *Scientific Reports*, *11*, 15459.

Lee, H.-S. (2015). Impact of maternal diet on the epigenome during in utero life and the developmental programming of diseases in childhood and adulthood. *Nutrients*, *7*, 9492–9507.

Leslie, E. J. (2020). Embracing human genetics: A primer for developmental biologists. *Development*, *147*.

Levine, A. J. (2020). p53: 800 million years of evolution and 40 years of discovery. *Nature Reviews. Cancer*, *20*, 471–480.

Link, N., & Bellen, H. J. (2020). Using Drosophila to drive the diagnosis and understand the mechanisms of rare human diseases. *Development*, *147*, dev191411.

Lo, H. F., Hong, M., & Krauss, R. S. (2021). Concepts in multifactorial etiology of developmental disorders: Gene-gene and gene-environment interactions in holoprosencephaly. *Frontiers in Cell and Development Biology*, *9*, 795194.

Lorente, S. O., Rodrigues, J. C., Jiménez Jiménez, C., Joyce-Menekse, M., Rodrigues, C., Croft, S. L., et al. (2004). Novel azasterols as potential agents for treatment of leishmaniasis and trypanosomiasis. *Antimicrobial Agents and Chemotherapy*, *48*, 2937–2950.

Lovely, C., Rampersad, M., Fernandes, Y., & Eberhart, J. (2017). Gene-environment interactions in development and disease. *Wiley Interdisciplinary Reviews. Developmental Biology*, *6*.

Lupo, P. J., Mitchell, L. E., & Jenkins, M. M. (2019). Genome-wide association studies of structural birth defects: A review and commentary. *Birth Defects Research, 111*, 1329–1342.

Lynch, H. T., Snyder, C. L., Shaw, T. G., Heinen, C. D., & Hitchins, M. P. (2015). Milestones of Lynch syndrome: 1895-2015. *Nature Reviews. Cancer, 15*, 181–194.

Lyu, G., Zhang, C., Ling, T., Liu, R., Zong, L., Guan, Y., et al. (2018). Genome and epigenome analysis of monozygotic twins discordant for congenital heart disease. *BMC Genomics, 19*, 428.

Mabon, M. E., Scott, B. A., & Crowder, C. M. (2009). Divergent mechanisms controlling hypoxic sensitivity and lifespan by the DAF-2/insulin/IGF-receptor pathway. *PLoS One, 4*, e7937.

Majumdar, U., Yasuhara, J., & Garg, V. (2021). In vivo and in vitro genetic models of congenital heart disease. *Cold Spring Harbor Perspectives in Biology, 13*.

Marcogliese, P. C., Deal, S. L., Andrews, J., Harnish, J. M., Bhavana, V. H., Graves, H. K., et al. (2022). Drosophila functional screening of de novo variants in autism uncovers damaging variants and facilitates discovery of rare neurodevelopmental diseases. *Cell Reports, 38*, 110517.

Martin, E. M., & Fry, R. C. (2018). Environmental influences on the epigenome: Exposure-associated DNA methylation in human populations. *Annual Review of Public Health, 39*, 309–333.

Matsunaga, E., & Shiota, K. (1977). Holoprosencephaly in human embryos: Epidemiologic studies of 150 cases. *Teratology, 16*, 261–272.

McCarthy, N., Wetherill, L., Lovely, C. B., Swartz, M. E., Foroud, T. M., & Eberhart, J. K. (2013). Pdgfra protects against ethanol-induced craniofacial defects in a zebrafish model of FASD. *Development, 140*, 3254–3265.

Millacura, N., Pardo, R., Cifuentes, L., & Suazo, J. (2017). Effects of folic acid fortification on orofacial clefts prevalence: A meta-analysis. *Public Health Nutrition, 20*, 2260–2268.

Miller, M. B. (1991). Extrasystolic activity of the myocardium in pipe-rolling workers during their work. *Kardiologiia, 31*, 64–66.

Modafferi, S., Zhong, X., Kleensang, A., Murata, Y., Fagiani, F., Pamies, D., et al. (2021). Gene-environment interactions in developmental neurotoxicity: A case study of synergy between Chlorpyrifos and CHD8 knockout in human brainspheres. *Environmental Health Perspectives, 129*, 77001.

Moreau, J. L. M., Kesteven, S., Martin, E. M. M. A., Lau, K. S., Yam, M. X., O'Reilly, V. C., et al. (2019). Gene-environment interaction impacts on heart development and embryo survival. *Development, 146*.

Morozova, T. V., Shankar, V., MacPherson, R. A., Mackay, T. F. C., & Anholt, R. R. H. (2022). Modulation of the Drosophila transcriptome by developmental exposure to alcohol. *BMC Genomics, 23*, 347.

Murray, J. C. (2002). Gene/environment causes of cleft lip and/or palate. *Clinical Genetics, 61*, 248–256.

Nikolopoulou, E., Galea, G. L., Rolo, A., Greene, N. D., & Copp, A. J. (2017). Neural tube closure: Cellular, molecular and biomechanical mechanisms. *Development, 144*, 552–566.

Nishimura, Y., & Kurosawa, K. (2022). Analysis of gene-environment interactions related to developmental disorders. *Frontiers in Pharmacology, 13*, 863664.

Noma, K., Mizoguchi, Y., Tsumura, M., & Okada, S. (2022). Mendelian susceptibility to mycobacterial diseases: State of the art. *Clinical Microbiology and Infection*.

Ojelade, S. A., Jia, T., Rodan, A. R., Chenyang, T., Kadrmas, J. L., Cattrell, A., et al. (2015). Rsu1 regulates ethanol consumption in Drosophila and humans. *Proceedings of the National Academy of Sciences of the United States of America, 112*, E4085–E4093.

Park, H. M., Woo, H., Jung, S. J., Jung, K. W., Shin, H. R., & Shin, A. (2016). Colorectal cancer incidence in 5 Asian countries by subsite: An analysis of Cancer Incidence in Five Continents (1998-2007). *Cancer Epidemiology, 45*, 65–70.

Parker, S. E., Mai, C. T., Canfield, M. A., Rickard, R., Wang, Y., Meyer, R. E., et al. (2010). Updated national birth prevalence estimates for selected birth defects in the United States, 2004–2006. *Birth Defects Research. Part A, Clinical and Molecular Teratology, 88*, 1008–1016.

Prevention, C.f.D.C.a. (2008). Update on overall prevalence of major birth defects—Atlanta. *Georgia, 1978–2005*, 1–5.

Prevention, C.f.D.C.a. (2017). *National vital statistics report, Deaths: Final data for 2015.*

Rager, J. E., Bangma, J., Carberry, C., Chao, A., Grossman, J., Lu, K., et al. (2020). Review of the environmental prenatal exposome and its relationship to maternal and fetal health. *Reproductive Toxicology, 98*, 1–12.

Rayon, T., Stamataki, D., Perez-Carrasco, R., Garcia-Perez, L., Barrington, C., Melchionda, M., et al. (2020). Species-specific pace of development is associated with differences in protein stability. *Science, 369.*

Rodin, R. E., & Walsh, C. A. (2018). Somatic mutation in pediatric neurological diseases. *Pediatric Neurology, 87*, 20–22.

Roessler, E., Belloni, E., Gaudenz, K., Jay, P., Berta, P., Scherer, S. W., et al. (1996). Mutations in the human Sonic Hedgehog gene cause holoprosencephaly. *Nature Genetics, 14*, 357–360.

Roessler, E., Hu, P., & Muenke, M. (2018). Holoprosencephaly in the genomics era. *American Journal of Medical Genetics. Part C, Seminars in Medical Genetics, 178*, 165–174.

Rogers, J. M., Francis, B. M., Sulik, K. K., Alles, A. J., Massaro, E. J., Zucker, R. M., et al. (1994). Cell death and cell cycle perturbation in the developmental toxicity of the demethylating agent, 5-aza-2'-deoxycytidine. *Teratology, 50*, 332–339.

Sarkar, S., & Feany, M. B. (2021). Precision medicine on the fly: Using Drosophila to decipher gene-environment interactions in Parkinson's disease. *Toxicological Sciences, 182*, 159–167.

Shahbazi, M. N., & Zernicka-Goetz, M. (2018). Deconstructing and reconstructing the mouse and human early embryo. *Nature Cell Biology, 20*, 878–887.

Shenefelt, R. E. (1972). Morphogenesis of malformations in hamsters caused by retinoic acid: Relation to dose and stage at treatment. *Teratology, 5*, 103–118.

Shi, H., O'Reilly, V. C., Moreau, J. L., Bewes, T. R., Yam, M. X., Chapman, B. E., et al. (2016). Gestational stress induces the unfolded protein response, resulting in heart defects. *Development, 143*, 2561–2572.

Shiota, K. (2021). A life-table analysis of the intrauterine fate of malformed human embryos and fetuses. *Birth Defects Research, 113*, 623–632.

Sidik, A., Dixon, G., Buckley, D. M., Kirby, H. G., Sun, S., & Eberhart, J. K. (2021). Exposure to ethanol leads to midfacial hypoplasia in a zebrafish model of FASD via indirect interactions with the Shh pathway. *BMC Biology, 19*, 134.

Sparrow, D. B. (2022). Better communication between experts is needed to solve the environmental origins of birth defects. *BioEssays, 44*, e2100241.

Stelzer, Y., Shivalila, C. S., Soldner, F., Markoulaki, S., & Jaenisch, R. (2015). Tracing dynamic changes of DNA methylation at single-cell resolution. *Cell, 163*, 218–229.

Swartz, M. E., Lovely, C. B., McCarthy, N., Kuka, T., & Eberhart, J. K. (2020). Novel ethanol-sensitive mutants identified in an F3 forward genetic screen. *Alcoholism, Clinical and Experimental Research, 44*, 56–65.

Tai, C. G., Graff, R. E., Liu, J., Passarelli, M. N., Mefford, J. A., Shaw, G. M., et al. (2015). Detecting gene-environment interactions in human birth defects: Study designs and statistical methods. *Birth Defects Research. Part A, Clinical and Molecular Teratology, 103*, 692–702.

Testa, J. R., Cheung, M., Pei, J., Below, J. E., Tan, Y., Sementino, E., et al. (2011). Germline BAP1 mutations predispose to malignant mesothelioma. *Nature Genetics*, *43*, 1022–1025.
Timberlake, A. T., Choi, J., Zaidi, S., Lu, Q., Nelson-Williams, C., Brooks, E. D., et al. (2016). Two locus inheritance of non-syndromic midline craniosynostosis via rare. *eLife*, *5*.
Totsuka, Y., Watanabe, M., & Lin, Y. (2021). New horizons of DNA adductome for exploring environmental causes of cancer. *Cancer Science*, *112*, 7–15.
Toufaily, M. H., Westgate, M. N., Lin, A. E., & Holmes, L. B. (2018). Causes of congenital malformations. *Birth Defects Research*, *110*, 87–91.
Walton, S. L., Singh, R. R., Little, M. H., Bowles, J., Li, J., & Moritz, K. M. (2018). Prolonged prenatal hypoxia selectively disrupts collecting duct patterning and postnatal function in male mouse offspring. *The Journal of Physiology*, *596*, 5873–5889.
Wang, X. X., Jia, H. J., Lv, Y. R., Sun, H. H., Wei, X. L., Tan, J. Y., et al. (2021). A luciferase-EGFP reporter system for the evaluation of DNA methylation in mammalian cells. *Molecular Biology*, *55*, 742–751.
Wang, J., Lu, J., Mook, R., Zhang, M., Zhao, S., Barak, L., et al. (2012). The insecticide synergist piperonyl butoxide inhibits Hedgehog signaling: Assessing chemical risks. *Toxicological Sciences*, *128*, 517–523.
Wang, T., Pehrsson, E. C., Purushotham, D., Li, D., Zhuo, X., Zhang, B., et al. (2018). The NIEHS TaRGET II consortium and environmental epigenomics. *Nature Biotechnology*, *36*, 225–227.
Wang, Z., Walker, G. W., Muir, D. C. G., & Nagatani-Yoshida, K. (2020). Toward a global understanding of chemical pollution: A first comprehensive analysis of national and regional chemical inventories. *Environmental Science & Technology*, *54*, 2575–2584.
Ward, E. M., Schulte, P. A., Bayard, S., Blair, A., Brandt-Rauf, P., Butler, M. A., et al. (2003). Priorities for development of research methods in occupational cancer. *Environmental Health Perspectives*, *111*, 1–12.
Webber, D. M., MacLeod, S. L., Bamshad, M. J., Shaw, G. M., Finnell, R. H., Shete, S. S., et al. (2015). Developments in our understanding of the genetic basis of birth defects. *Birth Defects Research. Part A, Clinical and Molecular Teratology*, *103*, 680–691.
Wei, W., Liu, L., Chen, J., Jin, K., Jiang, F., Liu, F., et al. (2010). Racial differences in MLH1 and MSH2 mutation: An analysis of yellow race and white race based on the InSiGHT database. *Journal of Bioinformatics and Computational Biology*, *8*(Suppl. 1), 111–125.
Weinhouse, C., Truong, L., Meyer, J. N., & Allard, P. (2018). Caenorhabditis elegans as an emerging model system in environmental epigenetics. *Environmental and Molecular Mutagenesis*, *59*, 560–575.
Weksberg, R., Shuman, C., Caluseriu, O., Smith, A. C., Fei, Y. L., Nishikawa, J., et al. (2002). Discordant KCNQ1OT1 imprinting in sets of monozygotic twins discordant for Beckwith-Wiedemann syndrome. *Human Molecular Genetics*, *11*, 1317–1325.
Wiley, D. S., Redfield, S. E., & Zon, L. I. (2017). Chemical screening in zebrafish for novel biological and therapeutic discovery. *Methods in Cell Biology*, *138*, 651–679.
Wu, S., Powers, S., Zhu, W., & Hannun, Y. A. (2016). Substantial contribution of extrinsic risk factors to cancer development. *Nature*, *529*, 43–47.
Wu, S., Zhu, W., Thompson, P., & Hannun, Y. A. (2018). Evaluating intrinsic and non-intrinsic cancer risk factors. *Nature Communications*, *9*, 3490.
Xiong, K. M., Peterson, R. E., & Heideman, W. (2008). Aryl hydrocarbon receptor-mediated down-regulation of sox9b causes jaw malformation in zebrafish embryos. *Molecular Pharmacology*, *74*, 1544–1553.
Xu, Z., Lie, R. T., Wilcox, A. J., Saugstad, O. D., & Taylor, J. A. (2019). A comparison of DNA methylation in newborn blood samples from infants with and without orofacial clefts. *Clinical Epigenetics*, *11*, 40.

Yoon, B., Yeung, P., Santistevan, N., Bluhm, L. E., Kawasaki, K., Kueper, J., et al. (2022). Zebrafish models of alx-linked frontonasal dysplasia reveal a role for Alx1 and Alx3 in the anterior segment and vasculature of the developing eye. *Biology Open*, *11*.
Young, J. I., Slifer, S., Hecht, J. T., & Blanton, S. H. (2021). DNA methylation variation is identified in monozygotic twins discordant for non-syndromic cleft lip and palate. *Frontiers in Cell and Development Biology*, *9*, 656865.
Zhang, C., Ball, J., Panzica-Kelly, J., & Augustine-Rauch, K. (2016). In vitro developmental toxicology screens: A report on the progress of the methodology and future applications. *Chemical Research in Toxicology*, *29*, 534–544.
Zhang, Z., Chen, Y., Wu, Y., Hao, Y., Zhao, X., Wang, X., et al. (2020). A twin-pair analysis indicates congenital scoliosis is associated with allele-specific methylation in the SVIL gene. *Molecular Medicine Reports*, *22*, 2093–2100.
Zhou, Y., Sinnathamby, V., Yu, Y., Sikora, L., Johnson, C. Y., Mossey, P., et al. (2020). Folate intake, markers of folate status and oral clefts: An updated set of systematic reviews and meta-analyses. *Birth Defects Research*, *112*, 1699–1719.

CHAPTER TWO

Transgenerational inheritance and its modulation by environmental cues

Roxane Verdikt[a,*], Abigail A. Armstrong[b], and Patrick Allard[a,c,*]
[a]Institute for Society and Genetics, University of California, Los Angeles, Los Angeles, CA, United States
[b]Department of Obstetrics/Gynecology and Division of Reproductive Endocrinology and Infertility, University of California, Los Angeles, CA, United States
[c]Molecular Biology Institute, University of California, Los Angeles, Los Angeles, CA, United States
*Corresponding authors: e-mail address: rverdikt@ucla.edu; pallard@ucla.edu

Contents

Abstract

The epigenome plays an important role in shaping phenotypes. However, whether the environment can alter an organism's phenotype across several generations through epigenetic remodeling in the germline is still a highly debated topic. In this chapter, we briefly review the mechanisms of epigenetic inheritance and their connection with germline development before highlighting specific developmental windows of susceptibility to environmental cues. We further discuss the evidence of transgenerational inheritance to a range of different environmental cues, both epidemiological in humans and experimental in rodent models. Doing so, we pinpoint the current challenges in

Current Topics in Developmental Biology, Volume 152
ISSN 0070-2153
https://doi.org/10.1016/bs.ctdb.2022.10.002

demonstrating transgenerational inheritance to environmental cues and offer insight in how recent technological advances may help deciphering the epigenetic mechanisms at play. Together, we draw a detailed picture of how our environment can influence our epigenomes, ultimately reshaping our phenotypes, in an extended theory of inheritance.

1. Introduction

The definition and associated understanding of "epigenetics" have dramatically shifted since the 1940's coining of the term *epigenotype* by Waddington (Waddington, 2012). Nearly 60 years later, the currently most accepted definition of epigenetics was put forth by Wu and Morris (Wu & Morris, 2001). They proposed that epigenetics should be defined as: "the study of changes in gene function that are mitotically and/or meiotically heritable and that do not entail a change in DNA sequence." As evidenced by the extensive correspondence associated with this proposal (Wu & Morris, 2001), the introduction of the concept of heritability in epigenetics was quite remarkable. Indeed, while epigenetic mechanisms operate on a short timescale for gene regulation, the potentially long-lasting nature of some epigenetic modifications has garnered significant attention for its ability to explain biological phenomena that unfold across several generations.

Contrasting with the relative stability of epigenetic marks, the epigenome is remarkably malleable and sensitive to environmental influences. Thus, epigenetics is intimately tied to the research into the Developmental Origins of Health and Disease (DOHaD), which aims to uncover the mechanisms by which early life (pre- or peri-natal) exposures may condition the health of an individual over their life course (Barrett, 2017; Bianco-Miotto, Craig, Gasser, van Dijk, & Ozanne, 2017; Ong, Lin, & Holbrook, 2015). In the context of an in utero (i.e., prenatal) exposure directly altering the health outcomes of the next generation, this phenomenon would call for an intergenerational role for the epigenome. Even more remarkable is the wealth of evidence, in a variety of species, indicating that many different types of environmental cues elicit epigenetic changes and phenotypic alterations over several generations. In many instances, these health outcomes are detected in generations beyond the point at which any directly exposed cells remain. This mechanism of epigenetic inheritance is termed "transgenerational" and is the most intriguing since it implies a transfer of biological information by germ cells independently of DNA sequences, yet, acting on gene expression and functioning in parallel with sequence-based regulation. Perhaps unsurprisingly, transgenerational epigenetic inheritance (TEI) has attracted both significant research interest and controversy when applied to humans (Breton et al., 2021).

In this chapter, we explore the molecular mechanisms underlying TEI. We first review the epigenetic mechanisms of inheritance, then discuss their involvement in germ cells development. We then focus on the evidence for TEI, but also highlight the knowledge gaps that require significant research investment. We end this chapter with forward-thinking approaches that we propose to deploy to address the remaining black boxes in the field.

2. Epigenetic mechanisms of inheritance

Epigenetic mechanisms make use of reversible modifications of chromatin states that impact gene expression (Bird, 2007; Murr, 2010). Such modifications include chemical changes of the DNA itself, in particular DNA methylation, changes in histones post-translational modifications, changes in nucleosomes position and composition, non-coding RNA (ncRNA)-mediated modifications of the chromatin structure and changes in high-order chromatin structure within the nucleus (Fig. 1). Importantly, epigenetic states are heritable and ensure the maintenance and long-term stability of gene expression across cell division and during development (Almouzni & Cedar, 2016). Here, we briefly review known mechanisms of epigenetic inheritance.

In mammals, DNA methylation mainly consists in the addition of a methyl group onto cytosines (5mC) within the context of CpG dinucleotides (Bird, 2002; Goll & Bestor, 2005). Dense regions of CpGs – termed CpG islands (CGIs) – are associated with the majority of human and mouse genome promoters and their methylation constitutes a major mechanism for stable repression of gene expression (Haberle & Stark, 2018; Zhu, Wang, & Qian, 2016). The DNA methyltransferase DNMT1, recruited by the ubiquitin E3 ligase UHRF1, has been shown to reproduce DNA methylation patterns during DNA replication and repair, ensuring faithful maintenance of methylation profiles (Nishiyama et al., 2020; Sharif et al., 2007). DNMT3A and DNMT3B have long been considered as "de novo" methyltransferases for their capacity to establish new patterns of DNA methylation (Okano, Bell, Haber, & Li, 1999). However, the distinction between maintenance and de novo DNA methyltransferases has been challenged by observations of their cooperation in establishing DNA methylation patterns (Haggerty et al., 2021; Lyko, 2018).

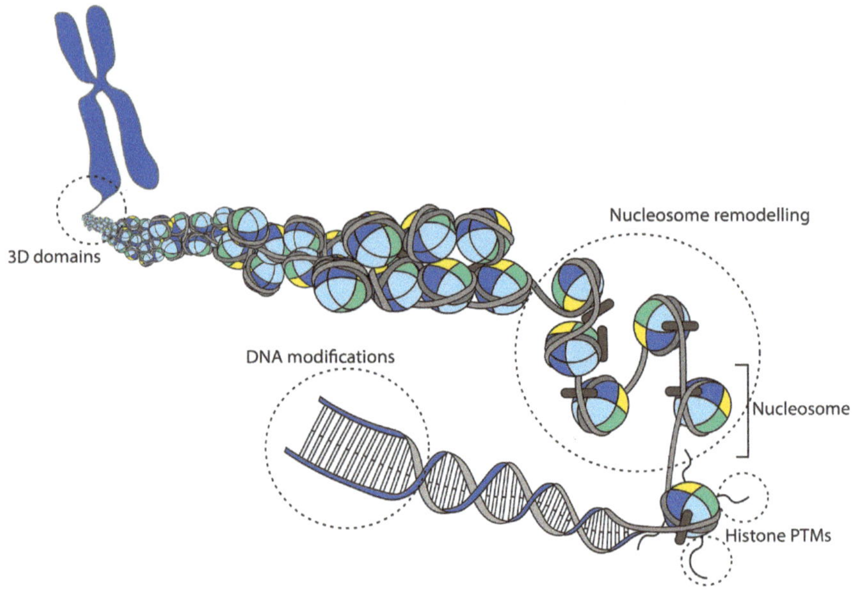

Fig. 1 Epigenetic Regulation of Chromatin Structure. The nucleosome is the basic structural unit of chromatin and consists in 146 base pairs of DNA tightly packaged around an octameric protein core of histones. Epigenetic mechanisms include the positioning and remodeling of nucleosomes, post-translational modifications of histones, chemical modifications of DNA itself and higher organization of three-dimensional chromatin domains.

In addition to changes in the DNA sequence itself, changes in nucleosome position and composition, in particular on the amino-terminal tails of histones, have also been shown to participate in epigenetic memory of phenotypes (Bannister & Kouzarides, 2011; Venkatesh & Workman, 2015). Multiple post-translational modifications (PTMs) of histone tails have been identified, including methylation, acetylation, ubiquitination, sumoylation, and this repertoire is in perpetual expansion with the discovery of crotonylation (Fang et al., 2021; Tan et al., 2011), malonylation, succinylation, and glutarylation (Sabari, Zhang, Allis, & Zhao, 2017). Histone PTMs can directly regulate chromatin compaction but can also indirectly regulate gene expression through the recruitment of chromatin-modifying enzymes, which has led to the notion of a histone code (Strahl & Allis, 2000; Tan et al., 2011). Indeed, specific genomic regions are occupied by histones displaying different PTMs and these epigenetic signatures are predictive of transcriptional states. For instance, transcriptionally-active enhancers are enriched in a subset of histone marks, including histone H3 lysine

27 acetylation (H3K27ac) and histone H3 lysine 4 monomethylation (H3K4me1) (Heintzman et al., 2007; Hon, Hawkins, & Ren, 2009). Maintenance of histone marks is regulated by the controlled recruitment of histone modifiers, principally during DNA replication (Shan, Fang, & Jia, 2021).

Non-coding RNAs regulate epigenomes by recruiting chromatin-modifying complexes (Holoch & Moazed, 2015). Regulatory ncRNAs can be classically classified according to their size, with transcripts longer than 200 nt termed long ncRNAs (lncRNAs) and smaller transcripts including microRNAs (miRNAs) or piwi-interacting RNAs (piRNAs) (Kaikkonen, Lam, & Glass, 2011). Small ncRNAs-mediated inheritance of epigenetic states is a hallmark in *Schizosaccharomyces pombe*, *Drosophila* and *Caenorhabditis elegans (*Heyt & Thakur*, 2021)*. In mammals, piRNAs and sperm-borne ncRNAs are the most well-understood small ncRNAs in the inheritance of epigenetic states (Chen, Yan, & Duan, 2016; Yan, 2014). In addition, lncRNAs are important regulators of epigenetic states and serves as hub for the concerted recruitment of multiple types of chromatin modifiers that are relevant for epigenetic inheritance (Legoff, D'Cruz, Tevosian, & Smagulova, 2019).

Finally, the three-dimensional organization of chromatin domains within the nucleus has been shown to be an important epigenetic mechanism (Pombo & Dillon, 2015). Indeed, depending on its transcriptional competence, chromatin dynamically transits between higher-order organization forms within different subnuclear compartments (Pombo & Dillon, 2015). Technical advances in the recent years have provided tools to analyze chromatin conformation but the three-dimensional control of epigenetic inheritance still needs to be better defined (Zheng & Xie, 2019).

3. Substrates for epigenetic inheritance

Germ cells are at the origin of new organisms and ensure a continuous link across generations and species (Lesch & Page, 2012). The germ cell lineage is responsible for the faithful propagation of genetic and epigenetic information to the next generation. Thus, for TEI to occur, the germline information needs to be altered and these changes retained throughout development. In this section, we will briefly review the development of germ cells and their epigenetic dynamics. We highlight key windows of susceptibility to environmental cues. We will further discuss how environmentally-induced epigenetic changes can be sustained during the intense epigenetic remodeling undergone by the germline.

3.1 Epigenetic dynamics during germ cells development

Germ cells are specified early in the developing embryo, in the perigastrulation epiblast, around embryonic day 7.5 (E7.5) in the mouse (Magnúsdóttir & Surani, 2014; Saitou & Yamaji, 2012) (Fig. 2). The founding population of mouse primordial germ cells (mPGCs) consists of about 40 cells and arise following specification cues—WNT3 signaling and bone morphogenetic proteins (BMP4, BMP2 and BMP8b)—from the extraembryonic ectoderm and visceral endoderm. Specification results in the induction of germline transcriptional programs, regulated by the lineage-determining transcription factors PRDM14, BLIMP1/PRDM1 and TFAP2C (Tang, Kobayashi, Irie, Dietmann, & Surani, 2016). From E9.5, mPGCs start migrating through the hindgut endoderm towards the genital ridge. The colonization of the genital ridge is accompanied by cell proliferation and an intense wave of epigenetic remodeling, which creates a naïve epigenome (Tang et al., 2016). In particular, mPGCs undergo a global and extensive loss of DNA methylation to reach a level of 5% of global DNA methylation (Guibert, Forné, & Weber, 2012; Hajkova et al., 2002; Seisenberger et al., 2012) (Fig. 2). Histone reprogramming and changes in histone post-translational modifications have been proposed to be functionally coordinated to preserve genomic stability and safeguard the genome from DNA demethylation and spurious transcription (Ancelin et al., 2006; Kurimoto et al., 2015; Seki et al., 2007). Indeed, mPGCs present a progressive reduction in global levels of H3K9me2 followed by an increase in global levels of H3K27me3, an increase of open domains marked by H3K4me1 and a global decrease in H2A/H4R3me2s (Ancelin et al., 2006; Kurimoto et al., 2015; Nagano et al., 2022; Ohta et al., 2017; Seki et al., 2007). In addition, it has been recently shown that the three-dimensional organization of the chromatin within the nucleus also preserves mPGCs from pervasive transcription, with an augmented chromatin insulation mediated by CTCF (Nagano et al., 2022).

At their arrival in the developing somatic gonad, by E11.5, mPGCs are on the onset of gametogenesis, during which sex differentiation and meiosis will occur, ultimately generating the functional gametes: oocytes or spermatozoa (Larose et al., 2019). During this time, mPGCs undergo licensing, a developmental transition regulated by the RNA-binding protein DAZL, that enables the cells to acquire the competence to respond to feminizing or masculinizing cues produced by the fetal ovary or testis (Gill, Hu, Lin, & Page, 2011). Subsequently, mPGCs will progress either to spermatogenesis or oogenesis depending on the sex of the embryo and epigenetics marks will be re-established in a sex-specific manner (Fraser & Lin, 2016; Kota & Feil, 2010; Sasaki & Matsui, 2008).

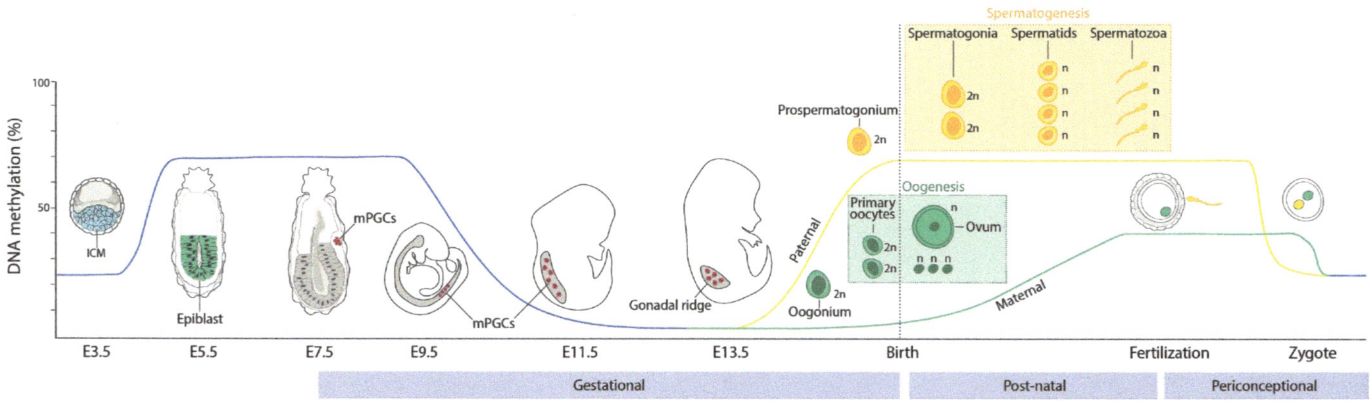

Fig. 2 Epigenetic Remodeling during germline development offer unique exposure windows for inheritance. Variations in DNA methylation levels (%) are shown during germline development in the mouse. Spermatogenesis and oogenesis are summarized and depicted in respective frames. Windows of susceptibility for epigenetic memory to environmental exposures are shown for the gestational, post-natal and periconceptional periods, respectively. 2n = diploid, n = haploid.

Oogenesis, the process of formation of mature female gametes, begins with differentiation of female PGCs into oogonia around E13.5 (Sánchez & Smitz, 2012) (Fig. 2). Oogonia proliferate by mitosis to form a pool of primary oocytes that arrest at the prophase stage of the first meiotic division until after birth when they are assembled into primordial follicles. Upon puberty, a small subset of oocytes will resume meiosis and undergo meiotic maturation concomitant with the process of ovulation prior to fertilization. During oogenesis, epigenetic marks are distinctively re-established on the genome (Sendžikaitė & Kelsey, 2019; Tomizawa, Nowacka-Woszuk, & Kelsey, 2012). At the later stages of their development, after birth, oocytes undergo de novo DNA methylation (Hiura, Obata, Komiyama, Shirai, & Kono, 2006). Mature oocytes are lowly methylated (40%) and present a unique bimodal DNA methylation pattern, with gene bodies of actively-transcribed genes being mostly hypermethylated (>75%) and large-scale genomic domains being hypomethylated (<25%) termed partially-methylated domains (PMDs) (Kobayashi et al., 2012; Veselovska et al., 2015) (Fig. 2). Mechanistically, de novo DNA methylation is mainly mediated by the DNA methyltransferase DNMT3A, guided by DNMT3L (Bourc'his, Xu, Lin, Bollman, & Bestor, 2001; Kaneda et al., 2004; Lucifero et al., 2007). Regions destined for DNA methylation have been shown to present specific histone signatures (reduced H3K4me2/3 and increased H3K36me3), indicating the importance of crosstalk in epigenetic reprogramming (Stewart et al., 2015; Xu et al., 2019). Furthermore, the oocyte PMDs are also associated with non-canonical distributions of histone modifications. Indeed, H3K4me3, a mark typically enriched at active promoters was found on intergenic domains where it accumulates following the activity of the H3K4 methyltransferases MLL2 and SETD1 and correlates with genome-wide transcriptional silencing (Andreu-Vieyra et al., 2010; Dahl et al., 2016; Hanna et al., 2018; Sha et al., 2020; Yu et al., 2017). Accumulation of the repressive mark H3K27me3 has also been reported on the oocyte PMDs but its functional relevance appears less clear (Inoue, Jiang, Lu, Suzuki, & Zhang, 2017; Zheng et al., 2016). A recent study highlighted that H2AK119ub1 was pervasively distributed in oocyte PMDs and that the co-existence of H3K27me3 and H2AK119ub1 at promoters was associated with gene silencing (Mei et al., 2021), further pointing to the importance of mechanistic crosstalk in oocytes epigenetic reprogramming.

Spermatogenesis, the process of formation of mature male germ cells, begins with differentiation of male PGCs into prospermatogonia/spermatogonial stem cells (SSCs) around E13.5 (Fayomi & Orwig, 2018;

Griswold, 2016) (Fig. 2). Prospermatogonia arrest at a mitotic G1/G0 phase until puberty during which mitosis and spermatogenesis resumes. Spermatogenesis takes about 34.5 days to complete in the mouse and is classically divided in three major phases: mitotic divisions of prospermatogonia into spermatogonia, two sets of meiotic divisions of spermatogonia to generate the haploid spermatids and spermiogenesis during which spermatids are transformed into spermatozoa (Fig. 2). Contrary to oocytes, spermatozoa DNA is highly methylated (>80%) and de novo DNA methylation is initiated before birth in the prospermatogonia between E15.5 and E18.5 (Ben Maamar, Beck, Nilsson, McCarrey, & Skinner, 2022; Kubo et al., 2015) (Fig. 2). DNA methylation in spermatozoa is evenly distributed, with the exception of the presence of PMDs in intergenic regions (Kubo et al., 2015; Shirane, Miura, Ito, & Lorincz, 2020). Mechanistically, DNMT3A guided by DNMT3L have also been shown to be important for spermatozoa DNA methylation (Bourc'his et al., 2001; Kato et al., 2007; Kobayashi et al., 2013). As for oocytes, DNA methylation and histone post-translation modifications are functionally linked, with domains destined for DNA methylation marked by H3K36me2 and H3K27me3 (Nagano et al., 2022; Shirane et al., 2020). Importantly, spermatogenesis is characterized by histone-to--protamine transition. While spermatogonia DNA is initially wrapped around canonical histones, these are progressively replaced first by testis-specific histone variants then by protamines during spermiogenesis (Bao & Bedford, 2016). Protamines are small arginine-rich nuclear proteins that allow strong DNA binding and tight packaging into the compact spermatozoa head. About 1% of the mature mouse spermatozoa genome retains canonical histones that present post-translational modifications transmitted to the early zygote and persisting in the early embryo (Brykczynska et al., 2010; Hammoud et al., 2009), suggesting a role for these canonical histones in epigenetic inheritance (Siklenka et al., 2015; Van Der Heijden et al., 2008). Finally, spermatozoa also encode a large proportion of small and long ncRNAs, which roles in sperm epigenetic remodeling remain to be determined (Jodar, 2019), but that have nevertheless been linked to epigenetic inheritance in the offspring (Chen, Yan, & Duan, 2016).

Following fertilization, another extensive epigenetic remodeling occurs asymmetrically to remove parental epigenetic marks in each parental genomes before fusion (Okada & Yamaguchi, 2017). Soon after the entry of the spermatozoon in the oocyte, the paternal genome is demethylated via Tet3-mediated oxidation and protamines are exchanged for the maternally-provided histones that eventually acquire post-translational modifications (Okada & Yamaguchi, 2017). In contrast, the maternal genome remains largely protected from demethylation through the H3K9me2-mediated recruitment of Dppa3/PGC7/STELLA (Nakamura et al., 2012). Certain maternal histone marks, such as H3K36me3, are erased prior zygotic genome activation (ZGA), whereas other marks such as the non-canonically-distributed H3K4me3 are redistributed to canonical promoters upon ZGA (Xia & Xie, 2020). This has led some authors to categorize histone remodeling in the early embryo based on functions: epigenetic marks that are required for gametogenesis, for transcriptional control both in the gametes and in the embryo or "passenger" marks that appear dispensable for the gametes but that are nevertheless maintained and could play a role later in embryogenesis (Xia & Xie, 2020). Of note, although early epigenetic reprogramming erases much of the parental epigenomes, certain loci called imprinting control regions (ICRs) conserve epigenetic features that are parentally-inherited and maintained in the offspring (Tucci et al., 2019). Understanding the mechanisms of epigenetic memory on ICRs is relevant for deciphering the mechanisms of epigenetic inheritance of environmental cues. ZGA marks a key step in early development during which the zygote acquires its transcriptional independency and establishes its own epigenome. As the embryo develops towards the blastocyst stage, active histone marks (H3K4me3 and H3K27ac) are established on promoters and enhancers, respectively, whereas DNA methylation and repressive histone marks are not restored until implantation, which could be functionally relevant for totipotency (Xia & Xie, 2020).

The developmental model for germ cells detailed above corresponds to the one of the mouse and is mostly conserved in mammals. The resolution of the epigenetic remodeling occurring during the germline development in humans has been challenged by technical and ethical constraints. However the development of in vitro germ cell models and low-input epigenetic interrogation techniques has enable to demonstrate that, in general, DNA methylation patterns are relatively conserved between mouse and humans (Hanna, Demond, & Kelsey, 2018). This further supports the use of mouse as a model organism for epigenetic studies in the germline.

3.2 Windows of susceptibility

Germ cell development is uniquely regulated by waves of epigenetic erasures and re-establishment. In particular, two major events of epigenetic reprogramming occur during mammalian development: in PGCs, serving to erase established patterns and restore pluripotency potential, and in the early zygote, serving to erase the gametic epigenomes and regain totipotency. For environmentally-induced changes in the epigenome to be passed across generations, they need to be maintained through these epigenetic events. In particular, three temporal windows of susceptibility to environmental cues can be theoretically defined during: (1) gestation, (2) post-natal development of gametes and (3) around conception (Fig. 2). Here we will further detail these temporal windows of susceptibility and the proposed mechanisms of epigenetic inheritance to environmental cues in the germline. We focus on such mechanisms that directly impact germ cells epigenomes, however, change in somatic cells, particularly in gonads, can indirectly alter the epigenome of germ cells (Bline, Le Goff, & Allard, 2020).

During gestation, maternal environmental exposure could affect PGCs' epigenome remodeling. DNA methylation, the most studied epigenetic mark, is globally reduced to the lowest-reported levels in mammalian epigenomes and would, therefore, not be a likely medium for transgenerational inheritance. However, certain loci corresponding to transposable elements have been identified as methylation escapees. Indeed, young long-terminal repeat (LTR) retrotransposons, known as endogenous retroviruses, resist programed global demethylation during PGC reprogramming. For instance, the intracisternal A-particle (IAP) in the mouse and Alu and HERVK elements in humans have been shown to display higher DNA methylation that the rest of the epigenome (Gkountela et al., 2015; Guo et al., 2015; Ohta et al., 2017; Seisenberger et al., 2012). Retrotransposon-associated inheritance of epialleles is a well-described phenomenon, particularly in the context of the extensively-studied agouti viable yellow (A^{vy}) epiallele. In agouti mice, an IAP retrotransposon element is inserted upstream of the *agouti* gene (Duhl, Vrieling, Miller, Wolff, & Barsh, 1994). Changes in the IAP LTR methylation status drive phenotypical changes, with unmethylated LTRs promoting ubiquitous expression of *agouti* (Duhl et al., *1994)*. These A^{vy} mice present a phenotype of yellow fur, obesity, glucose intolerance and increased susceptibility to tumors development (Duhl et al., 1994). Hence, epigenetic changes in endogenous retroviruses can drive inheritance of phenotypes in mammals, however, if the endovirome can serve as hotspots for epigenetic memory in the context transgenerational inheritance remains to be determined (Blake & Watson, 2016; Friedli & Trono, 2015; van Otterdijk & Michels, 2016; Zeng et al., 2022). To do so, the epigenetic mechanisms responsible for endogenous retroviruses silencing during normal development need to be better understood before analyzing the effect of epigenetic enzymes recruitment upon altered environmental conditions.

The sex-specific reestablishment of epigenetic marks on gametes occurs in part after birth and during puberty, with differences between oocytes and spermatozoa. Hence, epigenetic changes in gametes postnatally, following exposures to environmental cues, has been proposed as a mechanism of transgenerational inheritance. Multiple studies have, in fact, reported the existence of aberrant differentially-methylated regions (DMRs) in paternal epigenomes in response to environmental cues (see Section 3). However, it is unlikely that phenotypic changes in the offspring are mediated solely by DNA methylation, considering the importance of epigenetic crosstalk in spermatogenesis. Rather, it is proposed that cumulative effects of dysregulation epigenetic patterns in the earlier development are sustained and exacerbated in transcriptional regulatory networks (Kremsky & Corces, 2020). In this sense, changes in sperm small ncRNAs, that has garnered much attention recently as a medium for epigenetic inheritance, could also participate in this global change (Chen, Yan, & Duan, 2016). Compared to the breadth of data on paternal-mediated epigenetic inheritance, much less is known on maternal-mediated inheritance, mainly due to the paucity of biological material for epigenetic studies. The advent of new techniques for chromatin interrogation and for modeling oocytes development in vitro will bring to light mechanisms of maternal-mediated inheritance in response to environmental cues (see Section 4).

Finally, the periconceptional environment could affect early embryo reprogramming and constitutes another window of susceptibility for TEI. In human, this specific period has been much studied for assisted reproductive technologies (ART), with accumulating evidence suggesting that ART may influence developmental trajectories in the offspring (Fleming et al., 2018). Indeed, despite advances in in vitro culture of embryos, developmental conditions remain sub-optimal (Watkins, Lucas, & Fleming, 2010), indicating the importance of the maternal reproductive tract for proper development and its sensitivity to environmental cues. In hypofertile conditions, it remains unclear what can be attributed to the periconceptional environment or to the parental gametes themselves. However, studies in mouse models have shown how variations in culture media components impact the developmental potential of the embryos (Watkins et al., 2010). In fact, the vast majority of data emerging from periconceptional sensitivity to the environment has been made in the context of maternal dietary environment (Fleming et al., 2018). In humans, maternal undernutrition was shown to be associated with subsequent adult cardiometabolic and neurological dysfunction in the offspring (Velazquez, Fleming, & Watkins, 2019). Mechanistically, metabolic dysfunction are linked with epigenetic mechanisms since several metabolites act as cofactors for epigenetic reactions (Boon, Silveira, & Mostoslavsky, 2020; Li, Egervari, Wang, Berger, & Lu, 2018). However, more studies are needed to define how the mammalian embryo responds to its immediate environment and whether epigenetic memory of periconceptional events are maintained throughout adulthood (Novakovic et al., 2019). This is further complexified by the existence of multiple simultaneous mechanisms of inheritance to environmental cues and confounding factors, particularly in social animals.

4. Evidence for environmentally-induced epigenetic inheritance in mammals

Exposure to environmental cues have the capacity to alter epigenomes and phenotypes. However, TEI postulates that an epigenetic memory of exposure is transmitted to the offspring through the germline without continuous exposure. For maternal inheritance, true transgenerational effects are detectable only at the third generation (F3). Indeed, for exposure in pregnant females (P0), the embryo (F1) is also exposed, as well as its fetal germ cells (F2). Thus, effects without continuous exposures do not occur before the F3 generation (Fig. 3). For paternal inheritance, the F2 is the first generation wherein transgenerational effects can be examined (Fig. 3). While TEI is well accepted in plants (Hauser, Aufsatz, Jonak, & Luschnig, 2011) and invertebrate animals, such as *Caenorhabditis elegans* (Baugh & Day, 2020), the existence or significance of TEI in mammals has been much disputed considering the extensive germline epigenetic reprogramming known to occur. In humans, in particular, evidence for transgenerational epigenetic inheritance to environmental cues is limited (Horsthemke, 2018; Svanes et al., 2021). This is because human studies tend to suffer from heterogeneous populations, reproducibility, recall- and selection- bias among other confounding factors (Marczylo, Jacobs, & Gant, 2016). Despite these conceptual challenges, several studies have provided evidence for at least intergenerational inheritance. Here, we compiled existing evidence of environmentally-induced transgenerational epigenetic inheritance in rodents in response to a range of environmental cues from manufactured chemicals, metals, diet to drugs (Table 1). For each environmental cue, we present the available data in humans and highlight the need for a better understanding of the molecular mechanisms of transgenerational

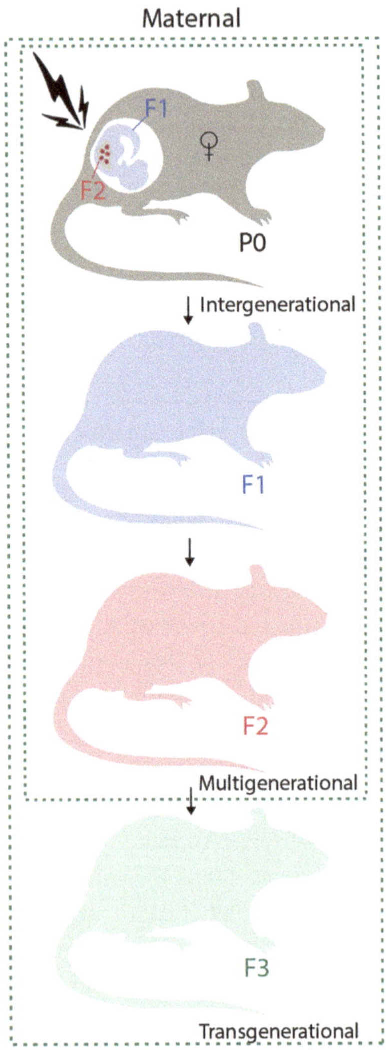

Fig. 3 Modes of Inheritance following Environmental Exposure. Exposure in gestational females (P0) also exposes the embryo (F1) and its germ cells (F2). Therefore, only the third generation is the one where the observed phenotype is not due to direct exposure and can truly be transgenerational. For post-natal exposure, as presented under the paternal scheme, exposure affects the F1, hence, the second generation is the one where transgenerational phenotypes can be observed.

versus intergenerational epigenetic inheritance. Of note, we focus here on environmentally-induced TEI and we do not discuss the epigenetic inheritance of behavioral phenotypes that has been covered elsewhere (Bale, 2014; Jawaid & Mansuy, 2019).

Table 1 Evidence of transgenerational epigenetic inheritance to environmental cues in rodents.

Animal Model	Environmental cue	Offspring phenotype	Epigenetic mark	Mode of exposure	References
Rat, Mouse	Manufactured chemical (Vinclozolin)	Impaired male fertility, increased adult onset disease	DNA methylation	Maternal (gestational)	Anway, Cupp, Uzumcu, and Skinner (2005); Anway, Memon, Uzumcu, and Skinner (2006); Anway, Leathers, and Skinner (2006); Guerrero-Bosagna, Settles, Lucker, and Skinner (2010); Guerrero-Bosagna et al. (2012)
Rat, Mouse	Manufactured chemical (Dioxin)	Impaired female fertility, increased adult onset disease	DNA methylation	Maternal (gestational)	Bruner-Tran and Osteen (2011); Manikkam, Tracey, Guerrero-Bosagna, and Skinner (2012)
Rat, Mouse	Manufactured chemical (BPA)	Impaired female fertility, increased adult onset disease	DNA methylation	Maternal (gestational) Paternal (pre-conceptional)	Manikkam, Tracey, Guerrero-Bosagna, and Skinner (2013); Ziv-Gal, Wang, Zhou, and Flaws (2015); Saidur Rahman, Pang, Ryu, Park, and Pang (2020)
Mouse	Manufactured chemical (Atrazine)	Impaired male fertility	H3K4me	Maternal (gestational)	Hao et al. (2016)
Mouse	Metal (Arsenic)	Impaired male fertility	DNA methylation	Maternal (gestational)	Yin et al. (2021)

Continued

Table 1 Evidence of transgenerational epigenetic inheritance to environmental cues in rodents.—cont'd

Animal Model	Environmental cue	Offspring phenotype	Epigenetic mark	Mode of exposure	References
Rat	Metal (Arsenic)	Impaired fertility	DNA methylation	Maternal (pre-conceptional) and Paternal (pre-conceptional)	Nava-Rivera et al. (2021)
Mouse	Diet (HFD)	Increased male body weight, impaired insulin sensitivity	DNA methylation	Maternal (gestational)	Sarker et al. (2018)
Rat	Diet (HFD)	Reduced birth-weight, female increased resistance to HFD-induced weight gain	Small RNAs	Paternal (pre-conceptional)	de Castro et al. (2016)
Mouse	Diet (HFD)	Increased obesity, insulin resistance	DNA methylation and small RNAs	Paternal (pre-conceptional)	Fullston et al. (2013); Wei et al. (2014); Chen et al. (2016)
Rat	Drugs (Alcohol)	Altered hypothalamic functions	DNA methylation	Maternal (gestational)	Govorko, Bekdash, Zhang, and Sarkar (2012)
Mouse	Drugs (Alcohol)	Altered neocortical development and behaviors	DNA methylation	Maternal (gestational)	Abbott, Rohac, Bottom, Patadia, and Huffman (2018); Bottom, Kozanian, Rohac, Erickson, and Huffman (2022)
Mouse	Drugs (Nicotine)	Reversal learning deficit	DNA methylation	Paternal (pre-conceptional)	McCarthy et al. (2018)

4.1 Manufactured chemicals

The first demonstration of transgenerational inheritance of an environmentally-induced phenotype in rodents was made nearly 20 years ago from studying the agricultural fungicide vinclozolin (Anway et al., 2005). Transient gestational exposure in female rats resulted in impaired sperm motility and reduced fertility in the male offspring up to the F4 (Anway et al., 2005; Anway, Memon, et al., 2006). Female descendants also developed transgenerational phenotypes up to the F4, including tumors and kidney disease (Anway, Leathers, & Skinner, 2006). Mechanistically, a methylated DNA immunoprecipitation followed by a promoter tilling microarray (MeDIP-chip) showed the presence of differentially-demethylated regions (DMRs) in F3 spermatozoa (Guerrero-Bosagna et al., 2010, 2012). However, the stability of the generated epimutations, or their relationship with known genetic abnormalities induced by vinclozolin were not assessed. Nevertheless, a following study showed that gestational vinclozolin exposure significantly altered F3 sperm-borne small ncRNAs that correlated with the DMRs and the vinclozolin transgenerational phenotypes (Schuster, Skinner, & Yan, 2016), suggesting a concerted vinclozolin-induced epigenetic transgenerational inheritance phenomenon.

Since these pioneering observations, exposure to multiple other manufactured chemicals, belonging to the large class of endocrine-disrupting chemicals (EDCs), have been shown to be inherited transgenerationally in rodents. However, the epigenetic mechanisms at play have not always been identified. Despite this, the question of EDCs-mediated epigenetic remodeling in the germline and phenotypic inheritance across generations needs to be urgently addressed. Indeed, EDCs are ubiquitous in our environment and because they directly interfere with the endocrine hormonal system, EDCs have been linked with multiple disorders in humans, including pregnancy complications, genital malformations, and cancer (Xin, Susiarjo, & Bartolomei, 2015).

Diethylstilbestrol (DES), a nonsteroidal estrogen, was initially prescribed in the 1950s with the goal of reducing miscarriage (Fénichel, Brucker-Davis, & Chevalier, 2015; Titus et al., 2019). The medication was taken off market in the 1970s after discovering cases of vaginal clear cell adenocarcinoma and breast cancer in the offspring born to women who were gestationally exposed to DES (Fénichel et al., 2015). The cohort of children resulting from DES-exposed women were also found to develop genital tract malformations, including hypospadias and cryptorchidism in males and uterine abnormalities in females (Fénichel et al., 2015). Additionally, both sexes were more likely to develop non-reproductive diseases including diabetes, hypercholesterolemia, hypertension and osteoporosis (Fénichel et al., 2015). Beyond intergenerational effects from mother to offspring, several studies have confirmed that DES has transgenerational effects up to the third generation (Fénichel et al., 2015; Titus et al., 2019). In particular, a study demonstrated that women from the third generation following DES gestational exposure developed higher rates of menstrual irregularity, preterm delivery and vaginal epithelial changes compared to the offspring of unexposed women (Titus et al., 2019). In addition, men from the third generation following DES gestational exposure had a higher likelihood of congenital anomalies, including hypospadias and heart defects (Fénichel et al., 2015). Without any chromosomal or genetic alterations, these offspring of DES-exposed women provide evidence for transmission of epigenetic inheritance across generations. Although the molecular mechanisms of transgenerational inheritance remain unclear, DES exposure may epigenetically alter estrogen receptors and impact the role of sex steroids in differentiating the genital tract during an important fetal exposure window (Fénichel et al., 2015; Titus et al., 2019).

Dioxin/TCDD (2,3,7,8-tetrachlorodibenzo-p-dioxin) is a by-product of industrial combustion processes and gestational exposure of this EDC has been shown to be transgenerationally inherited up to the F3, with an observed phenotype of reduced female fertility (Bruner-Tran & Osteen, 2011). Further studies in rats showed a similar phenotype in the F3 and that the F3 sperm epigenome was altered, with the identification of several DMRs (Manikkam et al., 2012). Again, the functional relevance of these DMRs, and their conservation across rodent models, need to be addressed to properly establish the mechanisms of TEI in response to dioxin. Dioxin exposure in humans is widespread and occurs primarily via food ingestion or contact with contaminated soil, air, and water (Travis & Hattemer-Frey, 1991). Studies in an Italian human cohort, following the explosion of a trichlorophenol manufacturing plant and the release of up to 30 kg of TCDD, showed that men whose mothers were exposed in utero to dioxin had impaired semen quality (reduction in spermatozoa concentration, count and quality) (Mocarelli et al., 2011). In a Russian cohort, whole genome bisulfite sequencing (WGBS) showed the presence of spermatozoa DMRs associated with serum TCDD concentrations (Pilsner et al., 2018). Together, these studies show that dioxin provokes spermatozoa epimutations and intergenerational phenotypes in humans, although impact across a larger number of generations has not been studied.

The plastic monomer bisphenol A (BPA) is another widespread EDC that is readily measured in humans following exposure to food and beverage containers (Vandenberg, Hauser, Marcus, Olea, & Welshons, 2007). BPA directly impacts human metabolic and reproductive organs (Rochester, 2013). However, epidemiological studies tend to support that, although BPA can cause fetal death and growth impairment during infancy at high dose, the risks for low-dose exposure are negligible (Ranjit, Siefert, & Padmanabhan, 2010). Contrarily to the evident paucity of research in humans, several studies in rodents have addressed the consequences of in-utero exposure to BPA across generations. BPA exposure has been associated with transgenerationally-acquired phenotypes of reduced fertility and adult onset disease, correlating with spermatozoa DMRs serving as BPA epimutations (Manikkam et al., 2013; Ziv-Gal et al., 2015). In a recent study focused on paternal inheritance of BPA exposure in mice, P0 males were exposed during the entire period of sexual maturity/puberty and phenotypes were analyzed in subsequent generations (Saidur Rahman et al., 2020). Both F2 and F3 males presented reduced fertility and a global increase in DNA methylation in spermatozoa. This study highlights that EDCs provoke transgenerational phenotypes that are both maternally and paternally inherited. Considering the different role of DNA methylation in oogenesis and spermatogenesis, further studies will need to focus on a better understanding of EDCs-mediated epigenetic remodeling in germ cells from P0 exposed parents.

In this regard, epigenetic crosstalk between histone PTMs and DNA methylation might be relevant in ensuring transgenerational inheritance to EDCs. For instance, gestational exposure to the herbicide atrazine was associated with reduced spermatozoa number in male F3 due to altered spermiogenesis (Hao et al., 2016). This phenotype was associated with a transgenerational global decrease in H3K4me3 level in the F3 (Hao et al., 2016). Occupancy of H3K4me3 was studied by chromatin immunoprecipitation followed by sequencing (ChIP-seq) and integration with transcriptomic data demonstrated that the epigenetic status of promoters of genes associated with germ cell differentiation were transgenerationally altered. Additionally, differential expression of lncRNAs in the F3 was also observed (Hao et al., 2016), further pointing to more potential mechanistic cross-talk in transgenerational inheritance of atrazine.

4.2 Metals

Toxic metals are ubiquitous in our environment and exposure has been associated with serious adverse health outcomes in human populations, such as cancer, cardiovascular diseases, and neurological disorders (Ray, Yosim, & Fry, 2014). With accumulating data showing that environmental toxic metals can impact epigenomes (Ray et al., 2014), their transgenerational inheritance has started to be studied in more depth.

Arsenic is a ubiquitous naturally-occurring toxic substance in the environment (Chung, Do, & Hong, 2014). Arsenic compounds dissolve readily in water; thus, consumption of contaminated water is a global route of exposure (Chung et al., 2014). Arsenic exposure is associated with a constellation of negative health effects impacting all major organs and organ systems and is a major carcinogen (Argos et al., 2010; Smith et al., 1992). For instance, in a Mexican human cohort, elevated arsenic levels in water correlated with hypermethylation in tumor-suppressor genes, pointing towards an epigenetic component in arsenic carcinogenic capacity (Smeester et al., 2011). Transgenerational phenotypes in response to arsenic exposure have been reported in *Caenorhabditis elegans* (Yu & Liao, 2016), and in zebrafish (Hernández-Sánchez, Valles, & Bardullas, 2022). In humans, a study in a Bengali cohort demonstrated that arsenic exposure in early pregnancy altered the cord blood methylome of the offspring, with more pronounced effects in male than in female newborns (Broberg et al., 2014). The identified DNA methylation changes were associated with cancer-related genes, which would suggest a sex specificity in arsenic carcinogenic capacity (Broberg et al., 2014). In agreement with epidemiological studies in humans, intergenerational effects of parental arsenic exposure have been reported in the mouse (Gong et al., 2021; Yin et al., 2021) and in the rat (Nava-Rivera et al., 2021). The first study on transgenerational inheritance of arsenic exposure in the mouse identified that pre-conceptional paternal exposure was associated with reduced adiposity in the F2 and increased adiposity in the F3 (Gong et al., 2021). Epigenetic mechanisms were not assessed but their adaptative changes were proposed to be responsible for the opposed phenotypes between the F2 and the F3 generations (Gong et al., 2021). More recently, another study in a mouse model of maternal gestational exposure to arsenic showed that the male F3 had impaired fertility associated with a decrease in the methylation levels in two specific imprinted genes (Yin et al., 2021). Finally, a study of pre-conceptional exposure to arsenic in both male and female rats showed impaired reproductive functions in the F2/F3 that were associated with a global increase in DNA methylation in gonads (Nava-Rivera et al., 2021). Thus, while there is evidence of phenotypic transgenerational inheritance of arsenic exposure, the role of DNA methylation in this phenomenon is unclear. This is representative of the unsatisfactory understanding of the effects of arsenic exposure on DNA methylation. Indeed, arsenic exposure provokes both global DNA hypomethylation, due to reduction in the levels of the universal methyl donor S-Adenosyl Methionine SAM, and hypermethylation of specific loci during carcinogenesis (Reichard & Puga, 2010; Ren et al., 2011). More fundamental studies are therefore needed to resolve the opposite effects of arsenic on DNA methylation (genome-wide vs locus-specific). Furthermore, SAM is also the methyl donor for histone methylation; thus, epigenetic crosstalks in arsenic TEI should be investigated.

The number of studies on transgenerational inheritance of metal exposure in rodents is limited, despite their wide distribution in our environment, their reported epigenetic inheritance in other animals (Carvan et al., 2017; Mu et al., 2021; Vandegehuchte, Lemière, Vanhaecke, Vanden Berghe, & Janssen, 2010) and epidemiological studies in humans (Maloney, Bayon, Zawia, & Lahiri, 2018; Sen et al., 2015). Cadmium and mercury co-administration in gestational mice was associated with reduced glucose tolerance up to the male F4 generation (Camsari et al., 2019). Lead gestational exposure in mice provoked growth stunt in the F3 generation (Sobolewski et al., 2020). However, in both studies, potential epigenetic mechanisms for transgenerational inheritance were not studied. Overall, mercury, cadmium, lead, selenium, and aluminum are all prevalent environmental toxic metals and their potential TEI in Mammals needs to be urgently examined.

4.3 Diet

Early long-term epidemiological studies and some empirical observations were the first testimonies that parental nutrition affects the offspring health (Guo, Luo, & Lin, 2020; Soubry, 2015). In recent years, studies in animal models have shed light on how changes in parental diets can modulate germ cells metabolism and epigenomes and induce phenotypic changes in the offspring, particularly the development of metabolic diseases. Here, we discuss how high-fat diet, caloric restriction or diet composition can be transgenerationally-inherited in the offspring upon epigenetic changes in the germline.

Increasing prevalence of obesity has become a major global health concern. A critical question is whether obesity in parents can influence the development of metabolic diseases over several generations in the offspring. Modeling of obesity in rodent models, using high-fat diet (HFD), has revealed the existence of transgenerational epigenetic inheritance of parental diet, both maternally and paternally. Maternal HFD during pregnancy is known to increase the rates of obesity and metabolic syndrome in the first and second generations (Kaspar, Hastreiter, Irmler, Hrabé de Angelis, & Beckers, 2020; Wu & Suzuki, 2006). However, maternal HFD can also be transmitted transgenerationally up to the third generation. Indeed, the F3 female offspring of mice fed with HFD before pregnancy showed increased body size and improved glucose tolerance, that were linked with dynamic patterns in imprinted genes expression (Dunn & Bale, 2011). More recently, the male F3 offspring of mothers perinatally fed with a HFD were shown to present an increased body weight and an impaired sensitivity to insulin (Sarker et al., 2018). These phenotypes were correlated with DNA methylation changes and signature DMRs in F2 spermatozoa (Sarker et al., 2018). Interestingly, observations regarding the phenotypes in the third generation vary between sexes (Sarker et al., 2018) and pre-conceptional versus gestational HFD also provoked different transgenerational phenotypes in the two above-mentioned studies. This indicates that multiple epigenetic mechanisms might be at play in maternal inheritance of HFD and more research is needed (Kaspar et al., 2020). Most studies on transgenerational inheritance of HFD have focused on the paternal route. The first study to address this question showed that female F1 descendent of male rat fed with HFD presented an impaired glucose-insulin homeostasis (Ng et al., 2010). This intergenerational phenotype for paternal inheritance of HFD in the rat was later shown to be transgenerationally transmitted by spermatozoa up to the second generation (de Castro et al., 2016). Indeed, HFD was associated with changes in F1 spermatozoa small RNAs consistent with their metabolic functions and the F2 phenotype (de Castro et al., 2016). Several studies have substantiated these findings in mice and shown that HFD provokes spermatozoa epigenome changes, in terms of DNA methylation and altered small RNA profiles (Chen, Yan, Cao, et al., 2016; Fullston et al., 2013; Wei et al., 2014). Since spermatozoa DNA methylation changes in response to HFD could not completely recapitulate the transcriptional outcomes in the offspring (Wei et al., 2014), attention is now shifting to the understanding of concerted epigenetic reprogramming in spermatozoa following HFD. Another open question is whether dietary changes can restore spermatozoa epigenomes after weight loss.

Inadequate protein intake is one example of malnutrition and can be modeled by feeding rodents with low-protein diet. One study found that the F1 generation of male mice fed with a low-protein diet had reduced

DNA methylation levels in the liver (Carone et al., 2010). These epigenetic signatures correlated with increased expression of genes involved in lipid and cholesterol biosynthesis (Carone et al., 2010). However, a later study with the same mice exposure paradigm showed that this intergenerational phenotype was not associated with changes in spermatozoa methylome (Shea et al., 2015), pointing towards the existence of other epigenetic mechanisms such as histone PTMs or small RNAs (Yoshida et al., 2020). In addition, a rat maternal low-protein diet also resulted in altered DNA methylation in the liver of male pups, due to decreased expression of the DNA methylation machinery (Gong, Pan, & Chen, 2010). These studies indicate that low-protein diet results at least in intergenerational phenotypes in rodents. Epidemiological studies in humans show how maternal low protein diet is associated with the lean variant of type 2 diabetes (Vipin, Blesson, & Yallampalli, 2022).Considering the worldwide increase in vegetarian and vegan diets, which are typically associated with low protein consumption (Vipin et al., 2022), more data will be needed to determine whether low protein diet associate with inter- and transgenerational epigenetic inheritance of metabolic disorders.

Undernutrition in early life may affect adult health. The famine period in the Netherlands during World War II, known as the Dutch Hunger Winter of 1944–1945, has allowed to study the intergenerational effect of parental nutrition on the metabolic status of grandchildren in humans (Lumey, Stein, & Susser, 2011). In this historical cohort, famine exposure during pregnancy (daily average of 900 kca/day or less) was associated with an adverse metabolic profile (suboptimal glucose handling, higher body mass index, etc.) and a higher risk for developing schizophrenia in the F1 offspring (Lumey et al., 2011). Adult offspring (F2) of gestationally exposed F1 fathers, but not mothers, were also found to present higher body mass index, suggesting a paternal intergenerational inheritance over two generations (Veenendaal et al., 2013). Subsequently, an methylome-wide approach by reduced representation bisulfite sequencing (RRBS) in the blood of F1 individuals demonstrated the presence of DMRs associated with the phenotypic outcome of early-life exposure to famine (Tobi et al., 2014). Studies in rodents support that gestational undernutrition in mice provokes intergenerational developmental reprogramming, with the apparition of DMRs in the F1 spermatozoa methylome (Radford et al., 2014) (Radford et al., 2014). Male mice undernutrition also resulted in spermatozoa global demethylation and an offspring with reduced postnatal weight (McPherson et al., 2016). However, the only evidence of TEI of undernutrition was found in the rat (Nowacka-Woszuk, Szczerbal, Malinowska, & Chmurzynska, 2018). Indeed, the F3 fetuses of pregnant females fed with a restricted diet presented lower hepatic levels of histone deacetylases (HDAC1 and SIN3A) but the increased H3 acetylation was not statistically significant (Nowacka-Woszuk et al., 2018). Altogether, more studies will be needed to pinpoint the epigenetic mechanisms of inter- and transgenerational inheritance of under- and malnutrition.

The impact of diet onto developmental processes needs to be placed in the larger context of the role of metabolism in early development and of the interplay between metabolism and epigenetics (Dahan, Lu, Nguyen, Kennedy, & Teitell, 2019; Verdikt & Allard, 2021). In this regard, another aspect of nutrition that has been explored is the change in diet composition, in particular, how micro-nutrient deficiency can impact gametes epigenomes and alter the offspring development. Folate is a metabolite that serves as a methyl-donor and its levels influence the availability of SAM (Clare, Brassington, Kwong, & Sinclair, 2019). Female mice carrying mutations in genes involved in the folate metabolism gave rise to an offspring with congenital malformations and tissue-specific loss of DNA methylation (Padmanabhan et al., 2013), demonstrating the importance of folate metabolism in maternal epigenetic inheritance. Folic acid supplementation is often recommended in early pregnancy and studies in humans indicate that changes in maternal plasma folate levels are associated with epigenome-wide DNA methylation variations in newborn cord blood (Caffrey et al., 2018; Joubert et al., 2016). In a recent study in the mouse, paternal inheritance of folate deficiency was shown to occur through dysregulation of H3K4me3 in spermatozoa (Lismer et al., 2021). Importantly, regions with altered H3K4me3 profiles in spermatozoa corresponded to developmental loci, were retained in the zygote and overlapped with differential zygotic gene expression (Lismer et al., 2021). Further research will need to determine whether folate deficiency-induced histone retention interplays with DNA methylation in transgenerational epigenetic inheritance. Another open question is whether the epigenetic modes of inheritance of folate deficiency vary in a sex-specific manner, considering the different role of chromatin methylation in gametogenesis.

4.4 Substances use and abuse

The evidence for epigenetic modifications following exposure to drugs of abuse, and their consequences on rewiring neuronal circuitry to maintain addiction, is well documented (Nestler, 2014). In recent years, several studies have started to address whether drugs of abuse can alter gametes epigenomes and the consequences for transgenerational inheritance (Wanner, Colwell, & Faulk, 2019; Yohn, Bartolomei, & Blendy, 2015; Zeid & Gould, 2020). Inheritance of drug abuse is particularly confounded by genetic and non-genetic (social, cultural, and environmental) factors (Yohn et al., 2015). Hence, modeling parental inheritance to drug abuse in rodent models provides mechanistic bases, i.e., the identification of epigenetic biomarkers, to study the relevance of TEI of drug abuse in human populations.

Ethyl alcohol (ethanol) effects are mediated through interference with several neurotransmitter systems (Chastain, 2006). Alcohol exposure provokes epigenetic and transcriptional changes in the brain, particularly in regions critical to addictive behaviors (Berkel & Pandey, 2017). Mechanistically, alcohol was shown to reduce the production of SAM provoking global hypomethylation in rat liver (Lu & Mato, 2005) as well as cause histone hyperacetylation by directly providing the acetyl groups through ethanol metabolism (Mews et al., 2019), although a tissue-specific understanding of epigenetic remodeling following alcohol exposure is currently lacking. In addition, alcohol exposure provokes genome-wide alteration in DNA methylation patterns in germ cells and several occurrences of intergenerational inheritance of these changes have been found both in rodent models and in human cohorts (Yohn et al., 2015). The risks of maternal alcohol abuse during pregnancy are well-established; however, gestational exposure to alcohol (prenatal ethanol exposure, PrEE) also provokes phenotypic and epigenetic changes across generations. For instance, in a rat model for fetal alcohol exposure, male from the F3 generation presented significant reduced expression of the proopiomelanocortin gene in the hypothalamus, concomitant with increased DNA methylation on its CGIs (Govorko et al., 2012). The fact that F3 male but not female progeny presented these changes support a paternal inheritance, although epigenome-wide studies in the brain are needed to confirm the observation. A similar paternal inheritance of gestational exposure to ethanol was shown in the mouse, where the F3 offspring presented an altered development of the neocortex, associated with a decreased expression of all three DNA methyltransferases and a global neocortical hypomethylation (Abbott et al., 2018). A study in the same transgenerational mouse model for fetal alcohol exposure confirmed that the developmental changes were associated with altered behavioral phenotypes (Bottom et al., 2022). Overall, these studies highlight the importance of paternal inheritance in the transmission of alcohol exposure. Considering recent works showing that chronic ethanol exposure alters spermatozoa small ncRNAs, and their nucleoside modifications (Rompala et al., 2018), further works will need to assess the contribution of pre-conceptional ethanol exposure to TEI of alcohol abuse.

The abuse of opioids, including the illegal drug heroin and pain relievers such as morphine, fentanyl, codeine, and oxycodone, has emerged as a major health threat over the last decade (Gilardi, Augsburger, & Thomas, 2018). Prevalence of opioid medication signifies that exposure in pregnant mothers has increased (Gilardi et al., 2018). While accumulating data evidence that gestational exposure provokes intergenerational phenotypic and epigenetic changes (Gilardi et al., 2018; Zeid & Gould, 2020), there are currently no studies examining epigenetic transgenerational effects of opioids beyond the F1 generation. This is true for another illicit drug: cocaine. Indeed, epigenetic reprogramming in male germ cells has been demonstrated in response to cocaine exposure (González et al., 2020; Swinford-Jackson et al., 2022) and intergenerational inheritance has been reported (He, Lidow, & Lidow, 2006; Swinford-Jackson et al., 2022; Vassoler, White, Schmidt, Sadri-Vakili, & Christopher, 2013). However, future studies will need to address whether pre-conceptional or gestational cocaine exposure can be transmitted through epigenetic remodeling of germ cells across generations.

With the growing use of medical and recreational cannabis, consumption of cannabis products has dramatically increased in recent years, including by expecting mothers (Carlier, Huestis, Zaami, Pichini, & Busardò, 2020; Volkow, Han, Compton, & McCance-Katz, 2019). More than 100 phytocannabinoids have been isolated from *Cannabis sativa*, with (−)-trans-Δ9-tetrahydrocannabinol (Δ9-THC) being the predominant phytocannabinoid responsible for the psychoactivity of cannabis (Andre, Hausman, & Guerriero, 2016). As a result, the level of Δ9-THC in cannabis has reached up to 20% of total compounds (Chandra et al., 2019). In recent years, several studies have shown that effects of gestational Δ9-THC exposure are intergenerationally inherited in rodents. Indeed, the descendants of rats exposed to Δ9-THC present functional alterations in brain regions controlling compulsive and goal-directed behaviors (Prini et al., 2018; Szutorisz et al., 2014; Vassoler, Johnson, & Byrnes, 2013; Watson et al., 2015). Furthermore, Δ9-THC also has a clearly described impact on the epigenome, especially on DNA methylation and histone PTMs both in the brain (Prini et al., 2018; Prini, Penna, Sciuccati, Alberio, & Rubino, 2017; Szutorisz & Hurd, 2018; Watson et al., 2015) and in spermatozoa (Murphy et al., 2018; Osborne et al., 2020; Schrott & Murphy, 2020). In particular, one study focusing on cross-comparison of spermatozoa methylome in Δ9-THC-exposed rats and human cannabis users showed that similar pathways were altered (Murphy et al., 2018). These DMRs could further serve as biomarkers when investigating the existence of TEI for cannabis in Mammals.

Finally, accumulating data indicate intergenerational epigenetic inheritance following exposure to tobacco smoke (Hammer et al., 2018). In particular, prior evidence has demonstrated that maternal smoking contributes to asthma and poor lung function in the offspring (Magnus et al., 2015; Rehan, Liu, Sakurai, & Torday, 2013). Whether transgenerational inheritance of asthma occurs in humans is more conflicting. Extensive population-based studies have found a significant correlation between a grandmother's smoking exposure and asthma in grandchildren, independent of the mother's smoking exposure (Li, Langholz, Salam, & Gilliland, 2005; Magnus et al., 2015). Mechanistically, the transgenerational effects of smoking on asthma susceptibility could arise from differential DNA methylation in fetal oocytes (Li et al., 2005). In this regard, newborn cord blood samples from the Norwegian Mother and Child Cohort Study have confirmed the existence of altered methylation patterns in newborns born to women who smoked (Joubert et al., 2012). Additionally, a Swedish Birth Cohort study observed higher DNA methylation of insulin-like growth factor 2 (IGF2) in newborns whose mothers smoked; this was correlated to the paternally expressed IGF2 with male offspring preferentially affected (Krauss-Etschmann, Meyer, Dehmel, & Hylkema, 2015). Thus, epigenetic mechanisms may play a role in explaining the increasing incidence of asthma in the offspring of smoking mothers. However, whether epigenetic changes in the germline arise from nicotine, the active component of tobacco, or other chemical constituents of tobacco is difficult to assess retroactively (Zeid & Gould, 2020). Studies in rodents have shown that nicotine can remodel the chromatin through DNA methylation and changes in histone PTMs (Levine et al., 2011; Toledo-Rodriguez et al., 2010). Intergenerational epigenetic inheritance of nicotine exposure has been reported in mice (Dai et al., 2017). However, to date, few studies have addressed the mechanisms of epigenetic inheritance to nicotine exposure in rodents further than the second generation. The F2 offspring of male mice exposed to nicotine displayed behavioral changes that might be due to the observed changes in P0 spermatozoa methylome (McCarthy et al., 2018). This warrants further research to understand the epigenetic mechanisms of paternal inheritance to nicotine exposure. Given the rise in the use of smokeless tobacco products, electronic cigarettes in particular

(McCarthy et al., 2020), another open question is whether nicotine and other components, such as flavoring agents, synergize in the epigenetic remodeling of germ cells. Data in a mouse intergenerational model for electronic cigarette showed that co-exposure to nicotine and saccharin in the P0 provoked memory deficits in the F1 whereas individual exposure to either substance alone did not (McCarthy et al., 2020). Overall, as for other drugs of abuse, research on TEI of nicotine is still in its early stage. An additional difficulty for nicotine, as well as cannabis and opioids, is to adapt the dose and delivery mode in rodent models as to mimic the best the reported uses in human populations.

5. Harnessing in vitro models for gene-environment interaction studies

Despite multiple lines of evidence for transgenerational inheritance in rodents, our understanding of the epigenetic mechanisms involved in phenotypic inheritance across multiple generations remains limited. Furthermore, the existence of epigenetic inheritance of phenotypes further than the second generation in humans remains controversial. Here we discuss the current limitations in studies on the impact of the environment on germs cells. We further highlight the potential of in vitro germ cell models and in vitro gametogenesis for TEI studies.

5.1 Current limitations to TEI studies in mammals

Beyond epigenetic inheritance, studies on the impact on the environment on subsequent generations must account for other mechanisms that may be interrelated, including genetic changes and non-genetic changes on the individual, ecological, cultural, or social levels (Danchin et al., 2011). Traditional Mendelian inheritance postulates that phenotypes are inherited based on the transmission of DNA across generations and that abnormal DNA sequences lead to abnormal phenotypes (Senaldi & Smith-Raska, 2020). Epigenetic inheritance builds on this genetic model of inheritance and adds that the chromatin structure of gametes, not just their DNA sequence, also contributes to the offspring phenotype. However, an under-appreciated dimension in inheritance studies is how genetic and epigenetic information are interrelated. For instance, DNA sequence variations affect DNA methylation, not just in CpG loci, but genome-wide (Hannon et al., 2018; McRae et al., 2014). In fact, genetic heritability was shown to be the major cause of similarity in DNA methylation levels in a human cohort

(McRae et al., 2014), although the mechanisms at play or the contribution of non-genetic inheritance were not assessed. Another dimension is the evolution of epigenomes over time. Indeed, aging is associated with changes in the chromatin structure both globally and locally (Kane & Sinclair, 2019). Hence, aging should be taken into account when studying interactions between the environment and epigenomes (Barrere-Cain & Allard, 2020; Bertozzi et al., 2021). Finally, non-genetic factors also contribute to epigenetic inheritance of phenotypes. In social animals, the inherited environment is shaped by social status, cultural norms and acquired behaviors (Danchin et al., 2011). For instance, in humans, socioeconomic status, political orientation or religious practices may influence a child's environment, with consequences on their epigenome and phenotype (Danchin et al., 2011). Overall, the epigenome is shaped by its interaction with the genome and the inherited environment. These confounding factors should be considered in the context of an extended theory of inheritance. In addition, studies aiming at proving TEI in animal models must account for genetic biases (by controlling the genetic background) and acknowledge the extent of non-genetic inheritance.

Another major issue in TEI studies is a technical one. Indeed, while epigenetic studies in male germ cells can be easily performed, considering the relatively large amount of material and accessibility, the small number of oocytes in vivo prevent detailed molecular analyses. In fact, the only epigenetic evidence of inheritance of phenotypes across generations following environmental exposure in rodents are paternal, mostly associated with detectable changes in spermatozoa methylome and ncRNAs profiles (Table 1). The development of new low-input chromatin analyses has allowed to further our understanding of the epigenetic mechanisms regulating the homeostatic development of oocytes (Skene & Henikoff, 2017; Xu & Xie, 2018). It is therefore likely that the same techniques will reveal molecular mechanisms of maternal inheritance of environmentally-induced phenotypes.

5.2 In vitro germ cell models

In parallel to the development of low-input chromatin techniques, epigenetic inheritance studies can benefit from in vitro models for germ cells development. Here, we briefly summarize the existing models and highlight how in vitro gametogenesis can be harnessed in studying the impact of the environment on germ cells epigenome.

The generation of functional mature and fertile gametes in vitro requires the reconstitution of the multi-stepped development of germ cells in vivo. Mouse embryonic stem cells (mESCs) are pluripotent stem cells derived from the inner cell mass of blastocysts and thus constitute an immortalized representation of the preimplantation embryo (Evans & Kaufman, 1981; Nichols & Smith, 2009). In early forays into in vitro gametogenesis, direct derivations of mPGCs from mESCs were found to be poorly efficient, with the induced cells not contributing to healthy offspring (Hayashi, Kawaguchi, Durcova-Hills, & Imai, 2018). The first successful reconstitution of the fetal mouse germ cell specification pathway in vitro was made over 10 years ago. In this system, the so-called mouse PGC-like cells (mPGCLCs) are obtained in a two-step differentiation from mESCs to a mouse epiblast-like intermediate (mEpiLCs), therefore mimicking the stepwise differentiation of mPGCs (Hayashi et al., 2018; Hayashi, Ohta, Kurimoto, Aramaki, & Saitou, 2011). Mouse PGCLCs have been found to accurately recapitulate early developmental processes and epigenetic remodeling of mouse PGCs (Hayashi et al., 2011; Ohta et al., 2017; von Meyenn et al., 2016). In particular, using transgenic ESCs harboring fluorescent reporters for germline markers, properly-specified mPGCLCs can be isolated by flow cytometry (Ohinata, Sano, Shigeta, Yamanaka, & Saitou, 2008). The scalability of this PGCLCs differentiation protocol allows, in turn, the rapid generation of a large number of cells, making of it a powerful system for the interrogation of underlying transcriptional and epigenetic mechanisms. In this regard, we have recently explored the developmental trajectory of mPGCLCs following exposure to BPA and showed a dose-dependent effect on mPGCLCs proliferation in XX but not XY cells (Ooi, Jiang, Kang, & Allard, 2021). Furthermore, mPGCLCs transcriptome revealed a dysregulation of X-linked genes and retrotransposons (Ooi et al., 2021). These results represent the first characterization of the effect of a toxicant exposure on fetal germ cells, highlighting their potential for TEI studies.

In the original mPGCLCs protocol, XY cells were shown to properly undergo spermatogenesis following transplantation in the seminiferous tubules of neonatal mice, whereas XX cells aggregated with female gonadal somatic cells in reconstituted ovaries also underwent maturation to oocytes (Hayashi et al., 2011). Since then, multiple studies have attempted to recreate the entire process of gametogenesis in culture. Today, in vitro full reconstitution of mouse spermatogenesis and oogenesis is a reality (Hikabe et al., 2016; Komeya, Sato, & Ogawa, 2018; Luo & Yu, 2022), with the micro-environment of the gonads and communication with somatic cells being replicated in organoids (Yoshino et al., 2021). However, transposition to human gametogenesis has been hindered by ethical and technical limitations (Luo & Yu, 2022). For instance, human PGCLCs (hPGCLCs) can be obtained from human induced-pluripotent stem cells (hiPSCs), however, proper induction cannot be confirmed by the production of fertile offspring as done in the mouse, for obvious ethical reasons (Luo & Yu, 2022). Furthermore, development of germ cells require cocultures of hPGCLCs with human embryonic somatic gonadal cells, which again, cannot be easily obtained for ethical and technical reasons (Luo & Yu, 2022). With the advent of organoid research (Yoshino et al., 2021), reconstituted gonads will provide a platform to study the molecular mechanisms of mammalian germ cells differentiation. These in vitro gametogenesis systems can also be harnessed for TEI studies, by allowing the expansion of initial material for downstream epigenetic interrogation techniques.

6. Conclusion and future challenges

Epigenetic mechanisms represent a formidable tool for organisms to evolve with their environment. Whether transgenerational inheritance of environmental cues across multiple generations occurs in Mammals is still a very much debated topic. Despite the epidemiological evidence in humans and phenotypical evidence in rodent models that we have described, research on the molecular mechanisms of TEI is still in its infancy and multiple open questions remain.

Future studies will need to define what are the epigenetic mechanisms of transgenerational inheritance. We have shown that oftentimes, changes in male germ cells methylome are pointed as a mode of inheritance. However, we have also discussed how DNA methylation and other epigenetic mechanisms, such as histone PTMs, for instance, are coordinated in homeostatic development of germ cells. Hence, understanding how environmental cues alter epigenetic crosstalks in the germline will be critical for TEI studies. Similarly, deciphering the role of sex specificity in epigenetic remodeling in response to the environment will shed light on conflicting results regarding maternal versus paternal modes of inheritance.

Another dimension is an ethical and practical one: if we are to prove epigenetic remodeling in human germ cells and in human embryos, we will need to rely on models. Most studies have been performed in rodent models, which, ultimately do not recapitulate fully the extent of epigenetic remodeling occurring in humans. In addition to these animal models, we have highlighted the relevance of using in vitro gametogenesis to unravel the molecular mechanisms of epigenetic inheritance in response to environmental changes.

Overall, while the study of transgenerational inheritance in response to changes in our environment is still in its early stages, the development of new tools, techniques and models opens the way for exciting future research that may reshape once more our definition of "epigenetics" in an extended theory of inheritance.

References

Abbott, C. W., Rohac, D. J., Bottom, R. T., Patadia, S., & Huffman, K. J. (2018). Prenatal ethanol exposure and neocortical development: A transgenerational model of FASD. *Cerebral Cortex*, *28*, 2908–2921. https://doi.org/10.1093/cercor/bhx168.

Almouzni, G., & Cedar, H. (2016). Maintenance of epigenetic information. *Cold Spring Harbor Perspectives in Biology*, *8*, 1–23. https://doi.org/10.1101/cshperspect.a019372.

Ancelin, K., Lange, U. C., Hajkova, P., Schneider, R., Bannister, A. J., Kouzarides, T., et al. (2006). Blimp1 associates with Prmt5 and directs histone arginine methylation in mouse germ cells. *Nature Cell Biology*, *8*, 623–630. https://doi.org/10.1038/ncb1413.

Andre, C. M., Hausman, J.-F., & Guerriero, G. (2016). Cannabis sativa: The Plant of the Thousand and one Molecules. *Frontiers in Plant Science*, *7*, 19. https://doi.org/10.3389/fpls.2016.00019.

Andreu-Vieyra, C. V., Chen, R., Agno, J. E., Glaser, S., Anastassiadis, K., Stewart Francis, A., et al. (2010). MLL2 is required in oocytes for bulk histone 3 lysine 4 trimethylation and transcriptional silencing. *PLoS Biology*, *8*, 53–54. https://doi.org/10.1371/journal.pbio.1000453.

Anway, M. D., Cupp, A. S., Uzumcu, M., & Skinner, M. K. (2005). Epigenetic transgenerational actions of endocrine disruptors and male fertility. *Reproductive Toxicology*, *31*, 1466–1469. https://doi.org/10.1016/j.reprotox.2010.10.012.

Anway, M. D., Leathers, C., & Skinner, M. K. (2006). Endocrine disruptor vinclozolin induced epigenetic transgenerational adult-onset disease. *Endocrinology*, *147*, 5515–5523. https://doi.org/10.1210/en.2006-0640.

Anway, M. D., Memon, M. A., Uzumcu, M., & Skinner, M. K. (2006). Transgenerational effect of the endocrine disruptor vinclozolin on male spermatogenesis. *Journal of Andrology*, *27*, 868–879. https://doi.org/10.2164/jandrol.106.000349.

Argos, M., Kalra, T., Rathouz, P. J., Chen, Y., Pierce, B., Parvez, F., et al. (2010). Arsenic exposure from drinking water, and all-cause and chronic-disease mortalities in Bangladesh (HEALS): A prospective cohort study. *The Lancet.*, *376*, 252–258. https://doi.org/10.1016/S0140-6736(10)60481-3.

Bale, T. L. (2014). Lifetime stress experience: Transgenerational epigenetics and germ cell programming. *Dialogues in Clinical Neuroscience*, *16*, 297–305.

Bannister, A. J., & Kouzarides, T. (2011). Regulation of chromatin by histone modifications. *Cell Research*, *21*, 381–395. https://doi.org/10.1038/cr.2011.22.

Bao, J., & Bedford, M. T. (2016). Epigenetic regulation of the histone-to-protamine transition during spermiogenesis. *Reproduction*, *151*, R55–R70. https://doi.org/10.1530/REP-15-0562.

Barrere-Cain, R., & Allard, P. (2020). An understudied dimension: Why age needs to be considered when studying epigenetic-environment interactions. *Epigenet Insights.*, *13*. https://doi.org/10.1177/2516865720947014. 2516865720947014.

Barrett, J. R. (2017). Programming the future: Epigenetics in the context of DOHaD. *Environmental Health Perspectives*, A72. https://doi.org/10.1289/ehp.125-A72.

Baugh, L. R., & Day, T. (2020). Nongenetic inheritance and multigenerational plasticity in the nematode C. elegans. *eLife.*, 1–13.

Ben Maamar, M., Beck, D., Nilsson, E., McCarrey, J. R., & Skinner, M. K. (2022). Developmental alterations in DNA methylation during gametogenesis from primordial germ cells to sperm. *iScience.*, *25*, 1–29. https://doi.org/10.1016/j.isci.2022.103786.

Berkel, T. D. M., & Pandey, S. C. (2017). Emerging role of epigenetic mechanisms in alcohol addiction. *Alcoholism: Clinical and Experimental Research*, *41*, 666–680. https://doi.org/10.1111/acer.13338.

Bertozzi, T. M., Becker, J. L., Blake, G. E. T., Bansal, A., Nguyen, D. K., Fernandez-Twinn, D. S., et al. (2021). Variably methylated retrotransposons are refractory to a range of environmental perturbations. *Nature Genetics*, *53*, 1233–1242. https://doi.org/10.1038/s41588-021-00898-9.

Bianco-Miotto, T., Craig, J. M., Gasser, Y. P., van Dijk, S. J., & Ozanne, S. E. (2017). Epigenetics and DOHaD: From basics to birth and beyond. *Journal of Developmental Origins of Health and Disease*, *8*, 513–519. https://doi.org/10.1017/S2040174417000733.

Bird, A. (2002). DNA methylation patterns and epigenetic memory. *Genes & Development*, *16*, 6–21. https://doi.org/10.1101/gad.947102.

Bird, A. (2007). Perceptions of epigenetics. *Nature*, *447*, 396–398. https://doi.org/10.1038/nature05913.

Blake, G. E. T., & Watson, E. D. (2016). Unravelling the complex mechanisms of transgenerational epigenetic inheritance. *Current Opinion in Chemical Biology*, *33*, 101–107. https://doi.org/10.1016/j.cbpa.2016.06.008.

Bline, A. P., Le Goff, A., & Allard, P. (2020). What is lost in the weismann barrier? *Journal of Developmental Biology*, 8. https://doi.org/10.3390/jdb8040035.

Boon, R., Silveira, G. G., & Mostoslavsky, R. (2020). Nuclear metabolism and the regulation of the epigenome. *Nature Metabolism*, *2*, 1190–1203. https://doi.org/10.1038/s42255-020-00285-4.

Bottom, R. T., Kozanian, O. O., Rohac, D. J., Erickson, M. A., & Huffman, K. J. (2022). Transgenerational effects of prenatal ethanol exposure in prepubescent mice. *Frontiers in Cell and Development Biology*, *10*, 812429. https://doi.org/10.3389/fcell.2022.812429.

Bourc'his, D., Xu, G. L., Lin, C. S., Bollman, B., & Bestor, T. (2001). Dnmt3L and the establishment of maternal genomic imprints. *Science*, *294*, 2536–2539. https://doi.org/10.1126/science.1065848.

Breton, C. V., Landon, R., Kahn, L. G., et al. (2021). Exploring the evidence for epigenetic regulation of environmental influences on child health across generations. *Communications Biology*, *4*, 769. https://doi.org/10.1038/s42003-021-02316-6.

Broberg, K., Ahmed, S., Engström, K., Hossain, M. B., Jurkovic Mlakar, S., Bottai, M., et al. (2014). Arsenic exposure in early pregnancy alters genome-wide DNA methylation in cord blood, particularly in boys. *Journal of Developmental Origins of Health and Disease*, *5*, 288–298. https://doi.org/10.1017/S2040174414000221.

Bruner-Tran, K. L., & Osteen, K. G. (2011). Developmental exposure to TCDD reduces fertility and negatively affects pregnancy outcomes across multiple generations. *Reproductive Toxicology*, *31*, 344–350. https://doi.org/10.1016/j.reprotox.2010.10.003.

Brykczynska, U., Hisano, M., Erkek, S., Ramos, L., Oakeley, E. J., Roloff, T. C., et al. (2010). Repressive and active histone methylation mark distinct promoters in human and mouse spermatozoa. *Nature Structural and Molecular Biology*, *17*, 679–687. https://doi.org/10.1038/nsmb.1821.

Caffrey, A., Irwin, R. E., McNulty, H., Strain, J. J., Lees-Murdock, D. J., McNulty, B. A., et al. (2018). Gene-specific DNA methylation in newborns in response to folic acid supplementation during the second and third trimesters of pregnancy: Epigenetic analysis from a randomized controlled trial. *American Journal of Clinical Nutrition, 107*, 566–575. https://doi.org/10.1093/ajcn/nqx069.

Camsari, C., Folger, J. K., Rajput, S. K., McGee, D., Latham, K. E., & Smith, G. W. (2019). Transgenerational effects of periconception heavy metal administration on adipose weight and glucose homeostasis in mice at maturity. *Toxicological Sciences, 168*, 610–619. https://doi.org/10.1093/toxsci/kfz008.

Carlier, J., Huestis, M. A., Zaami, S., Pichini, S., & Busardò, F. P. (2020). Monitoring perinatal exposure to Cannabis and synthetic cannabinoids. *Therapeutic Drug Monitoring, 42*, 194–204. https://doi.org/10.1097/FTD.0000000000000667.

Carone, B. R., Fauquier, L., Habib, N., Shea, J. M., Hart, C. E., Li, R., et al. (2010). Paternally induced transgenerational environmental reprogramming of metabolic gene expression in mammals. *Cell, 143*, 1084–1096. https://doi.org/10.1016/j.cell.2010.12.008.

Carvan, M. J., Kalluvila, T. A., Klingler, R. H., Larson, J. K., Pickens, M., Mora-Zamorano, F. X., et al. (2017). Mercury-induced epigenetic transgenerational inheritance of abnormal neurobehavior is correlated with sperm epimutations in zebrafish. *PLoS One, 12*, 1–26. https://doi.org/10.1371/journal.pone.0176155.

Chandra, S., Radwan, M. M., Majumdar, C. G., Church, J. C., Freeman, T. P., & ElSohly, M. A. (2019). New trends in cannabis potency in USA and Europe during the last decade (2008–2017). *European Archives of Psychiatry and Clinical Neuroscience, 269*, 5–15. https://doi.org/10.1007/s00406-019-00983-5.

Chastain, G. (2006). Alcohol, neurotransmitter systems, and behavior. *The Journal of General Psychology, 133*, 329–335. https://doi.org/10.3200/GENP.133.4.329-335.

Chen, Q., Yan, M., Cao, Z., Li, X., Zhang, Y., Shi, J., et al. (2016). Sperm tsRNAs contribute to intergenerational inheritance of an acquired metabolic disorder. *Science, 351*, 397–400. https://doi.org/10.1126/science.aad7977.

Chen, Q., Yan, W., & Duan, E. (2016). Epigenetic inheritance of acquired traits through sperm RNAs and sperm RNA modifications. *Nature Publishing Group, 17*, 733–743. https://doi.org/10.1038/nrg.2016.106.

Chung, J. Y., Do, Y. S., & Hong, Y. S. (2014). Environmental source of arsenic exposure. *Journal of Preventive Medicine and Public Health, 47*, 253–257. https://doi.org/10.3961/jpmph.14.036.

Clare, C. E., Brassington, A. H., Kwong, W. Y., & Sinclair, K. D. (2019). One-carbon metabolism: Linking nutritional biochemistry to epigenetic programming of long-term development. *Annual Review of Animal Biosciences, 7*, 263–287. https://doi.org/10.1146/annurev-animal-020518-115206.

Dahan, P., Lu, V., Nguyen, R. M. T., Kennedy, S. A. L., & Teitell, M. A. (2019). Metabolism in pluripotency : Both driver and passenger ? *The Journal of Biological Chemistry, 294*, 5420–5429. https://doi.org/10.1074/jbc.TM117.000832.

Dahl, J. A., Jung, I., Aanes, H., Greggains, G. D., Manaf, A., Lerdrup, M., et al. (2016). Broad histone H3K4me3 domains in mouse oocytes modulate maternal-to-zygotic transition. *Nature, 537*, 548–552. https://doi.org/10.1038/nature19360.

Dai, J., Wang, Z., Xu, W., Zhang, M., Zhu, Z., Zhao, X., et al. (2017). Paternal nicotine exposure defines different behavior in subsequent generation via hyper-methylation of mmu-miR-15b. *Scientific Reports, 7*, 1–15. https://doi.org/10.1038/s41598-017-07920-3.

Danchin, É., Charmantier, A., Champagne, F. A., Mesoudi, A., Pujol, B., & Blanchet, S. (2011). Beyond DNA: Integrating inclusive inheritance into an extended theory of evolution. *Nature Reviews. Genetics, 12*, 475–486. https://doi.org/10.1038/nrg3028.

de Castro, B. T., Ingerslev, L. R., Alm, P. S., Versteyhe, S., Massart, J., Rasmussen, M., et al. (2016). High-fat diet reprograms the epigenome of rat spermatozoa and transgenerationally affects metabolism of the offspring. *Molecular Metabolism., 5*, 184–197. https://doi.org/10.1016/j.molmet.2015.12.002.

Duhl, D. M. J., Vrieling, H., Miller, K. A., Wolff, G. L., & Barsh, G. S. (1994). Neomorphic agouti mutations in obese yellow mice. *Nature*, *8*, 59–65.

Dunn, G. A., & Bale, T. L. (2011). Maternal high-fat diet effects on third-generation female body size via the paternal lineage. *Endocrinology*, *152*, 2228–2236. https://doi.org/10.1210/en.2010-1461.

Evans, M. J., & Kaufman, M. H. (1981). Establishment in culture of pluripotential cells from mouse embryos. *Nature*, *292*, 154–156.

Fang, Y., Xu, X., Ding, J., Yang, L., Doan, M. T., Karmaus, P. W. F., et al. (2021). Histone crotonylation promotes mesoendodermal commitment of human embryonic stem cells. *Cell Stem Cell*, 1–16. https://doi.org/10.1016/j.stem.2020.12.009.

Fayomi, A. P., & Orwig, K. E. (2018). Spermatogonial stem cells and spermatogenesis in mice, monkeys and men. *Stem Cell Research*, *29*, 207–214. https://doi.org/10.1016/j.scr.2018.04.009.

Fénichel, P., Brucker-Davis, F., & Chevalier, N. (2015). The history of Distilbène® (diethylstilbestrol) told to grandchildren-the transgenerational effect. *Annales d'Endocrinologie*, *76*, 253–259. https://doi.org/10.1016/j.ando.2015.03.008.

Fleming, T. P., Watkins, A. J., Velazquez, M. A., Mathers, J. C., Prentice, A. M., Stephenson, J., et al. (2018). Origins of lifetime health around the time of conception: Causes and consequences. *The Lancet.*, *391*, 1842–1852. https://doi.org/10.1016/S0140-6736(18)30312-X.

Fraser, R., & Lin, C. J. (2016). Epigenetic reprogramming of the zygote in mice and men: On your marks, get set, go! *Reproduction*, *152*, R211–R222. https://doi.org/10.1530/REP-16-0376.

Friedli, M., & Trono, D. (2015). The developmental control of transposable elements and the evolution of higher species. *Annual Review of Cell and Developmental Biology*, *31*, 429–451. https://doi.org/10.1146/annurev-cellbio-100814-125514.

Fullston, T., Teague, E. M. C. O., Palmer, N. O., Deblasio, M. J., Mitchell, M., Corbett, M., et al. (2013). Paternal obesity initiates metabolic disturbances in two generations of mice with incomplete penetrance to the F2 generation and alters the transcriptional profile of testis and sperm microRNA content. *FASEB Journal*, *27*, 4226–4243. https://doi.org/10.1096/fj.12-224048.

Gilardi, F., Augsburger, M., & Thomas, A. (2018). Will widespread synthetic opioid consumption induce epigenetic consequences in future generations? *Frontiers in Pharmacology*, *9*, 1–9. https://doi.org/10.3389/fphar.2018.00702.

Gill, M. E., Hu, Y. C., Lin, Y., & Page, D. C. (2011). Licensing of gametogenesis, dependent on RNA binding protein DAZL, as a gateway to sexual differentiation of fetal germ cells. *Proceedings of the National Academy of Sciences of the United States of America*, *108*, 7443–7448. https://doi.org/10.1073/pnas.1104501108.

Gkountela, S., Zhang, K. X., Shafiq, T. A., Liao, W. W., Hargan-Calvopiña, J., Chen, P. Y., et al. (2015). DNA demethylation dynamics in the human prenatal germline. *Cell*, *161*, 1425–1436. https://doi.org/10.1016/j.cell.2015.05.012.

Goll, M. G., & Bestor, T. H. (2005). Eukaryotic cytosine methyltransferases. *Annual Review of Biochemistry*, *74*, 481–514. https://doi.org/10.1146/annurev.biochem.74.010904.153721.

Gong, L., Pan, Y. X., & Chen, H. (2010). Gestational low protein diet in the rat mediates Igf2 gene expression in male offspring via altered hepatic DNA methylation. *Epigenetics*, *5*, 619–626. https://doi.org/10.4161/epi.5.7.12882.

Gong, Y., Xue, Y., Li, X., Zhang, Z., Zhou, W., Marcolongo, P., et al. (2021). Inter- and transgenerational effects of paternal exposure to inorganic arsenic.Pdf. Advanced. *Science*, *8*, 1–14. https://doi.org/10.1002/advs.202002715.

González, B., Gancedo, S. N., Garazatua, S. A. J., Roldán, E., Vitullo, A. D., & González, C. R. (2020). Dopamine receptor D1 contributes to cocaine epigenetic reprogramming of histone modifications in male germ cells. *Frontiers in Cell and Development Biology*, *8*, 1–9. https://doi.org/10.3389/fcell.2020.00216.

Govorko, D., Bekdash, R. A., Zhang, C., & Sarkar, D. K. (2012). Male germline transmits fetal alcohol adverse effect on hypothalamic proopiomelanocortin gene across generations. *Biological Psychiatry*, *72*, 378–388. https://doi.org/10.1016/j.biopsych.2012.04.006.
Griswold, M. D. (2016). Spermatogenesis: The commitment to meiosis. *Physiological Reviews*, *96*, 1–17. https://doi.org/10.1152/physrev.00013.2015.
Guerrero-Bosagna, C., Covert, T. R., Haque, M. M., Settles, M., Nilsson, E. E., Anway, M. D., et al. (2012). Epigenetic transgenerational inheritance of vinclozolin induced mouse adult onset disease and associated sperm epigenome biomarkers. *Reproductive Toxicology*, *34*, 694–707. https://doi.org/10.1016/j.reprotox.2012.09.005.
Guerrero-Bosagna, C., Settles, M., Lucker, B., & Skinner, M. K. (2010). Epigenetic transgenerational actions of vinclozolin on promoter regions of the sperm epigenome. *PLoS One*, 5. https://doi.org/10.1371/journal.pone.0013100.
Guibert, S., Forné, T., & Weber, M. (2012). Global profiling of DNA methylation erasure in mouse primordial germ cells. *Genome Research*, *22*, 633–641. https://doi.org/10.1101/gr.130997.111.partments.
Guo, T., Luo, F., & Lin, Q. (2020). You are affected by what your parents eat: Diet, epigenetics, transgeneration and intergeneration. *Trends in Food Science and Technology*, *100*, 248–261. https://doi.org/10.1016/j.tifs.2020.04.021.
Guo, F., Yan, L., Guo, H., Li, L., Hu, B., Zhao, Y., et al. (2015). The transcriptome and DNA methylome landscapes of human primordial germ cells. *Cell*, *161*, 1437–1452. https://doi.org/10.1016/j.cell.2015.05.015.
Haberle, V., & Stark, A. (2018). Eukaryotic core promoters and the functional basis of transcription initiation. *Nature Reviews. Molecular Cell Biology*, *19*, 621–637. https://doi.org/10.1038/s41580-018-0028-8.
Haggerty, C., Kretzmer, H., Riemenschneider, C., Kumar, A. S., Mattei, A. L., Bailly, N., et al. (2021). Dnmt1 has de novo activity targeted to transposable elements. *Nature Structural and Molecular Biology*. https://doi.org/10.1038/s41594-021-00603-8.
Hajkova, P., Erhardt, S., Lane, N., Haaf, T., El-Maarri, O., Reik, W., et al. (2002). Epigenetic reprogramming in mouse primordial germ cells. *Mechanisms of Development*, *117*, 15–23. Available: https://www.ncbi.nlm.nih.gov/pubmed/12204247.
Hammer, B., Wagner, C., Divac Rankov, A., Reuter, S., Bartel, S., Hylkema, M. N., et al. (2018). In utero exposure to cigarette smoke and effects across generations: A conference of animals on asthma. *Clinical and Experimental Allergy*, *48*, 1378–1390. https://doi.org/10.1111/cea.13283.
Hammoud, S. S., Nix, D. A., Zhang, H., Purwar, J., Carrell, D. T., & Cairns, B. R. (2009). Distinctive chromatin in human sperm packages genes for embryo development. *Nature*, *460*, 473–478. https://doi.org/10.1038/nature08162.
Hanna, C. W., Demond, H., & Kelsey, G. (2018). Epigenetic regulation in development: Is the mouse a good model for the human? *Human Reproduction Update*, *24*, 556–576. https://doi.org/10.1093/humupd/dmy021.
Hanna, C. W., Taudt, A., Huang, J., Gahurova, L., Kranz, A., Andrews, S., et al. (2018). MLL2 conveys transcription-independent H3K4 trimethylation in oocytes. *Nature Structural and Molecular Biology*, *25*, 73–82. https://doi.org/10.1038/s41594-017-0013-5.
Hannon, E., Knox, O., Sugden, K., Burrage, J., Wong, C. C. Y., Belsky, D. W., et al. (2018). Characterizing genetic and environmental influences on variable DNA methylation using monozygotic and dizygotic twins. *PLoS Genetics*, *14*, 1–27. https://doi.org/10.1371/journal.pgen.1007544.
Hao, C., Gely-Pernot, A., Kervarrec, C., Boudjema, M., Becker, E., Khil, P., et al. (2016). Exposure to the widely used herbicide atrazine results in deregulation of global tissue-specific RNA transcription in the third generation and is associated with a global decrease of histone trimethylation in mice. *Nucleic Acids Research*, *44*, 9784–9802. https://doi.org/10.1093/nar/gkw840.

Hauser, M. T., Aufsatz, W., Jonak, C., & Luschnig, C. (2011). Transgenerational epigenetic inheritance in plants. *Biochimica et Biophysica Acta, Gene Regulatory Mechanisms, 1809*, 459–468. https://doi.org/10.1016/j.bbagrm.2011.03.007.

Hayashi, M., Kawaguchi, T., Durcova-Hills, G., & Imai, H. (2018). Generation of germ cells from pluripotent stem cells in mammals. *Reproductive Medicine and Biology., 17*, 107–114. https://doi.org/10.1002/rmb2.12077.

Hayashi, K., Ohta, H., Kurimoto, K., Aramaki, S., & Saitou, M. (2011). Reconstitution of the mouse germ cell specification pathway in culture by pluripotent stem cells. *Cell, 146*, 519–532. https://doi.org/10.1016/j.cell.2011.06.052.

He, F., Lidow, I. A., & Lidow, M. S. (2006). Consequences of paternal cocaine exposure in mice. *Neurotoxicology and Teratology, 28*, 198–209. https://doi.org/10.1016/j.ntt.2005.12.003.

Heintzman, N., Stuart, R., Hon, G., Fu, Y., Ching, C., Hawkins, R. D., et al. (2007). Distinct and predictive chromatin signatures of transcriptional promoters and enhancers in the human genome. *Nature Genetics, 39*, 311–318. https://doi.org/10.1038/ng1966.

Hernández-Sánchez, J., Valles, S., & Bardullas, U. (2022). Linking arsenic, DNA methylation biomarkers, and transgenerational neurotoxicity: Modeling in zebrafish. Biomarkers. *Toxicology*, 1–24. https://doi.org/10.1007/978-3-030-87225-0_59-1.

Heyt, K. M., & Thakur, J. (2021). Regulation of epigenetic processes by non-coding RNAs. *Nucleus, 64*, 285–301. https://doi.org/10.1007/s13237-021-00372-1.

Hikabe, O., Hamazaki, N., Nagamatsu, G., Obata, Y., Hirao, Y., Hamada, N., et al. (2016). Reconstitution in vitro of the entire cycle of the mouse female germ line. *Nature, 539*, 299–303. https://doi.org/10.1038/nature20104.

Hiura, H., Obata, Y., Komiyama, J., Shirai, M., & Kono, T. (2006). Oocyte growth-dependent progression of maternal imprinting in mice. *Genes to Cells, 11*, 353–361. https://doi.org/10.1111/j.1365-2443.2006.00943.x.

Holoch, D., & Moazed, D. (2015). RNA-mediated epigenetic regulation of gene expression. *Nature Reviews. Genetics, 16*, 71–84. https://doi.org/10.1038/nrg3863.

Hon, G. C., Hawkins, R. D., & Ren, B. (2009). Predictive chromatin signatures in the mammalian genome. *Human Molecular Genetics, 18*, 12–14. https://doi.org/10.1093/hmg/ddp409.

Horsthemke, B. (2018). A critical view on transgenerational epigenetic inheritance in humans. *Nature Communications, 9*, 1–4. https://doi.org/10.1038/s41467-018-05445-5.

Inoue, A., Jiang, L., Lu, F., Suzuki, T., & Zhang, Y. (2017). Maternal H3K27me3 controls DNA methylation-independent imprinting. *Nature, 547*, 419–424. https://doi.org/10.1038/nature23262.

Jawaid, A., & Mansuy, I. M. (2019). Inter- and transgenerational inheritance of behavioral phenotypes. *Current Opinion in Behavioral Sciences, 25*, 96–101. https://doi.org/10.1016/j.cobeha.2018.12.004.

Jodar, M. (2019). Sperm and seminal plasma RNAs: What roles do they play beyond fertilization? *Reproduction, 158*, R113–R123.

Joubert, B. R., Den Dekker, H. T., Felix, J. F., Bohlin, J., Ligthart, S., Beckett, E., et al. (2016). Maternal plasma folate impacts differential DNA methylation in an epigenome-wide meta-analysis of newborns. Nature. *Communications*, 7. https://doi.org/10.1038/ncomms10577.

Joubert, B. R., Håberg, S. E., Nilsen, R. M., Wang, X., Vollset, S. E., Murphy, S. K., et al. (2012). 450K epigenome-wide scan identifies differential DNA methylation in newborns related to maternal smoking during pregnancy. *Environmental Health Perspectives, 120*, 1425–1431. https://doi.org/10.1289/ehp.1205412.

Kaikkonen, M. U., Lam, M. T. Y., & Glass, C. K. (2011). Non-coding RNAs as regulators of gene expression and epigenetics. *Cardiovascular Research, 90*, 430–440. https://doi.org/10.1093/cvr/cvr097.

Kane, A. E., & Sinclair, D. A. (2019). Epigenetic changes during aging and their reprogramming potential. *Critical Reviews in Biochemistry and Molecular Biology*, *54*, 61–83. https://doi.org/10.1080/10409238.2019.1570075.

Kaneda, M., Okano, M., Hata, K., Sado, T., Tsujimoto, H., Li, E., et al. (2004). Essential role for de novo DNA methyltransferase Dnmt3a in paternal and maternal imprinting. *Nature*, *429*, 900–903. https://doi.org/10.1038/nature02633.

Kaspar, D., Hastreiter, S., Irmler, M., Hrabé de Angelis, M., & Beckers, J. (2020). Nutrition and its role in epigenetic inheritance of obesity and diabetes across generations. *Mammalian Genome*, *31*, 119–133. https://doi.org/10.1007/s00335-020-09839-z.

Kato, Y., Kaneda, M., Hata, K., Kumaki, K., Hisano, M., Kohara, Y., et al. (2007). Role of the Dnmt3 family in de novo methylation of imprinted and repetitive sequences during male germ cell development in the mouse. *Human Molecular Genetics*, *16*, 2272–2280. https://doi.org/10.1093/hmg/ddm179.

Kobayashi, H., Sakurai, T., Imai, M., Takahashi, N., Fukuda, A., Yayoi, O., et al. (2012). Contribution of intragenic DNA methylation in mouse gametic DNA methylomes to establish oocyte-specific heritable marks. *PLoS Genetics*, 8. https://doi.org/10.1371/journal.pgen.1002440.

Kobayashi, H., Sakurai, T., Miura, F., Imai, M., Mochiduki, K., Yanagisawa, E., et al. (2013). High-resolution DNA methylome analysis of primordial germ cells identifies gender-specific reprogramming in mice. *Genome Research*, *23*, 616–627. https://doi.org/10.1101/gr.148023.112.Freely.

Komeya, M., Sato, T., & Ogawa, T. (2018). In vitro spermatogenesis: A century-long research journey still half way. *Reproductive Medicine and Biology.*, *17*, 407–420. https://doi.org/10.1002/rmb2.12225.

Kota, S. K., & Feil, R. (2010). Epigenetic transitions in germ cell development and meiosis. *Developmental Cell*, *19*, 675–686. https://doi.org/10.1016/j.devcel.2010.10.009.

Krauss-Etschmann, S., Meyer, K. F., Dehmel, S., & Hylkema, M. N. (2015). Inter-and trans-generational epigenetic inheritance: Evidence in asthma and COPD? *Clinical Epigenetics*, 7, 1–13. https://doi.org/10.1186/s13148-015-0085-1.

Kremsky, I., & Corces, V. G. (2020). Protection from DNA re-methylation by transcription factors in primordial germ cells and pre-implantation embryos can explain trans-generational epigenetic inheritance. *Genome Biology*, *21*, 118. https://doi.org/10.1186/s13059-020-02036-w.

Kubo, N., Toh, H., Shirane, K., Shirakawa, T., Kobayashi, H., Sato, T., et al. (2015). DNA methylation and gene expression dynamics during spermatogonial stem cell differentiation in the early postnatal mouse testis. *BMC Genomics*, *16*, 1–17. https://doi.org/10.1186/s12864-015-1833-5.

Kurimoto, K., Yabuta, Y., Hayashi, K., Ohta, H., Kiyonari, H., Mitani, T., et al. (2015). Quantitative dynamics of chromatin remodeling during germ cell specification from mouse embryonic stem cells. *Cell Stem Cell*, *16*, 517–532. https://doi.org/10.1016/j.stem.2015.03.002.

Larose, H., Shami, A. N., Abbott, H., Manske, G., Lei, L., & Hammoud, S. S. (2019). Gametogenesis: A journey from inception to conception. In *Current topics in developmental biology* (pp. 257–310). Elsevier Inc. https://doi.org/10.1016/bs.ctdb.2018.12.006.

Legoff, L., D'Cruz, S. C., Tevosian, S., & Smagulova, F. (2019). Long noncoding RNA and epigenetic inheritance. Transgenerational. *Epigenetics*, 213–248. https://doi.org/10.1016/b978-0-12-816363-4.00010-9.

Lesch, B. J., & Page, D. C. (2012). Genetics of germ cell development. *Nature Reviews. Genetics*. https://doi.org/10.1038/nrg3294.

Levine, A., Huang, Y. Y., Drisaldi, B., Griffin, E. A., Pollak, D. D., Xu, S., et al. (2011). Molecular mechanism for a gateway drug: Epigenetic changes initiated by nicotine prime gene expression by cocaine. *Science Translational Medicine*, *3*, 1–10. https://doi.org/10.1126/scitranslmed.3003062.

Li, X., Egervari, G., Wang, Y., Berger, S. L., & Lu, Z. (2018). Regulation of chromatin and gene expression by metabolic enzymes and metabolites. *Nature Reviews. Molecular Cell Biology*, *19*, 563–578. https://doi.org/10.1038/s41580-018-0029-7.

Li, Y.-F., Langholz, B., Salam, M. T., & Gilliland, F. D. (2005). Maternal and grandmaternal smoking patterns are associated with early childhood asthma. *Chest*, *127*, 1232–1241. https://doi.org/10.1016/s0012-3692(15)34472-x.

Lismer, A., Dumeaux, V., Lafleur, C., Lambrot, R., Brind'Amour, J., Lorincz, M. C., et al. (2021). Histone H3 lysine 4 trimethylation in sperm is transmitted to the embryo and associated with diet-induced phenotypes in the offspring. *Developmental Cell*, *56*, 671–686.e6. https://doi.org/10.1016/j.devcel.2021.01.014.

Lu, S. C., & Mato, J. M. (2005). Role of methionine adenosyltransferase and S-adenosylmethionine in alcohol-associated liver cancer. *Alcohol*, *35*, 227–234. https://doi.org/10.1016/j.alcohol.2005.03.011.

Lucifero, D., La Salle, S., Bourc'his, D., Martel, J., Bestor, T. H., & Trasler, J. M. (2007). Coordinate regulation of DNA methyltransferase expression during oogenesis. *BMC Developmental Biology*, 7, 1–14. https://doi.org/10.1186/1471-213X-7-36.

Lumey, L. H., Stein, A. D., & Susser, E. (2011). Prenatal famine and adult health L.H. *Annual Review of Public Health*, *32*, 1–26. https://doi.org/10.1146/annurev-publhealth-031210-101230.

Luo, Y., & Yu, Y. (2022). Research advances in gametogenesis and embryogenesis using pluripotent stem cells. *Frontiers in Cell and Development Biology*, *9*, 1–14. https://doi.org/10.3389/fcell.2021.801468.

Lyko, F. (2018). The DNA methyltransferase family: A versatile toolkit for epigenetic regulation. *Nature Reviews. Genetics*, *19*, 81–92. https://doi.org/10.1038/nrg.2017.80.

Magnus, M. C., Håberg, S. E., Karlstad, Ø., Nafstad, P., London, S. J., & Nystad, W. (2015). Grandmother's smoking when pregnant with the mother and asthma in the grandchild: The Norwegian mother and child cohort study. *Thorax*, *70*, 237–243. https://doi.org/10.1136/thoraxjnl-2014-206438.

Magnúsdóttir, E., & Surani, M. A. (2014). How to make a primordial germ cell. *Development (Cambridge).*, *141*, 245–252. https://doi.org/10.1242/dev.098269.

Maloney, B., Bayon, B. L., Zawia, N. H., & Lahiri, D. K. (2018). Latent consequences of early-life lead (pb) exposure and the future: Addressing the pb crisis. *Neurotoxicology*, *68*, 126–132. https://doi.org/10.1016/j.neuro.2018.06.016.

Manikkam, M., Tracey, R., Guerrero-Bosagna, C., & Skinner, M. K. (2012). Dioxin (TCDD) induces epigenetic transgenerational inheritance of adult onset disease and sperm epimutations. *PLoS One*, 7, e46249-1–15. https://doi.org/10.1371/journal.pone.0046249.

Manikkam, M., Tracey, R., Guerrero-Bosagna, C., & Skinner, M. K. (2013). Plastics derived endocrine disruptors (BPA, DEHP and DBP) induce epigenetic transgenerational inheritance of obesity, reproductive disease and sperm epimutations. *PLoS One*, 8. https://doi.org/10.1371/journal.pone.0055387.

Marczylo, E. L., Jacobs, M. N., & Gant, T. W. (2016). Environmentally induced epigenetic toxicity: Potential public health concerns. *Critical Reviews in Toxicology*, *46*, 676–700. https://doi.org/10.1080/10408444.2016.1175417.

McCarthy, D. M., Lowe, S. E., Morgan, T. J., Cannon, E. N., Biederman, J., Spencer, T. J., et al. (2020). Transgenerational transmission of behavioral phenotypes produced by exposure of male mice to saccharin and nicotine. *Scientific Reports*, *10*, 1–14. https://doi.org/10.1038/s41598-020-68883-6.

McCarthy, D. M., Morgan, T. J., Jr., Lowe, S. E., Williamson, M. J., Spencer, T. J., Biederman, J., et al. (2018). Nicotine exposure of male mice produces behavioral impairment in multiple generations of descendants. *PLoS Biology*, *16*, e2006497. https://doi.org/10.1371/journal.pbio.2006497.

McPherson, N. O., Fullston, T., Kang, W. X., Sandeman, L. Y., Corbett, M. A., Owens, J. A., et al. (2016). Paternal under-nutrition programs metabolic syndrome in offspring which can be reversed by antioxidant/vitamin food fortification in fathers. *Scientific Reports*, *6*, 1–14. https://doi.org/10.1038/srep27010.

McRae, A. F., Powell, J. E., Henders, A. K., Bowdler, L., Hemani, G., Shah, S., et al. (2014). Contribution of genetic variation to transgenerational inheritance of DNA methylation. *Genome Biology*, *15*, R73. https://doi.org/10.1186/gb-2014-15-5-r73.

Mei, H., Kozuka, C., Hayashi, R., Kumon, M., Koseki, H., & Inoue, A. (2021). H2AK119ub1 guides maternal inheritance and zygotic deposition of H3K27me3 in mouse embryos. *Nature Genetics*, *53*, 539–550. https://doi.org/10.1038/s41588-021-00820-3.

Mews, P., Egervari, G., Nativio, R., Sidoli, S., Donahue, G., Lombroso, S. I., et al. (2019). Alcohol metabolism contributes to brain histone acetylation. *Nature*, *574*, 717–721. https://doi.org/10.1038/s41586-019-1700-7.

Mocarelli, P., Gerthoux, P. M., Needham, L. L., Patterson, D. G., Limonta, G., Falbo, R., et al. (2011). Perinatal exposure to low doses of dioxin can permanently impair human semen quality. *Environmental Health Perspectives*, *119*, 713–718. https://doi.org/10.1289/ehp.1002134.

Mu, Y., Hu, X., Yang, P., Sun, L., Gu, W., & Zhang, M. (2021). The effects of cadmium on the development of Drosophila and its transgenerational inheritance effects. *Toxicology*, *462*, 1–10. https://doi.org/10.1016/j.tox.2021.152931.

Murphy, S. K., Itchon-Ramos, N., Visco, Z., Huang, Z., Grenier, C., Schrott, R., et al. (2018). Cannabinoid exposure and altered DNA methylation in rat and human sperm. *Epigenetics*, *13*, 1208–1221. https://doi.org/10.1080/15592294.2018.1554521.

Murr, R. (2010). Interplay between different epigenetic modifications and mechanisms. *Advances in Genetics*, *70*, 101–141. https://doi.org/10.1016/B978-0-12-380866-0.60005-8.

Nagano, M., Hu, B., Yokobayashi, S., Yamamura, A., Umemura, F., Coradin, M., et al. (2022). Nucleome programming is required for the foundation of totipotency in mammalian germline development. *The EMBO Journal.*, *41*, 1–34. https://doi.org/10.15252/embj.2022110600.

Nakamura, T., Liu, Y. J., Nakashima, H., Umehara, H., Inoue, K., Matoba, S., et al. (2012). PGC7 binds histone H3K9me2 to protect against conversion of 5mC to 5hmC in early embryos. *Nature*, *486*, 415–419. https://doi.org/10.1038/nature11093.

Nava-Rivera, L. E., Betancourt-Martínez, N. D., Lozoya-Martínez, R., Carranza-Rosales, P., Guzmán-Delgado, N. E., Carranza-Torres, I. E., et al. (2021). Transgenerational effects in DNA methylation, genotoxicity and reproductive phenotype by chronic arsenic exposure. *Scientific Reports*, *11*, 8276. https://doi.org/10.1038/s41598-021-87677-y.

Nestler, E. J. (2014). Epigenetic mechanisms of drug addiction. *Neuropharmacology*, *76*, 259–268. https://doi.org/10.1016/j.neuropharm.2013.04.004.

Ng, S. F., Lin, R. C. Y., Laybutt, D. R., Barres, R., Owens, J. A., & Morris, M. J. (2010). Chronic high-fat diet in fathers programs β 2-cell dysfunction in female rat offspring. *Nature*, *467*, 963–966. https://doi.org/10.1038/nature09491.

Nichols, J., & Smith, A. (2009). Naive and primed pluripotent states. *Cell Stem Cell*, *4*, 487–492. https://doi.org/10.1016/j.stem.2009.05.015.

Nishiyama, A., Mulholland, C. B., Bultmann, S., Kori, S., Endo, A., Saeki, Y., et al. (2020). Two distinct modes of DNMT1 recruitment ensure stable maintenance DNA methylation. *Nature Communications*, *11*, 1–7. https://doi.org/10.1038/s41467-020-15006-4.

Novakovic, B., Lewis, S., Halliday, J., Kennedy, J., Burgner, D. P., Czajko, A., et al. (2019). Assisted reproductive technologies are associated with limited epigenetic variation at birth that largely resolves by adulthood. Nature. *Communications*, 10. https://doi.org/10.1038/s41467-019-11929-9.

Nowacka-Woszuk, J., Szczerbal, I., Malinowska, A. M., & Chmurzynska, A. (2018). Transgenerational effects of prenatal restricted diet on gene expression and histone modifications in the rat. *PLoS One*, *13*, 1–14. https://doi.org/10.1371/journal.pone.0193464.

Ohinata, Y., Sano, M., Shigeta, M., Yamanaka, K., & Saitou, M. (2008). A comprehensive, non-invasive visualization of primordial germ cell development in mice by the Prdm1-mVenus and Dppa3-ECFP double transgenic reporter. *Reproduction*, *136*, 503–514. https://doi.org/10.1530/REP-08-0053.

Ohta, H., Kurimoto, K., Okamoto, I., Nakamura, T., Yabuta, Y., Miyauchi, H., et al. (2017). In vitro expansion of mouse primordial germ cell- like cells recapitulates an epigenetic blank slate. *The EMBO Journal.*, *36*, 1888–1907. https://doi.org/10.15252/embj.201695862.

Okada, Y., & Yamaguchi, K. (2017). Epigenetic modifications and reprogramming in paternal pronucleus: Sperm, preimplantation embryo, and beyond. *Cellular and Molecular Life Sciences*, *74*, 1957–1967. https://doi.org/10.1007/s00018-016-2447-z.

Okano, M., Bell, D. W., Haber, D. A., & Li, E. (1999). DNA methyltransferases Dnmt3a and Dnmt3b are essential for De novo methylation and mammalian development. *Cell*, *99*, 247–257.

Ong, M.-L., Lin, X., & Holbrook, J. D. (2015). Measuring epigenetics as the mediator of gene/environment interactions in DOHaD. *Journal of Developmental Origins of Health and Disease*, *6*, 10–16. https://doi.org/10.1017/S2040174414000506.

Ooi, S. K. T., Jiang, H., Kang, Y., & Allard, P. (2021). Examining the developmental trajectory of an in vitro model of mouse primordial germ cells following exposure to environmentally relevant bisphenol a levels. *Environmental Health Perspectives*, *129*, 97013. https://doi.org/10.1289/EHP8196.

Osborne, A. J., Pearson, J. F., Noble, A. J., Gemmell, N. J., Horwood, L. J., Boden, J. M., et al. (2020). Genome-wide DNA methylation analysis of heavy cannabis exposure in a New Zealand longitudinal cohort. *Translational Psychiatry*, *10*, 1–10. https://doi.org/10.1038/s41398-020-0800-3.

Padmanabhan, N., Jia, D., Geary-Joo, C., Wu, X., Ferguson-Smith, A. C., Fung, E., et al. (2013). Mutation in folate metabolism causes epigenetic instability and transgenerational effects on development. *Cell*, *155*, 81–93. https://doi.org/10.1016/j.cell.2013.09.002.

Pilsner, J. R., Shershebnev, A., Medvedeva, Y. A., Suvorov, A., Wu, H., Goltsov, A., et al. (2018). Peripubertal serum dioxin concentrations and subsequent sperm methylome profiles of young Russian adults. *Reproductive Toxicology*, *78*, 40–49. https://doi.org/10.1016/j.reprotox.2018.03.007.

Pombo, A., & Dillon, N. (2015). Three-dimensional genome architecture: Players and mechanisms. *Nature Reviews. Molecular Cell Biology*, *16*, 245–257. https://doi.org/10.1038/nrm3965.

Prini, P., Penna, F., Sciuccati, E., Alberio, T., & Rubino, T. (2017). Chronic Δ9-THC exposure differently affects histone modifications in the adolescent and adult rat brain. *International Journal of Molecular Sciences*, 18. https://doi.org/10.3390/ijms18102094.

Prini, P., Rusconi, F., Zamberletti, E., Gabaglio, M., Penna, F., Fasano, M., et al. (2018). Adolescent THC exposure in female rats leads to cognitive deficits through a mechanism involving chromatin modifications in the prefrontal cortex. *Journal of Psychiatry & Neuroscience*, *43*, 87–101. https://doi.org/10.1503/jpn.170082.

Radford, E. J., Ito, M., Shi, H., Corish, J. A., Yamazawa, K., Isganaitis, E., et al. (2014). In utero undernourishment perturbs the adult sperm methylome and intergenerational metabolism. *Science*, *345*, 1255903-1–8. https://doi.org/10.1126/science.1255903.

Ranjit, N., Siefert, K., & Padmanabhan, V. (2010). Bisphenol-a and disparities in birth outcomes: A review and directions for future research. *Journal of Perinatology*, *30*, 2–9. https://doi.org/10.1038/jp.2009.90.

Ray, P. D., Yosim, A., & Fry, R. C. (2014). Incorporating epigenetic data into the risk assessment process for the toxic metals arsenic, cadmium, chromium, lead, and mercury: Strategies and challenges. *Frontiers in Genetics*, *5*, 1–26. https://doi.org/10.3389/fgene.2014.00201.

Rehan, V. K., Liu, J., Sakurai, R., & Torday, J. S. (2013). Perinatal nicotine-induced transgenerational asthma. *American Journal of Physiology. Lung Cellular and Molecular Physiology*, *305*, L501–L507. https://doi.org/10.1152/ajplung.00078.2013.

Reichard, J. F., & Puga, A. (2010). Effects of arsenic exposure on DNA methylation and epigenetic gene regulation. *Epigenomics*, *2*, 87–104. https://doi.org/10.2217/epi.09.45.

Ren, X., McHale, C. M., Skibola, C. F., Smith, A. H., Smith, M. T., & Zhang, L. (2011). An emerging role for epigenetic dysregulation in arsenic toxicity and carcinogenesis. *Environmental Health Perspectives*, *119*, 11–19. https://doi.org/10.1289/ehp.1002114.

Rochester, J. R. (2013). Bisphenol a and human health: A review of the literature. *Reproductive Toxicology*, *42*, 132–155. https://doi.org/10.1016/j.reprotox.2013.08.008.

Rompala, G. R., Mounier, A., Wolfe, C. M., Lin, Q., Lefterov, I., & Homanics, G. E. (2018). Heavy chronic intermittent ethanol exposure alters small noncoding RNAs in mouse sperm and epididymosomes. *Frontiers in Genetics*, *9*, 1–14. https://doi.org/10.3389/fgene.2018.00032.

Sabari, B. R., Zhang, D., Allis, C. D., & Zhao, Y. (2017). Metabolic regulation of gene expression through histone acylations. *Nature Reviews. Molecular Cell Biology*, *18*, 90–101. https://doi.org/10.1038/nrm.2016.140.

Saidur Rahman, M. D., Pang, W. K., Ryu, D. Y., Park, Y. J., & Pang, M. G. (2020). Multigenerational and transgenerational impact of paternal bisphenol a exposure on male fertility in a mouse model. *Human Reproduction*, *35*, 1740–1752. https://doi.org/10.1093/humrep/deaa139.

Saitou, M., & Yamaji, M. (2012). Primordial germ cells in mice. *Cold Spring Harbor Perspectives in Biology*, 4. https://doi.org/10.1101/cshperspect.a008375.

Sánchez, F., & Smitz, J. (2012). Molecular control of oogenesis. *Biochimica et Biophysica Acta, Molecular Basis of Disease*, *1822*, 1896–1912. https://doi.org/10.1016/j.bbadis.2012.05.013.

Sarker, G., Berrens, R., von Arx, J., Pelczar, P., Reik, W., Wolfrum, C., et al. (2018). Transgenerational transmission of hedonic behaviors and metabolic phenotypes induced by maternal overnutrition. *Translational Psychiatry*, *8*, 1–16. https://doi.org/10.1038/s41398-018-0243-2.

Sasaki, H., & Matsui, Y. (2008). Epigenetic events in mammalian germ-cell development : Reprogramming and beyond. *Nature Reviews. Genetics*, 9. https://doi.org/10.1038/nrg2295.

Schrott, R., & Murphy, S. K. (2020). Cannabis use and the sperm epigenome: A budding concern? *Environmental Epigenetics.*, *6*, 1–10. https://doi.org/10.1093/eep/dvaa002.

Schuster, A., Skinner, M. K., & Yan, W. (2016). Ancestral vinclozolin exposure alters the epigenetic transgenerational inheritance of sperm small noncoding RNAs. *Environmental Epigenetics.*, *2*, 1–10. https://doi.org/10.1093/eep/dvw001.

Seisenberger, S., Andrews, S., Krueger, F., Arand, J., Walter, J., Santos, F., et al. (2012). The dynamics of genome-wide DNA methylation reprogramming in mouse primordial germ cells. *Molecular Cell*, *48*, 849–862. https://doi.org/10.1016/j.molcel.2012.11.001.

Seki, Y., Yamaji, M., Yabuta, Y., Sano, M., Shigeta, M., Matsui, Y., et al. (2007). Cellular dynamics associated with the genome-wide epigenetic reprogramming in migrating primordial germ cells in mice. *Development*, *134*, 2627–2638. https://doi.org/10.1242/dev.005611.

Sen, A., Heredia, N., Senut, M. C., Land, S., Hollocher, K., Lu, X., et al. (2015). Multigenerational epigenetic inheritance in humans: DNA methylation changes associated with maternal exposure to lead can be transmitted to the grandchildren. *Scientific Reports*, *5*, 1–10. https://doi.org/10.1038/srep14466.

Senaldi, L., & Smith-Raska, M. (2020). Evidence for germline non-genetic inheritance of human phenotypes and diseases. *Clinical Epigenetics*, *12*, 1–12. https://doi.org/10.1186/s13148-020-00929-y.

Sendžikaitė, G., & Kelsey, G. (2019). The role and mechanisms of DNA methylation in the oocyte. *Essays in Biochemistry*, *63*, 691–705. https://doi.org/10.1042/EBC20190043.

Sha, Q. Q., Jiang, Y., Yu, C., Xiang, Y., Dai, X. X., Jiang, J. C., et al. (2020). CFP1-dependent histone H3K4 trimethylation in murine oocytes facilitates ovarian follicle recruitment and ovulation in a cell-nonautonomous manner. *Cellular and Molecular Life Sciences*, 77, 2997–3012. https://doi.org/10.1007/s00018-019-03322-y.

Shan, C. M., Fang, Y., & Jia, S. (2021). Leaving histone unturned for epigenetic inheritance. *FEBS Journal*, 1–11. https://doi.org/10.1111/febs.16260.

Sharif, J., Muto, M., Takebayashi, S.-I., Suetake, I., Iwamatsu, A., Endo, T. A., et al. (2007). The SRA protein Np95 mediates epigenetic inheritance by recruiting Dnmt1 to methylated DNA. *Nature*, *450*, 908–912. https://doi.org/10.1038/nature06397.

Shea, J. M., Serra, R. W., Carone, B. R., Shulha, H. P., Kucukural, A., Ziller, M. J., et al. (2015). Genetic and epigenetic variation, but not diet, Shape the Sperm Methylome. *Developmental Cell*, *35*, 750–758. https://doi.org/10.1016/j.devcel.2015.11.024.

Shirane, K., Miura, F., Ito, T., & Lorincz, M. C. (2020). NSD1-deposited H3K36me2 directs de novo methylation in the mouse male germline and counteracts Polycomb-associated silencing. *Nature Genetics*, *52*, 1088–1098. https://doi.org/10.1038/s41588-020-0689-z.

Siklenka, K., Erkek, S., Godmann, M., Lambrot, R., McGraw, S., Lafleur, C., et al. (2015). Disruption of histone methylation in developing sperm impairs offspring health transgenerationally. *Science*, *350*, aab2006. https://doi.org/10.1126/science.aab2006.

Skene, P. J., & Henikoff, S. (2017). An efficient targeted nuclease strategy for high-resolution mapping of DNA binding sites. *eLife*, *6*, 1–35. https://doi.org/10.7554/eLife.21856.

Smeester, L., Rager, J. E., Bailey, K. A., Guan, X., Smith, N., García-Vargas, G., et al. (2011). Epigenetic changes in individuals with arsenicosis. *Chemical Research in Toxicology*, *24*, 165–167. https://doi.org/10.1021/tx1004419.

Smith, A. H., Hopenhayn-Rich, C., Bates, M. N., Goeden, H. M., Hertz-Picciotto, I., Duggan, H. M., et al. (1992). Cancer risks from arsenic in drinking water. *Environmental Health Perspectives*, *97*, 259–267. https://doi.org/10.1289/ehp.9297259.

Sobolewski, M., Abston, K., Conrad, K., Marvin, E., Harvey, K., Susiarjo, M., et al. (2020). Lineage-and sex-dependent behavioral and biochemical transgenerational consequences of developmental exposure to lead, prenatal stress, and combined lead and prenatal stress in mice. *Environmental Health Perspectives*, *128*, 1–14. https://doi.org/10.1289/EHP4977.

Soubry, A. (2015). Epigenetic inheritance and evolution: A paternal perspective on dietary influences. *Progress in Biophysics and Molecular Biology*, *118*, 79–85. https://doi.org/10.1016/j.pbiomolbio.2015.02.008.

Stewart, K. R., Veselovska, L., Kim, J., Huang, J., Saadeh, H., Tomizawa, S. I., et al. (2015). Dynamic changes in histone modifications precede de novo DNA methylation in oocytes. *Genes and Development*, *29*, 2449–2462. https://doi.org/10.1101/gad.271353.115.

Strahl, B. D., & Allis, D. C. (2000). The language of covalent histone modifications. *Nature*, *403*, 41–45. https://doi.org/10.1038/47412.

Svanes, C., Bertelsen, R. J., Accordini, S., Holloway, J. W., Júlíússon, P., Boateng, E., et al. (2021). Exposures during the prepuberty period and future offspring's health: Evidence from human cohort studies. *Biology of Reproduction*, *105*, 667–680. https://doi.org/10.1093/biolre/ioab158.

Swinford-Jackson, S. E., Fant, B., Wimmer, M. E., Chan, D., Knouse, M. C., Sarmiento, M., et al. (2022). Cocaine-induced changes in sperm Cdkn1a methylation are associated with cocaine resistance in male offspring. *The Journal of neuroscience : The Official Journal of the Society for Neuroscience.*, *42*, 2905–2916. https://doi.org/10.1523/JNEUROSCI.3172-20.2022.

Szutorisz, H., DiNieri, J. A., Sweet, E., Egervari, G., Michaelides, M., Carter, J. M., et al. (2014). Parental THC exposure leads to compulsive heroin-seeking and altered striatal synaptic plasticity in the subsequent generation. *Neuropsychopharmacology, 39*, 1315–1323. https://doi.org/10.1038/npp.2013.352.

Szutorisz, H., & Hurd, Y. L. (2018). High times for cannabis: Epigenetic imprint and its legacy on brain and behavior. *Neuroscience and Biobehavioral Reviews, 85*, 93–101. https://doi.org/10.1016/j.neubiorev.2017.05.011.

Tan, M., Luo, H., Lee, S., Jin, F., Yang, J. S., Montellier, E., et al. (2011). Identification of 67 histone marks and histone lysine crotonylation as a new type of histone modification. *Cell, 146*, 1016–1028. https://doi.org/10.1016/j.cell.2011.08.008.

Tang, W. W. C., Kobayashi, T., Irie, N., Dietmann, S., & Surani, M. A. (2016). Specification and epigenetic programming of the human germ line. *Nature Reviews. Genetics*. https://doi.org/10.1038/nrg.2016.88.

Titus, L., Hatch, E. E., Drake, K. M., Parker, S. E., Hyer, M., Palmer, J. R., et al. (2019). Reproductive and hormone-related outcomes in women whose mothers were exposed in utero to diethylstilbestrol (DES): A report from the US National Cancer Institute DES third generation study. *Reproductive Toxicology, 84*, 32–38. https://doi.org/10.1016/j.reprotox.2018.12.008.

Tobi, E. W., Goeman, J. J., Monajemi, R., Gu, H., Putter, H., Zhang, Y., et al. (2014). DNA methylation signatures link prenatal famine exposure to growth and metabolism. *Nature Communications, 5*, 5592. https://doi.org/10.1038/ncomms6592.

Toledo-Rodriguez, M., Lotfipour, S., Leonard, G., Perron, M., Richer, L., Veillette, S., et al. (2010). Maternal smoking during pregnancy is associated with epigenetic modifications of the brain-derived neurotrophic Factor-6 exon in adolescent offspring. *American Journal of Medical Genetics, 153B*, 1350–1354.

Tomizawa, S. I., Nowacka-Woszuk, J., & Kelsey, G. (2012). DNA methylation establishment during oocyte growth: Mechanisms and significance. *International Journal of Developmental Biology, 56*, 867–875. https://doi.org/10.1387/ijdb.120152gk.

Travis, C. C., & Hattemer-Frey, H. A. (1991). Human exposure to dioxin. *The Science of the Total Environment., 104*, 97–127.

Tucci, V., Isles, A. R., Kelsey, G., Ferguson-Smith, A. C., Bartolomei, M. S., Benvenisty, N., et al. (2019). Genomic imprinting and physiological processes in mammals. *Cell, 176*, 952–965. https://doi.org/10.1016/j.cell.2019.01.043.

Van Der Heijden, G. W., Ramos, L., Baart, E. B., Van Den Berg, I. M., Derijck, A. A. H. A., Van Der Vlag, J., et al. (2008). Sperm-derived histones contribute to zygotic chromatin in humans. *BMC Developmental Biology, 8*, 6–11. https://doi.org/10.1186/1471-213X-8-34.

van Otterdijk, S. D., & Michels, K. B. (2016). Transgenerational epigenetic inheritance in mammals: How good is the evidence? *FASEB Journal, 30*, 2457–2465. https://doi.org/10.1096/fj.201500083.

Vandegehuchte, M. B., Lemière, F., Vanhaecke, L., Vanden Berghe, W., & Janssen, C. R. (2010). Direct and transgenerational impact on Daphnia magna of chemicals with a known effect on DNA methylation. *Comparative Biochemistry and Physiology, Part C: Toxicology & Pharmacology, 151*, 278–285. https://doi.org/10.1016/j.cbpc.2009.11.007.

Vandenberg, L. N., Hauser, R., Marcus, M., Olea, N., & Welshons, W. V. (2007). Human exposure to bisphenol a (BPA). *Reproductive Toxicology, 24*, 139–177. https://doi.org/10.1016/j.reprotox.2007.07.010.

Vassoler, F. M., Johnson, N. L., & Byrnes, E. M. (2013). Female adolescent exposure to cannabinoids causes transgenerational effects on morphine sensitization in female offspring in the absence of in utero exposure. *Journal of Psychopharmacology, 27*, 1015–1022. https://doi.org/10.1177/0269881113503504.

Vassoler, F. M., White, S. L., Schmidt, H. D., Sadri-Vakili, G., & Christopher, P. R. (2013). Epigenetic inheritance of a cocaine-resistance phenotype. *Nature Neuroscience, 16*, 42–47. https://doi.org/10.1038/nn.3280.

Veenendaal, M. V. E., Painter, R. C., De Rooij, S. R., Bossuyt, P. M. M., Van Der Post, J. A. M., Gluckman, P. D., et al. (2013). Transgenerational effects of prenatal exposure to the 1944-45 Dutch famine. *BJOG : An International Journal of Obstetrics and Gynaecology*, *120*, 548–554. https://doi.org/10.1111/1471-0528.12136.

Velazquez, M. A., Fleming, T. P., & Watkins, A. J. (2019). Periconceptional environment and the developmental origins of disease. *Journal of Endocrinology*, *242*, T33–T49. https://doi.org/10.1530/JOE-18-0676.

Venkatesh, S., & Workman, J. L. (2015). Histone exchange, chromatin structure and the regulation of transcription. *Nature Reviews. Molecular Cell Biology*, *16*, 178–189. https://doi.org/10.1038/nrm3941.

Verdikt, R., & Allard, P. (2021). Metabolo-epigenetics: The interplay of metabolism and epigenetics during early germ cells development. *Biology of Reproduction*. https://doi.org/10.1093/biolre/ioab118.

Veselovska, L., Smallwood, S. A., Saadeh, H., Stewart, K. R., Krueger, F., Maupetit Méhouas, S., et al. (2015). Deep sequencing and de novo assembly of the mouse oocyte transcriptome define the contribution of transcription to the DNA methylation landscape. *Genome Biology*, *16*, 1–17. https://doi.org/10.1186/s13059-015-0769-z.

Vipin, V. A., Blesson, C. S., & Yallampalli, C. (2022). Maternal low protein diet and fetal programming of lean type 2 diabetes. *World Journal of Diabetes*, *13*, 185–202. https://doi.org/10.4239/wjd.v13.i3.185.

Volkow, N. D., Han, B., Compton, W. M., & McCance-Katz, E. F. (2019). Self-reported medical and nonmedical Cannabis use among pregnant women in the United States. *JAMA : The Journal of the American Medical Association*, *322*, 167–169. https://doi.org/10.1001/jama.2018.20391.

von Meyenn, F., Berrens, R. V., Andrews, S., Santos, F., Collier, A. J., Krueger, F., et al. (2016). Comparative principles of DNA methylation reprogramming during human and mouse in vitro primordial germ cell specification. *Developmental Cell*, *39*, 104–115. https://doi.org/10.1016/j.devcel.2016.09.015.

Waddington, C. H. (2012). The epigenotype. 1942. *International Journal of Epidemiology*, *41*, 10–13. https://doi.org/10.1093/ije/dyr184.

Wanner, N. M., Colwell, M. L., & Faulk, C. (2019). The epigenetic legacy of illicit drugs: Developmental exposures and late-life phenotypes. *Environmental Epigenetics.*, *5*, 1–10. https://doi.org/10.1093/eep/dvz022.

Watkins, A. J., Lucas, E. S., & Fleming, T. P. (2010). Impact of the periconceptional environment on the programming of adult disease. *Journal of Developmental Origins of Health and Disease*, *1*, 87–95. https://doi.org/10.1017/S2040174409990195.

Watson, C. T., Szutorisz, H., Garg, P., Martin, Q., Landry, J. A., Sharp, A. J., et al. (2015). Genome-wide DNA methylation profiling reveals epigenetic changes in the rat nucleus Accumbens associated with cross-generational effects of adolescent THC exposure. *Neuropsychopharmacology*, *40*, 2993–3005. https://doi.org/10.1038/npp.2015.155.

Wei, Y., Yang, C. R., Wei, Y. P., Zhao, Z. A., Hou, Y., Schatten, H., et al. (2014). Paternally induced transgenerational inheritance of susceptibility to diabetes in mammals. *Proceedings of the National Academy of Sciences of the United States of America*, *111*, 1873–1878. https://doi.org/10.1073/pnas.1321195111.

Wu, C., & Morris, J. R. (2001). Genes, genetics, and epigenetics: A correspondence. *Science*, *293*, 1103–1105. https://doi.org/10.1126/science.293.5532.1103.

Wu, Q., & Suzuki, M. (2006). Parental obesity and overweight affect the body-fat accumulation in the offspring: The possible effect of a high-fat diet through epigenetic inheritance. *Obesity Reviews*, 7, 201–208. https://doi.org/10.1111/j.1467-789X.2006.00232.x.

Xia, W., & Xie, W. (2020). Rebooting the epigenomes during mammalian early embryogenesis. *Stem Cell Reports*, *15*, 1158–1175. https://doi.org/10.1016/j.stemcr.2020.09.005.

Xin, F., Susiarjo, M., & Bartolomei, M. S. (2015). Multigenerational and transgenerational effects of endocrine disrupting chemicals: A role for altered epigenetic regulation? *Seminars in Cell and Developmental Biology*, *43*, 66–75. https://doi.org/10.1016/j.semcdb.2015.05.008.

Xu, Q., Xiang, Y., Wang, Q., Wang, L., Brind'Amour, J., Bogutz, A. B., et al. (2019). SETD2 regulates the maternal epigenome, genomic imprinting and embryonic development. *Nature Genetics*, *51*, 844–856. https://doi.org/10.1038/s41588-019-0398-7.

Xu, Q., & Xie, W. (2018). Epigenome in early mammalian development: Inheritance, reprogramming and establishment. *Trends in Cell Biology*, *28*, 237–253. https://doi.org/10.1016/j.tcb.2017.10.008.

Yan, W. (2014). Potential roles of noncoding RNAs in environmental epigenetic transgenerational inheritance. *Molecular and Cellular Endocrinology*, *398*, 24–30. https://doi.org/10.1016/j.mce.2014.09.008.

Yin, G., Xia, L., Hou, Y., Li, Y., Cao, D., Liu, Y., et al. (2021). Transgenerational male reproductive effect of prenatal arsenic exposure: Abnormal spermatogenesis with Igf2/H19 epigenetic alteration in CD1 mouse. *International Journal of Environmental Health Research*, 1–13. https://doi.org/10.1080/09603123.2020.1870668.

Yohn, N. L., Bartolomei, M. S., & Blendy, J. A. (2015). Multigenerational and transgenerational inheritance of drug exposure: The effects of alcohol, opiates, cocaine, marijuana, andnicotine. *Progress in Biophysics and Molecular Biology*, 21–33. https://doi.org/10.1016/j.pbiomolbio.2015.03.002.

Yoshida, K., Maekawa, T., Ly, N. H., Fujita, S. I., Muratani, M., Ando, M., et al. (2020). ATF7-dependent epigenetic changes are required for the intergenerational effect of a paternal low-protein diet. *Molecular Cell*, *78*, 445–458.e6. https://doi.org/10.1016/j.molcel.2020.02.028.

Yoshino, T., Suzuki, T., Nagamatsu, G., Yabukami, H., Ikegaya, M., Kishima, M., et al. (2021). Generation of ovarian follicles from mouse pluripotent stem cells. *Science*, *373*, eabe0237. https://doi.org/10.1126/science.abe0237.

Yu, C., Fan, X., Sha, Q. Q., Wang, H. H., Li, B. T., Dai, X. X., et al. (2017). CFP1 regulates histone H3K4 trimethylation and developmental potential in mouse oocytes. *Cell Reports*, *20*, 1161–1172. https://doi.org/10.1016/j.celrep.2017.07.011.

Yu, C. W., & Liao, V. H. C. (2016). Transgenerational reproductive effects of Arsenite are associated with H3K4 Dimethylation and SPR-5 downregulation in Caenorhabditis elegans. *Environmental Science & Technology*, *50*, 10673–10681. https://doi.org/10.1021/acs.est.6b02173.

Zeid, D., & Gould, T. J. (2020). Impact of nicotine, alcohol, and cocaine exposure on germline integrity and epigenome. *Neuropharmacology*, *173*, 1–11. https://doi.org/10.1016/j.neuropharm.2020.108127.

Zeng, L., Wang, M., Zhou, J., Wang, X., Zhang, Y., & Su, P. (2022). A hypothesis: Retrotransposons as a relay of epigenetic marks in intergenerational epigenetic inheritance. *Gene*, *817*, 1–11. https://doi.org/10.1016/j.gene.2022.146229.

Zheng, H., Huang, B., Zhang, B., Xiang, Y., Du, Z., Xu, Q., et al. (2016). Resetting epigenetic memory by reprogramming of histone modifications in mammals. *Molecular Cell*, *63*, 1066–1079. https://doi.org/10.1016/j.molcel.2016.08.032.

Zheng, H., & Xie, W. (2019). The role of 3D genome organization in development and cell differentiation. *Nature Reviews. Molecular Cell Biology*, *20*, 535–550. https://doi.org/10.1038/s41580-019-0132-4.

Zhu, H., Wang, G., & Qian, J. (2016). Transcription factors as readers and effectors of DNA methylation. *Nature Reviews. Genetics*, *17*, 551–565. https://doi.org/10.1038/nrg.2016.83.

Ziv-Gal, A., Wang, W., Zhou, C., & Flaws, J. A. (2015). The effects of in utero bisphenol a exposure on reproductive capacity in several generations of mice. *Toxicology and Applied Pharmacology*, *284*, 354–362. https://doi.org/10.1016/j.taap.2015.03.003.

CHAPTER THREE

Gene-alcohol interactions in birth defects

Joshua L. Everson[a,b,*] and Johann K. Eberhart[a,b]
[a]Department of Molecular Biosciences, School of Natural Sciences, University of Texas at Austin, Austin, TX, United States
[b]Waggoner Center for Alcohol and Addiction Research, School of Pharmacy, University of Texas at Austin, Austin, TX, United States
*Corresponding author: e-mail address: jleverson@utexas.edu; eberhart@austin.utexas.edu

Contents

Abstract

Most human birth defects are thought to result from complex interactions between combinations of genetic and environmental factors. This is true even for conditions that, at face value, may appear simple and straightforward, like fetal alcohol spectrum disorders (FASD). FASD describe the full range of structural and neurological disruptions that result from prenatal alcohol exposure. While FASD require alcohol exposure, evidence from human and animal model studies demonstrate that additional genetic and/or environmental factors can influence the embryo's susceptibility to alcohol. Only a limited number of alcohol interactions in birth defects have been identified, with many sensitizing genetic and environmental factors likely yet to be identified. Because of this, while unsatisfying, there is no definitively "safe" dose of alcohol for all pregnancies. Determining these other factors, as well as mechanistically characterizing known interactions, is critical for better understanding and preventing FASD and requires combined scrutiny of human and model organism studies.

Current Topics in Developmental Biology, Volume 152
ISSN 0070-2153
https://doi.org/10.1016/bs.ctdb.2022.10.003

Abbreviations

ADH	alcohol dehydrogenase
ALDH	acetaldehyde dehydrogenase
BCE	before common era
CBD	cannabidiol
CL/P	cleft lip with or without cleft palate
CNS	central nervous system
CP	cleft palate
DEAB	4-diethylaminobenzaldehyde
ENU	N-ethyl-N-nitrosourea
EPA	environmental protection agency
FAE	fetal alcohol effects
FAS	fetal alcohol syndrome
FASD	fetal alcohol spectrum disorders
GD	gestational day
GWAS	genome-wide association study
HPE	holoprosencephaly
NAC	N-acetyl Cysteine
NO	nitric oxide
OFC	orofacial clefting
PBO	piperonyl butoxide
pFAS	partial fetal alcohol syndrome
RA	retinoic acid
ROS	reactive oxygen species
SAG	smoothened agonist
Shh	Sonic hedgehog

1. Introduction

Humankind has grappled with the effects of alcohol since its discovery. Alcohol (ethanol) is a teratogen (an agent capable of causing congenital defects). Given the widespread use of alcohol, it is likely the most common environmental cause of human birth defects, and is certainly the most common preventable cause of human birth defects. Fetal alcohol spectrum disorders (FASD) describe the full constellation of congenital abnormalities and neurological deficits associated with embryonic or fetal exposure to alcohol (Riley, Infante, & Warren, 2011; Wozniak, Riley, & Charness, 2019). Prenatal alcohol exposure causes a wide range of embryonic/fetal abnormalities, including malformations of the brain, face, heart, and other organs, as well as short stature, which can cause lifelong, irreparable burdens for those affected (Caputo, Wood, & Jabbour, 2016; Coriale et al., 2013; Denny, Coles, & Blitz, 2017; Wilhoit, Scott, & Simecka, 2017). With recent advances in FASD diagnostic criteria, FASD are now understood to affect a shocking 1–8% of all children born in the United States, making it one of the most common human birth defects (May et al., 2009, 2014, 2020a, 2020b, 2020c).

This high prevalence cannot be explained simply as only a problem of alcoholism, as has been suggested in the past (Ulleland, 1972). Researchers have identified a worrying new trend of increased drinking, including binge drinking, while pregnant (Howard et al., 2022). This trend appears to be driven by a combination of factors, including inadequate or unreliable information from health professionals and/or a disregard of public health recommendations by patients (Hammer & Rapp, 2022). While difficult to truly calculate, it is likely that a majority of embryonic alcohol exposures are unintentional. Around half of all pregnancies are unintended (Finer & Zolna, 2016), and critical stages of embryogenesis, including formation of the brain and face (Sulik, 2005), occur prior to the average gestational age of pregnancy detection, 5.5 weeks (Branum & Ahrens, 2017). For example, a Norwegian study found that 89% of participants reported drinking prior to pregnancy (59% binge drinking), of which 85% changed their drinking behavior once their pregnancy was identified. This suggests that a large percentage of pregnancies were at risk for prenatal alcohol exposure during the critical early weeks of development.

While alcohol exposure is required for FASD, phenotypic outcomes following prenatal alcohol exposure are highly variable. Clearly, not all alcohol exposures result in FASD. This suggests additional factors, either genetic or environmental, can affect embryo susceptibility to alcohol. This also means that a dose of alcohol that is not deleterious for one embryo, could be harmful to another. Thus, there is no single dose of alcohol that is safe for all pregnancies. Instead, multiple lines of evidence from human and model organism studies demonstrate that genetic and environmental factors can alter alcohol toxicity. However, the identities and, moreover, the mechanisms of these myriad interactions with alcohol are mostly unknown, with few notable exceptions. In this review, we examine the history of alcohol-induced birth defects and the evidence for gene-alcohol and multifactorial interactions in alcohol-induced birth defects. We focus this chapter on interactions affecting the face and brain, as most studies on gene-alcohol interactions have focused on these structures.

2. FASD background and prevalence

Homo sapiens have assuredly been consuming alcohol since they first emerged on Earth. There is evidence of humans storing fermented beverages dating as far back as 7000 BCE (McGovern et al., 2004). The association between excessive alcohol consumption and the immediate negative effects on the drinker likely dates back at least equally as far. When humans understood the potential adverse effects of embryonic alcohol exposure is much less clear.

An understanding of the effects of alcohol exposure on a developing embryo in turn requires an understanding of embryogenesis, or development, or at a minimum that such a process happens. In Western writing, many give Aristotle (BCE 384-322) credit for the foundations of developmental biology (Wallingford, 2021). At this same time, or even earlier, Eastern cultures, and potentially cultures of ancient Mexico, also appreciated embryogenesis (Wallingford, 2021). Readers interested in the history of developmental biology are encouraged to read Dr. Wallingford's detailed review on the subject (Wallingford, 2021). Thus, cultures across the globe had the capacity to associate alcohol consumption and embryonic outcomes over two millennia ago. But, when did they?

Passages from the ancient Greek and Roman as well as *the Bible* have been used to suggest that people from these ages understood the potential for adverse outcomes following embryonic alcohol exposure. However, careful interpretations taking into account the culture at the time do not particularly support such claims. Rather, these passages relate birth outcomes to the state of being of the parents at the time of conception (Abel, 1997, 1999; Brown, Bland, Jonsson, & Greenshaw, 2019; Obladen, 2021). Notably, the idea that birth outcomes related to the mental state of one or both parents during conception was not specific to alcohol (Wallingford, 2021). Therefore, it is highly unlikely that these statements demonstrate an understanding of alcohol teratogenesis. While we cannot know with 100% certainty, it seems safest to presume that such statements relate to an understanding of the immediate consequences of heavy intoxication and not an effect on the developing embryo.

Part of the difficulty in determining when there was a general recognition that ethanol consumption could be teratogenic is disentangling the social stigma of excessive alcohol consumption from birth outcomes. Authors in the 1700s commented on the deleterious effects of gin (Warren, 2015). Hogarth's etching *Gin Lane* has been used by some as evidence for an understanding of the deleterious effects of drinking on offspring (Warner & Rosett, 1975). However, it is worth noting that the companion etching *Beer Street* also shows a large group of individuals consuming alcohol, but this time beer, with no deleterious effects. The real difference between the two etchings would appear to be class.

There are sporadic manuscripts reporting negative effects of alcohol in pregnant women through the 1800s. These manuscripts are also often rife with sexism, classism and eugenics. Additionally, it is worth bearing in mind, that these manuscripts were published during the growth and peak of the temperance movement, a social movement that promoted moderation or abstinence from alcohol. Thus, many individuals would believe that alcohol was detrimental in its entirety and in all aspects of life. For instance, a report to the British House of Commons in 1834 states "...the highest medical authorities, examined in great numbers before Your Committee, being uniform in their testimony that ardent spirits are absolutely poisonous to the human constitution; that in no case whatever are they necessary, or even useful, to persons in health; that they are always, in every case and to the smallest extent, deleterious, pernicious or destructive, according to the proportions in which they may be taken into the system..." (Great Britain. Parliament. House of commons. Select committee on inquiry into drunkenness & Buckingham, 1834). Thus, it seems imperative to consider whether or not such studies were simply observational.

In the late 1800s numerous publications comment on birth outcomes of mothers with alcoholism. In 1876, Haddon states that intemperate women often had spontaneous abortions (Haddon, 1876). This appears to be purely observational as no comparison group was analyzed and the author attributes the effect on uterine function as opposed to embryogenesis (although it should be noted that uterine function may well play a role in FASD). It is further worth noting that, while the author was discussing pregnancy outcomes in alcoholic women he states, without supporting evidence, that moderate alcohol consumption likely "differs only in degree, and not in kind," suggesting a temperance bias. In 1888, there appears to be debate as to whether paternal or maternal alcohol consumption has a greater effect on the intellectual outcomes of pregnancy. In his manuscript, Duncan pushes back against the idea that the male has a greater influence saying, "What is the comparative potency of alcoholic disease of the testicles and of the ovaries on population we have at present no definite means of even guessing, but it may be safely judged that that of the ovaries is paramount on account of the much greater part taken in reproduction by the female than by the male" (Duncan, 1888). Here, again, the effect of alcohol appears to be attributed to sex organs as opposed to a direct effect on the embryo. By 1899 it was noted that infant mortality (before age 2) and epilepsy was higher in children born to alcoholic mothers relative to the general population (Sullivan, 2011). This study suffered somewhat from a lack of a truly appropriate comparison group. However, it did conclude that these effects were due to the continued exposure of the developing embryo to alcohol and is perhaps the first strong evidence for what we now term FASD.

The early 1900s provides further careful study indicating the deleterious effects of prenatal alcohol exposure. For instance, in a 1906 conference on infant mortality, two papers reference alcoholism (National Conference on Infantile Mortality, 1906). Dr. Woodhead states that alcohol exerts "poisonous action" on all cells and that the embryo should be particularly sensitive to alcohol due to its proliferative nature. This appears to be the first attempt to gain systematic information on the effects of prenatal alcohol exposure as the author proposes a questionnaire, including questions on the quantity of alcohol consumption as well as hereditary diseases of the parents, a much needed control. However, the only birth outcome examined was suckling. Ballantyne discusses the importance of preventative medicine and prenatal development. In contrast to Dr. Woodhead, Dr. Ballantyne suggests that the effect of alcohol is not on the fetus itself: "...this effect is due rather to the changes in the maternal kidneys and in the placenta than to any direct influence which the drug has upon the fetal tissues themselves." Thus, while it was becoming clear that alcohol exposure could associate with deleterious birth outcomes at this time, the cause was not understood.

The first descriptions of what we now know of as fetal alcohol syndrome (FAS) would be made in the mid-1900s. Dr. Paul Lemoine is often credited with the first description of what would later be termed FAS in 1968 (translated from French to English in 2003) (Lemoine, Harousseau, Borteyru, & Menuet, 2003). Interestingly, and relevant to whether previous generations had appreciated alcohol as a teratogen, he stated in a later review that it took some time for him to determine that the syndrome he was seeing was due to alcohol, because in his medical training "...we were taught without any doubt that it [alcohol] had no teratogenic effects..." (Lemoine, 1997). It was a staff member at the hospital that informed him that the mothers of two children he was seeing were alcoholics (Lemoine, 1997). In addition to describing the syndrome, Lemoine followed a subset of children up to 16 years of age and noted that they did show catch-up growth but displayed psychomotor deficits and increased aggression (Lemoine et al., 2003). The manuscript is notable also as it out and out rejects the theory that excess alcohol use at the time of conception was the culprit (Lemoine et al., 2003).

While Lemoine, indeed, deserves credit for identifying FAS, he notes that Dr. Jaqueline Rouquette provided the first characterization of what would be FAS in her thesis in 1957 (Rouquette, 1957). While her thesis has been referenced in passing by a handful of reviews on FAS, no details have previously been provided, presumably because it was only published as a thesis and in French. Therefore, we provide some details to put Dr. Rouquette's work into proper historical perspective. Between the years 1954 and 1957, Dr. Rouquette examined 100 children of confirmed alcoholic parents admitted to the Paul Parquet Infant Hygiene Center, which provided convalescent care to children from 1 month to 3 years of age. She detailed whether it was the mother (n=28), father (n=42) or both parents (n=30) that were alcoholic. In addition to alcoholism alone, she performed analyses of additional comorbidities. She concluded that maternal alcoholism was the predominant factor in preterm birth and congenital birth defects. She also noted a characteristic face in these children. This included a short upper lip, flattened root of the nose, hypertelorism and microcephaly. While she only dealt with young children, she also noted that they exhibited nervous, aggressive and unsociable behaviors. Through the comparison of co-morbidities, she concluded that alcohol was the culprit, as such, Dr. Rouquette's work appears to be the first clear identification of FAS.

It would be a series of manuscripts from the University of Washington that culminated in the definition and acceptance of FAS. In an 18 month period from the beginning of 1968 to mid-1969, Ulleland found that children born to alcoholic mothers were small for gestational age and failed to thrive (Ulleland, 1972). Notably, she found that over 80% of infants born to alcoholic mothers were below the 10th percentile in birth weight (Ulleland, 1972). In 1973 Jones and colleagues published their findings of eight children across three ethnic groups born to alcoholic mothers that largely shared a set of growth, craniofacial, limb and cardiac defects (Jones, Smith, Ulleland, & Streissguth, 1973). In the same year, Jones and Smith reported three additional cases and formalized the name Fetal Alcohol Syndrome (FAS) (Jones & Smith, 1973).

Despite this, it was not until 1981 that a U.S. surgeon general advisory warning recommended that pregnant women avoid consumption of alcohol. Just a year later, the noted fetal alcohol effects researcher Henry Rosett stated in the Hearing before the Subcommittee on Alcoholism and Drug Abuse of the Committee on Labor and Human Resources, "The credibility of the strong evidence of the dangers to the fetus from heavy drinking and alcoholism is weakened through exaggeration of possible dangers from small amounts of alcohol" (United States. Congress. Senate. Committee on Labor and Human Resources. Subcommittee on Alcoholism and Drug Abuse, 1982). Dr. Rosett provided references for studies failing to find effects of moderate alcohol on birth outcomes (Hingson et al., 1982; Kaminski, Rumeau-Rouquette, & Schwartz, 1976; Tennes & Blackard, 1980).

While true that some studies found no association between moderate alcohol consumption and deleterious outcomes, others did. Indeed, the Tennes and Blackard study referenced by Dr. Rosett did demonstrate a weak but significant association of moderate alcohol consumption and premature delivery (Tennes & Blackard, 1980). Prior to Rosett's remarks, a study of 163 infants with confirmed alcohol exposure were evaluated and 11 were found to have features consistent with FAS (Hanson, Streissguth, & Smith, 1978). The mothers of six of these children reported consuming between 1 and 2 oz of absolute alcohol/day during pregnancy and only two were noted to be alcoholics (Hanson et al., 1978). Additionally, two manuscripts published side by side in *the Lancet* in 1980 also associated more moderate alcohol consumption patterns with miscarriage (Harlap & Shiono, 1980; Kline, Shrout, Stein, Susser, & Warburton, 1980). In a very large study of 32,019 women, Harlap and Shiono found that moderate alcohol consumption (1–2 drinks/day) during the second, but interestingly not the first, trimester associated with almost double the risk of miscarriage (Harlap & Shiono, 1980). Kline and colleagues found that consumption of alcohol 2–6 days/week more than doubled the chances of spontaneous abortion (Kline et al., 1980).

While Dr. Rosette's remarks likely were directed to even lower consumption levels than those in these studies, they do bring up a question of immense importance: What is a safe level of alcohol consumption during pregnancy? The unsatisfying answer is that we simply do not know. Clearly, birth outcomes following exposure vary widely and this is due to a large number of other factors, environmental and genetic, that impact the result of prenatal alcohol exposure. Until we can understand how alcohol can interact with the tens of thousands of genes in our body and the hundreds of thousands of chemicals in our environment, we will not have an answer to this question.

3. What are multifactorial interactions in birth defects?

Despite being the leading cause of infant mortality (Matthews, MacDorman, & Thoma, 2015), the vast majority (80%) of human birth defects do not have a known cause (Feldkamp, Carey, Byrne, Krikov, & Botto, 2017). This has led to the prevailing hypothesis that most birth defects do not have a single cause, but are instead the result of "multiple hits" (reviewed in (Beames & Lipinski, 2020; Brent, 2004; Krauss & Hong, 2016)). In this model, two or more insults intersect to affect embryogenesis. These insults can be either genetic, such as a mutation or polymorphism, or environmental, such as exposure to a teratogen like alcohol. As such, multifactorial interactions fall under three general categories: gene-gene interactions, gene-environment interactions, and co-environmental interactions (co-exposure to two or more toxicants). Thus, for alcohol related birth defects the relevant multifactorial interactions are gene-environment (specifically gene-alcohol) and co-environmental interactions (*e.g.*, alcohol + another toxicant).

Multifactorial interactions are not always direct. That is, the factors can impinge on the same genetic networks (such as a *Sonic Hedgehog* (*Shh*) mutant exposed to a Shh pathway inhibitor), but they can also be indirect, such as two mechanistically distinct perturbations that result in disruptions of common or related biological processes. Multifactorial interactions also can be either additive, where the effect of both factors together equals the cumulative effect of both factors alone, or synergistic, where the effect of both factors is greater than the cumulative effect of both factors alone. These synergistic interactions pose the most concern from the perspective of public health and risk assessment. Most developmental toxicity dose-response assays are performed on single chemicals in non-genetically predisposed subjects, while almost nothing is known about the effect of chemical mixtures or single exposures to genetically predisposed individuals (Krewski et al., 2010; Tice, Austin, Kavlock, & Bucher, 2013). Most chemical exposures occur in mixtures (Koppe et al., 2006; Mitro, Johnson, & Zota, 2015), and each person is born with millions of genetic variants, of which 250–300 have been found to be predicted loss-of-function mutations in annotated genes, with 50–100 of these variants being previously associated with a heritable diseases (Abecasis et al., 2010). Thus, reference doses of chemicals that pass regulatory assessment as single exposures could still be harmful when combined with genetic susceptibility or with other toxicants in mixtures.

4. Evidence for gene-alcohol interactions

While alcohol exposure has been definitely linked to birth defects, what's less understood is the basis for variability in phenotypic outcomes following an exposure. For example, not all alcohol exposures cause birth defects, even when dose, duration, and timing of exposure are controlled. That is, the same alcohol exposure can cause variable outcomes, with some resulting in malformations while others cause no apparent phenotypes. This apparent discordance could be explained by the presence of genetic and/or environmental modifiers. Several lines of evidence suggest embryonic and maternal genetics can influence embryonic susceptibility or resilience to alcohol. These include human genetic studies, especially on twins and siblings, strain differences in susceptibility in multiple species, and specific gene-environment interactions in both humans and research animals.

4.1 Human studies

One of the first pieces of evidence used to suggest a genetic component in FASD was the observation that children born to mothers who have previously given birth to a child with FASD are more likely to also have FASD. However, this observation could be explained by environment, *e.g.*, a common pattern of drinking during pregnancy, rather than an intrinsic factor like genetics.

4.1.1 Twin studies

The most direct evidence for a genetic component in human FASD is from twin and sibling association studies. In these studies, pairs of twins or siblings with confirmed prenatal alcohol exposure are assessed for FASD, and concordance rates (*i.e.*, whether the twins had the same or different phenotypic outcomes) are calculated. These studies are the gold standard for human association studies, as they enable comparisons between two individuals with near identical embryonic environments and exposures. In addition, comparison of monozygotic twins (100% shared genetic identity) and dizygotic twins (50% shared genetic identity) enables direct comparison of the amount of shared genetic relatedness to concordance in FASD outcomes. Siblings (50% shared genetic identity) and half-siblings (25% shared genetic identity) can also be analyzed, although here exposures are not identical. Together, this provides a sort of "dose-response" for genetic relatedness.

The first of these studies by Streissguth and Dehaene in 1993 provided clear support for a genetic component to the effect of prenatal alcohol exposure (Streissguth & Dehaene, 1993). The authors wondered why FASD outcomes can vary so widely, even in cases of known high alcohol exposure during the critical period for FAS, hypothesizing that genetics can modify embryo susceptibility. The authors compared the rate of concordance of FAS and the milder (antiquated term) fetal alcohol effects (FAE) between 16 pairs of either monozygotic (n = 5) or dizygotic (n = 11) twins with confirmed heavy alcohol exposure born in the United States or France. The pairwise concordance rate (*i.e.*, FAS diagnosis in both children) was 100% for monozygotic twins, but only 64% (7/11) in dizygotic twins. In addition, the authors noted 4/11 dizygotic twins had markedly opposing phenotypes, two pairs in which one child had FAS and the other FAE, as well as two pairs in which one child had FAE and the other was apparently unaffected (Streissguth & Dehaene, 1993). This provides clear evidence that even when ethanol dose and time are controlled, genetic differences can alter embryo susceptibility.

Building upon this study, a 2019 publication by Hemingway and colleagues supported and elaborated on the findings of Streissguth and Dehaene (Hemingway et al., 2018). In it, the authors included full (n = 27) and half (n = 9) siblings along with monozygotic (n = 9) and dizygotic (n = 39) twins, expanding on the effective "genetic relatedness dose-response" to the following: monozygotic twins (100% shared genetic identity), dizygotic twins (50%), full siblings (50%) and half siblings with a common birth mother (25%). As genetic relatedness increased pairwise FASD diagnosis concordance increased between siblings: monozygotic = 100%, dizygotic = 66%, full sibling = 41%, half sibling = 22% (Hemingway et al., 2018). Remarkably, this nearly perfectly recapitulates the findings of Streissguth and Dehaene's foundational work, despite being conducted nearly 30 years later using the new FASD diagnostic criteria, and with a different and much larger study cohort. Notably, the study also highlighted 4/39 twins whose FASD diagnoses were on opposite ends of the severity spectrum (partial fetal alcohol syndrome (pFAS) *vs.* neurodevelopmental disorder/alcohol exposed), meaning one child had FAS facial phenotypes with severe central nervous system (CNS) dysfunction, but their twin had only moderate CNS dysfunction without facial dysmorphology. No significant sex differences were observed in either phenotypic severity nor pairwise discordance. Using the following formula, the authors then calculated the percent heritability to estimate the importance of genetic factors *vs.* environmental factors using the following formula: [((dizygotic discordance—monozygotic discordance)/dizygotic discordance) *100%] for both the FASD diagnosis and FAS facial dysmorphology rank, finding both were 100% (Hemingway et al., 2018). This means embryo genotype fully explained the differential outcomes observed between twin pairs. These results finely demonstrate dose and timing of alcohol exposure alone cannot account for phenotypic variability seen in response to alcohol, and strongly support a model in which embryo genetics alters susceptibility to prenatal alcohol toxicity.

4.1.2 Human gene-disease association studies

Association of genotypic variants to FASD phenotypes has been limited but with notable successes. Genome-wide association studies (GWAS) studies on FASD (as with other individual birth defect syndromes) have proven difficult. This is due in large part to the relatively low prevalence of the syndrome, making enrollment of an adequate number of study participants difficult (Khokha, Mitchell, & Wallingford, 2017). This problem is compounded for genes in which the effect size for disease risk is small. Due to the pleiotropic nature of alcohol and the multifactorial gene-alcohol interactions that are predicted to give rise to FASD, study cohorts must be even larger to encompass enough children with FASD from the same etiologies (*i.e.*, with mutations in the same locus).

4.1.2.1 Alcohol metabolizing genes

For these reasons, the majority of FASD modifying genes identified in humans have been candidate genes, with members of the ethanol metabolism pathway being some of the most likely candidates. Alcohol (ethanol) is first metabolized into the toxic metabolite acetaldehyde by alcohol dehydrogenase (*ADH*) members, which is then metabolized into the non-toxic metabolite acetate by acetaldehyde dehydrogenase (*ALDH*) members. Both alcohol and acetaldehyde have been demonstrated to act as teratogens, so damage to the embryo can happen until this final metabolic step. In humans, polymorphisms in these gene families alter the rate of metabolism of either ethanol or acetaldehyde, which is predicted to alter the teratogenic potential of identical alcohol consumption patterns (reviewed in (Crabb, Matsumoto, Chang, & You, 2004; Warren & Li, 2005)). However, the results of these studies are inconsistent.

The first study by McCarver et al. examined 243 African American infant-mother pairs with known alcohol exposure during pregnancy for interactions between alcohol and polymorphisms in *ADHB1* (previously called *ADH2*) (McCarver, Thomasson, Martier, Sokol, & Li, 1997). The *ADHB1*3* allele encodes a fast-metabolizing allele and is very common in African Americans (25%) (Thomasson, Beard, & Li, 1995). Importantly, this study did not correlate genotypes to FAS facial phenotypes, but instead used as the endpoint the Bayley Scales of Infant Development Mental Index (MDI) at 1 year of age. The authors discovered that the *ADHB1*3* allele in both offspring or the mother appeared to protect the child from alcohol-induced deficits.

However, the opposite result was found 5 years later by Stoler et al. The study examined 404 pregnant women and 139 infants (Stoler, Ryan, & Holmes, 2002). The authors found that a higher frequency than expected (46%) of African American women enrolled in the study had the *AHDB1*3* allele. In contrast to the previous findings by McCarver et al., Stoler et al. observed maternal *ADHB1*3* mutations were over-represented in children with FAS facial phenotypes and growth restriction after adjusting for confounding variables (McCarver et al., 1997; Stoler et al., 2002). While the authors hypothesized this may be due to increased drinking in the mothers with *ADHB1*3* polymorphisms, studies have shown that *ADH1B*3* polymorphisms are negatively-associated with greater drinking in humans (Zaso et al., 2018). In addition, the fact that infant *ADHB1*3* was also over-represented in children with FAS facial phenotypes implies there could be another mechanism, such as increased acetaldehyde accumulation. However, as noted, another major difference between these studies was the endpoint measured—McCarver assessed a neurobehavioral endpoint, while Stoler assessed FAS facial phenotypes.

These differences in study endpoints were reconciled in a follow-up of the McCarver study in 2004. Das et al. photographed the faces of 247 African American infants (173 of which (70%) participated in the original McCarver et al. 1997 study) and their mothers to assess genotype-FAS facial phenotype correlations, as was done by Stoler and colleagues. After selecting only mothers who reported drinking prior to the first prenatal visit, the authors reported that all groups in which the mother or the child had at least one *ADH1B*3* had significant ($p < 0.05$), improvements in FAS facial phenotypes compared to infant-mother pairs with no alleles of *ADH1B*3* (Das, Cronk, Martier, Simpson, & McCarver, 2004). This result is consistent with *ADH1B*3* conferring protection. Thus, the differences between these studies is difficult to ascertain, but could relate to the accumulation of acetaldehyde.

A final study by Jacobson et al. in 2006 also analyzed *ADH1B*3* genotype-FAS facial phenotype correlations (Jacobson et al., 2006). 263 African American mother-child pairs were analyzed both as infants at 1 year and at 7.5 years old using a battery of neurobehavioral assays, as well as several morphological endpoints (including head circumference). Maternal *ADH1B*3* polymorphism was found to be associated with unaffected head circumference, consistent with a protective effect (Jacobson et al., 2006). In addition, the study confirmed McCarver et al.'s finding that maternal *ADH1B*3* is associated with improved Bayley MDI neurobehavior performance at 1 year (Jacobson et al., 2006; McCarver et al., 1997). In contrast, offspring *ADH1B*3* polymorphism did not appear protective for Bayley MDI, but did appear protective against reduced head circumference at 1 year (Jacobson et al., 2006).

Together these studies support a role for the fast metabolizing *ADH1B*3* polymorphism in protecting against the deleterious effects of prenatal alcohol exposure. However, there are conflicting results that suggest a more complicated or multiple mechanisms. Additionally, the allele is associated with decreased alcohol consumption, which confounds the association. Thus, while most results support a protective function for the maternal *ADH1B*3* polymorphism, the mechanism is not known.

Next, a study by Viljoen et al. examined a slow-metabolizing allele of *ADHB1*, *ADH1B*2* (Viljoen et al., 2001). The study included 56 children with FAS, their mothers, and 172 controls from the same region in South Africa. The participants were of a Khoisan-White mixed-ancestry. The authors found the *ADH1B*2* mutation was more common in control mothers and offspring, consistent with the mutation conferring protection from alcohol-induced malformations (Viljoen et al., 2001). In addition, *ADH1B*3* was examined, but the frequency of this polymorphism was low and not significant between groups. A notable difference between this study and the two previous is the ethnicity of the participants, which explains the low frequency of *ADH1B**3 polymorphisms.

One final study by Boyles et al. from 2010 examined the effect of polymorphisms in another ADH gene, *ADH1C* (formerly *ADH3*), on orofacial clefting (OFC) (Boyles et al., 2010). This study came on the heels of a clinical trial in France that reported a 1.70 increased odds ratio for OFC in children born to mothers who reported first trimester drinking, which was reduced to 0.71 or 0.63 for children with *ADH1C* single or dual allele polymorphisms, respectively (Chevrier et al., 2005). The larger study by Boyles et al. included 483 infants with OFC and 503 controls in

Norway. The authors had previously examined 573 infants with OFC and 763 controls born in Norway and found an increased odds ratio of 2.2 for cleft lip with or without cleft palate (CL/P) and 2.6 for cleft palate only (CP) in children born to mothers who reported first trimester binge drinking (>5 drinks per setting) (DeRoo, Wilcox, Drevon, & Lie, 2008). Odds ratios were increased to 3.2 for CL/P and 3.0 for CP with three or more first trimester binge episodes. Boyles et al. found an odds ratio for OFC of 2.6 in children born to mothers who reported first trimester binge drinking, but this increased odds ratio was only observed when the mother or child had the *ADH1C* haplotype associated with reduced alcohol metabolism (odds ratio = 3.0). Conversely, when both the mother and child had the *ADH1C* haplotype associated with increased alcohol metabolism the observed odds ratio for OFC was 0.9, consistent with this rapid metabolizing isozyme being protective against alcohol-induced OFC. This finding helps contextualize the positive association found by DeRoo et al., as most studies examining a connection between maternal drinking and OFC have been negative (meta-analyzed in (Yin, Li, Li, & Zou, 2019)). A high presence of *ADH1C* slow-metabolizer polymorphisms would explain this apparent discrepancy, as was seen by Boyles and colleagues.

The collective findings on the effects of *ADH* polymorphisms are somewhat difficult to reconcile. Both slow and fast metabolizing polymorphisms show a protective effect, so clearly the kinetics of ADH is not the entire story. ALDH is required to convert acetaldehyde to acetate, but these studies did not analyze this second enzyme. Additionally, the overall rate of metabolism is also controlled by the level of expression of these enzymes, meaning metabolism rates can vary even between individuals with the same genotype (O'Connor, Morzorati, Christian, & Li, 1998). *ADH* expression level can be altered by alcohol consumption (Badger et al., 2000). While it seems very likely that enzymes involved in metabolizing alcohol would impact the outcome of an alcohol exposure, the best predictive value likely requires information regarding genotype and epigenetic modifications at multiple loci.

4.1.2.2 Chromosome architecture genes

Two genes involved in chromosome architecture and/or cohesion, *SMC2* and *MLLT3,* have also been associated with alcohol-induced cleft palate (Beaty et al., 2011). Interestingly, several chromosome cohesion complex genes were recently associated with a brain malformation, holoprosencephaly (Kruszka et al., 2019), which shares many phenotypic and etiological characteristics with severe FAS (see Hedgehog Signaling section). The GWAS study included 550 case-parent trios of children with cleft palate to identify associations between cleft palate and SNPs either alone or in combination with one of three common environmental exposures (alcohol, smoking, or multivitamin supplementation). While no genome-wide SNP associations were identified when examined in isolation (*i.e.*, without consideration of maternal exposures), when considering cases with known maternal alcohol exposure, significant SNPs were found in *SMC2* and *MLLT3* (Beaty et al., 2011). These data support gene-alcohol interactions in the genesis of human birth defects.

Notably, a study in mouse also points to chromosome stability as an important modifier of prenatal alcohol-induced defects. *Fancd2* mutations in humans are associated with Fanconi anemia, a genetic disorder that causes genomic instability, leading to developmental defects and other abnormalities (Langevin, Crossan, Rosado, Arends, & Patel, 2011). Langevin et al. generated mice with dual mutations in *Fancd2* and the acetaldehyde metabolizing enzyme *Aldh2* (Langevin et al., 2011). Mice with mutations in these genes were exposed to 5 g/kg alcohol on gestational day (GD)7.5, targeting gastrulation (2–3 weeks in humans). Eye malformations and exencephaly were frequently observed in dual *Fancd2*; *Aldh2* mutants. More rarely, these malformations were also seen in *Fancd2* null; *Aldh2* heterozygotes. No eye malformations or exencephaly were observed in other genotypes, demonstrating Fancd2 and Aldh2 have protective roles in alcohol pathogenesis. This result suggests that there are likely to be complex multifactorial interactions in the genesis of FASD in humans.

4.2 Animal model studies

4.2.1 Strain differences

The generation of genetically- and phenotypically-distinct animal strains by animal breeders has provided a powerful tool for assessing the contribution of genetics in a disease's pathogenesis. These studies provide some of the best evidence for a genetic component to alcohol susceptibility, but so far have yielded few specific targets. Nevertheless, these foundational studies help explain the variability observed in FASD by suggesting genetics can strongly influence susceptibility. These include studies in mouse, chicken, and zebrafish.

4.2.1.1 Mouse

Ogawa et al. were the first to examine differences in alcohol susceptibility between two inbred mouse strains, C57BL/6 and DBA/2 (Ogawa, Kuwagata, Ruiz, & Zhou, 2005). These lines were chosen because of their wide use by alcohol researchers. However, the authors do not designate whether these are the 6J or 6N C57Bl6 substrain (an important distinction discussed below). While B6 mice are enthusiastic alcohol drinkers, DBA mice avoid alcohol. This makes comparison of these lines an attractive model for parsing the detrimental effects of alcohol. The authors used whole embryo culture to directly assess the sensitivity of each strain to alcohol-induced defects in conditions where embryo stage as well as ethanol dose and timing were controlled. The authors removed and cultured B6 and DBA embryos at GD8 (targeting primary neurulation, 3 weeks in humans) with or without 400 mg/dL of ethanol for 44 h. B6 mice were found to be more sensitive to alcohol-induced malformations than DBA mice. While optic, forebrain, and total defects did not differ between the two strains, alcohol specifically caused caudal neural tube and heart defects in B6 mice, while it specifically caused hindbrain and somite defects in DBA mice.

A follow-up study the same group compared ethanol sensitivity between C57BL/6N (B6), DBA/2 (DBA), and 129S6/SvEvTac (129S6) (Chen, Ozturk, Ni, Goodlett, & Zhou, 2011). The authors again used whole embryo culture to expose stage-matched embryos to a standardized ethanol dose and time of 88 mM ethanol for 6 h starting at GD8.25. Media was then replaced and embryos were developed for an additional 42 h in ethanol-free media. Again, the B6 strain was found to be most susceptible to alcohol-induced defects, with forebrain, midbrain, hindbrain, heart, optic vesical, caudal neural tube, and hindlimbs severely malformed or developmentally delayed. In contrast, only the forebrains and optic vesicles were malformed or delayed in DBA and 129S6 mouse embryos. Interestingly, these differential outcomes corresponded with higher levels of apoptosis (cleaved caspase 3) in B6 compared to DBA and 129S6 embryos.

A similar analysis was conducted to compare the alcohol sensitivity of five inbred mouse lines: Inbred Short-Sleep (ISS), C57BL/6J (B6), C3H/Ibg (C3H), A/Ibg (A), and 129S6/SvEvTac (129S6) (Downing, Balderrama-Durbin, Broncucia, Gilliam, & Johnson, 2009). Pregnant dams of each strain were administered 5.8 g/kg alcohol or control by oral gavage on GD9 (early pharengeal stage, 4 weeks in humans) and fetuses were examined at GD18 (just before birth). As such, this study could be confounded by differences in gestational timing or alcohol metabolism between the strains. In addition, it's unclear why the GD9 exposure was chosen rather than the more typical gastrulation/early neurulation-stage exposure. Nevertheless, the authors confirmed the previous finding, that 6J mice are most susceptible to alcohol-induced malformations. In addition, they found the 129S6 line was the most resistant to alcohol-induced defects, with the remaining lines, A, ISS, and C3H showing moderate ethanol sensitivity, ordered by decreasing sensitivity.

While C57BL/6 mice appear highly sensitive to ethanol across these studies, the C57BL/6 strain has more sensitive and less sensitive substrains (Green et al., 2007). Two closely related mouse lines, 6J (maintained by Jackson Labs) and 6N (maintained by Envigo (formerly Harlan Labs)) have been maintained separately since 1951, leading to genetic drift and substrain divergence. Pregnant 6J or 6N dams were exposed to 2.9 g/kg ethanol 4 h apart at GD7 to specifically target gastrulation and development of the brain and face. The authors found that the 6N substrain was less susceptible to alcohol-induced face defects than 6J embryos. Interestingly, the authors found 6J mice could be protected by pre-exposure to a peripheral benzodiazepine receptor (PBR) agonist PK-11195, which had previously been shown capable of protecting embryos from teratogen-induced defects (Charlap, Donahue, & Knudsen, 2003; O'Hara et al., 2003; O'Hara, Charlap, Craig, & Knudsen, 2002). In contrast, PK-11195 treatment increased ethanol sensitivity of 6N embryos. Thus, important genetic background differences between mouse strains as well as the 6N and 6J substrains must exist to explain these differences in ethanol sensitivity.

Two notable genomic variations have been identified through genome-wide comparison of the 6J and 6N substrains—mutation of Nicotinamide nucleotide transhydrogenase (*Nnt*) in 6J (Ronchi et al., 2013) and Crumbs Cell Polarity Complex Component 1 (*Crb1*) in 6N (Mattapallil et al., 2012). *Crb1* has a role in eye development, with the specific mutation *Rd8* linked to retinal degeneration. *Nnt* is a ubiquitously expressed gene involved in redox reactions, and therefore may mediate clearance of alcohol-induced reactive oxygen species (ROS). Transcriptomic analyses comparing microdissected anterior poles of GD7 embryos across the 6J and 6N substrains have identified 80 genes with basal differential expression (Boschen, Ptacek, Berginski, Simon, & Parnell, 2021). The authors found 67 genes with higher expression in 6J and 13 with lower expression in 6J.

As expected, one gene with lower expression in 6J was *Nnt*. These analyses confirm the reduction of *Nnt* expression in the 6J substrain, making differences in ROS clearance a possible mechanism explaining the differential ethanol sensitivity.

ROS have often been implicated in FASD pathogenesis (reviewed in (Brocardo, Gil-Mohapel, & Christie, 2011)). For example, maternal mutations that either increase or decrease superoxide dismutase (SOD) activity are protective or sensitizing in mice, respectively (Wentzel & Eriksson, 2006). Furthermore, interventions that reduce ROS can mitigate alcohol-induced malformations (reviewed in (Zhang, Wang, Li, & Peng, 2018)). For example, administration of the anti-oxidant N-Acetyl Cysteine (NAC) rescued alcohol-induced eye defects in mice exposed to two doses of 2.9 g/kg ethanol on GD7 (Parnell, Sulik, Dehart, & Chen, 2010). While 300 mg/kg NAC partially, but not significantly, reduced eye defects, 600 mg/kg NAC significantly reduced alcohol-induced eye defects, demonstrating a partial rescue of alcohol teratogenicity.

Our lab has used CRISPR-Cas9 to generate an *nnt* mutant zebrafish line to directly interrogate the role of *nnt* in modifying alcohol susceptibility (Mazumdar, unpublished). Consistent with the mouse substrain differences, *nnt* mutant zebrafish were sensitized to alcohol induced malformations at a normally sub-teratogenic 1% ethanol dose (in media). Unpublished data suggest *nnt* mutants accumulate more ROS in response to alcohol, leading to increased apoptosis in the brain and pharyngeal arch neural crest cells (Mazumdar, unpublished). In sum, these data support a model in which 6J mice (with *Nnt* mutation) have increased alcohol sensitivity due to a diminished capacity to tolerate alcohol-induced oxidative stress. The genes that modify ethanol teratogenicity across other mouse strains remain to be determined.

4.2.1.2 Chicken

Debelak and Smith examined 11 chick strains for differences in susceptibility for alcohol-induced neural crest apoptosis (measured by acridine orange uptake) (Debelak & Smith, 2000). The authors demonstrated that cell death was variable based on genetic strain between the lines, with three very sensitive (Babcock ISA, HyLine W98, Babcock B300/Hampshire Red cross (BxHR)), three moderately sensitive (Spafas, HyLine W36, Babcock B300), and five ethanol non-responsive lines (DeKalb White and Black, Shaver White and 2000, DcKalb White/Hampshire Red cross). More detailed interrogation of the two most sensitive lines (W98 and BxHR) compared to the most resistant line (DeKalb White) showed that no differences in embryo staging/timing of alcohol delivery, nor dose of ethanol reaching the embryo explained the differential vulnerability to alcohol (Debelak & Smith, 2000).

These authors published a second analysis in which nine of the 11 chicken strains were assessed for differences in craniofacial outcomes following the difference in cell death they had observed (Su, Debelak, Tessmer, Cartwright, & Smith, 2001). The authors observed three categories of facial malformations in the strains: midfacial flattening (BxHR, ISA-Babcock, HyLine W98, and HyLine W36 strains), facial expansion (Spafas and Babcock B300), or facial hypoplasia (DeKalb White and Black). However, level of cell death was not directly predictive of facial dysmorphology, suggesting more complex compensatory mechanisms exist (Su et al., 2001). Nevertheless, these studies provide excellent evidence for genetic differences being capable of altering an embryo's sensitivity to alcohol-induced malformations.

Interestingly, similar to mouse, substrains differences exist in chicken. Garic et al. further subtyped the highly sensitive W98 White Leghorn strain, into W98S (from Spencer, IA) and W98D (from Dallas Center, IA) substrains (Garic, Berres, & Smith, 2014). W98D was derived from W98S and separated by at least 10 generations. W98D was selectively bred for egg production. W98S and W98D embryos were exposed to an alcohol dose-response before apoptosis was measured (using acridine orange), confirming the differential sensitivity between the two lines. Next, gene expression differences were examined in the head folds of unexposed W98D and W98S embryos with RNAseq to examine possible genetic mechanisms underlying ethanol vulnerability. These studies identified proteasome and protein synthesis (*e.g.*, ribosome biogenesis) as well as oxidative phosphorylation as genetic networks with differential expression between the two substrains. The authors went on to assess each substrain's differentially expressed genes following alcohol exposure (Berres, Garic, Flentke, & Smith, 2017), identifying largely the same pathways that were basally differentially expressed between the substrains, including ribosome, oxidative phosphorylation, and spliceosome. Three ribosomal genes were functionally tested using zebrafish. Zebrafish injected with *rps3a, rpl5a* or *rpl11* morpholinos had increased alcohol mediated apoptosis and craniofacial malformations (Berres et al., 2017).

4.2.1.3 Zebrafish

Several analyses of differential sensitivity to ethanol have been performed in zebrafish. Loucks et al. examined three strains of zebrafish for differences in ethanol sensitivity. These included Ekkwill (EK) (ZFIN ID: ZDB-GENO-990520-2), AB (ZFIN ID: ZDB-GENO-960809-7), and Tuebingen (TU) (ZFIN ID: ZDB-GENO-990623-3) strains (Loucks & Carvan, 2004). Embryos were subjected to a dose-response of alcohol to observe differences in cell death (*via* acridine orange) or survival. The authors found that the EK strain was the most resistant to alcohol-induced embryo lethality, but surprisingly showed the highest cell death in response to alcohol. AB strain embryos showed moderate lethality and cell death. Finally, TU strain embryos had the highest embryo lethality but the lowest cell death. Notably, ethanol uptake between two of these strains, AB and TU has been shown to be the same (Lovely, Nobles, & Eberhart, 2014). However, the differences in cell death observed in this study were modest. For example, at 100 mM ethanol, dead cell counts in the head were EK = 2.28 ± 0.25, AB = 1.70 ± 0.20, and TU = 1.27 ± 0.13, deviating by only about 1 dead cell between groups. However, the differences in embryo survival were much more robust, with day six LD50 concentrations of EK = 214 mM, AB = 100 mM, and TU = 6.5 mM, clearly demonstrating differences in embryo susceptibility between these strains. In addition, differences in the specific craniofacial malformations observed differed by strain. Substrain differences in ethanol response in zebrafish have not been carefully analyzed, although such differences are known to exist based on copy number variant analyses (Brown et al., 2012).

Collectively, these data on strain differences in alcohol sensitivity support a model in which genetic factors influence ethanol vulnerability. In some instances, these differences have led to the identification of genes modulating the sensitivity to ethanol teratogenesis. More often, these differences allow researchers to study the effects of ethanol in a sensitized genetic background. The preponderance of our understanding of the genetic modulation of ethanol teratogenesis come from candidate gene and forward genetic approaches.

4.2.2 Forward genetics

In forward genetics, random mutations are generated and individual mutant lines are screened for a phenotype of interest. After identifying interesting phenotypes, the mutation is genetically mapped to identify the causative mutation. Due to the use of random mutagenesis, forward genetics is inherently unbiased, facilitating discovery of novel genes or novel functions of known genes. The downside of forward genetics, at least historically, has been the relatively arduous job of genetic mapping. However, improved methods of genetic mapping using whole genome sequencing and innovative transposable elements that result in fluorescence following mutagenic integrations (Clark et al., 2011; Obholzer et al., 2012), make this approach more time efficient.

Among vertebrate models, zebrafish are highly amenable to forward genetic screens given their external development and fecundity. Given the variety of FASD-related phenotypes, such screens could expand our understanding of the genetics modulating the outcomes of ethanol exposure. To date, one such screen of *N*-ethyl-*N*-nitrosourea (ENU)-mutagenized zebrafish has been published (Swartz, Lovely, McCarthy, Kuka, & Eberhart, 2020). From 126 mutagenized families, 5 mutations that disrupted craniofacial development were identified. The majority (4/5) only had abnormal phenotypes in the presence of ethanol. One mutation was genetically mapped and found to be a disruption in an uncharacterized gene (Swartz et al., 2020). This finding supports the utility of forward genetics in our understanding of FASD.

4.2.3 Reverse genetics, candidate-based studies

The other side of genetic screens is reverse genetics. Here, targeted mutagenesis is used to create loss-of-function mutants for known genes of interest. These studies benefit from being faster, but are more biased, as the researcher must already have genes of interest identified. Despite this, reverse genetic approaches have provided important insights into the genetics of ethanol teratogenicity. There are several examples of reverse genetics successfully identifying ethanol-sensitizing mutations, particularly in mouse and zebrafish.

4.2.3.1 Retinoic acid signaling

One of the first major developmental signaling pathways to be linked to modifying alcohol teratogenesis was retinoic acid (RA) signaling. RA signaling is critical in embryogenesis. RA, a metabolite of retinol (Vitamin A), is involved in patterning the anterior-posterior axis of the embryo, and RA signaling is perturbed by prenatal alcohol exposure (reviewed in (Fainsod, Bendelac-Kapon, & Shabtai, 2020; Shabtai & Fainsod, 2018)).

Analyses in both zebrafish and *Xenopus*, support a role for RA in ethanol teratogenesis. Marrs et al. found that exposing zebrafish embryos to 100 mM alcohol from 3 to 24 h post fertilization (hpf) (blastula through pharyngula) resulted in a variety of severe craniofacial defects (Marrs et al., 2010). However, co-treatment of embryos with the same dose of alcohol plus supplemented RA (1 nM) resulted in a remarkable rescue of alcohol-induced craniofacial defects. In *Xenopus* embryos, defects caused by high dose of alcohol (2% in media) could be phenocopied in embryos co-exposed to a low dose of alcohol (1% in media) combined with 20 μM 4-diethylaminobenzaldehyde (DEAB), an inhibitor of retinaldehyde dehydrogenase (Raldh). The authors then overexpressed *Raldh2 via* RNA microinjection and exposed these embryos to high dose alcohol. Consistent with previous work on direct RA supplementation, *Raldh2* overexpression rescued alcohol-induced defects (Kot-Leibovich & Fainsod, 2009). Together, these data suggest that susceptibility to alcohol-induced teratogenesis can be influenced by RA signaling status.

4.2.3.2 Hedgehog signaling pathway

Another genetic pathway with ample supportive evidence for a role in modifying ethanol susceptibility is the Sonic Hedgehog (Shh) signaling pathway. Shh signaling is required for brain and face development in vertebrates. Single allele mutations in the gene *SHH*, which encodes the critically important pathway ligand, or *GLI2*, encoding a downstream transcriptional activator of the pathway, are the most common causes of a severe midline brain malformation called holoprosencephaly (HPE). Another gene, *CDON*, is an important SHH co-receptor in the pathway. Interestingly, like FASD, HPE has a complex etiology that is thought to be caused by mixed genetic and environmental interactions (Addissie et al., 2020; Beames & Lipinski, 2020; Grinblat & Lipinski, 2019; Hong & Krauss, 2018; Roessler et al., 2003; Roessler & Muenke, 2010). Furthermore, some of the midline brain and face dysmorphologies observed in HPE are similar to malformations seen in FAS, including agenesis of the corpus callosum (Godin, Dehart, Parnell, O'Leary-Moore, & Sulik, 2011; Hong & Krauss, 2017; O'Leary-Moore, Parnell, Lipinski, & Sulik, 2011; Riley et al., 1995; Wozniak et al., 2009; Yang et al., 2011, 2012). Finally, alcohol has been shown to disrupt Shh signaling (Boschen, Ptacek, et al., 2021; Li et al., 2007). Thus, links between Shh signaling and alcohol in birth defects are extensive.

Similar to work with RA, several studies have rescued the effects of prenatal alcohol exposure by exogenously activating the Shh pathway. This has been accomplished using genetic mutants (*e.g., Ptch1* mutants) (Hong & Krauss, 2013), Smoothened Agonist (Burton et al., 2022), cholesterol (which indirectly augments Shh signaling (Li et al., 2007)), and retroviral infection (Ahlgren, Thakur, & Bronner-Fraser, 2002). Care must be taken when interpreting these results, as some approaches more accurately recapitulate normal biology than others. For example, *Ptch1* mutants or SAG treatment would be expected to more accurately maintain normal Shh paracrine signaling (Cordero et al., 2004; Eberhart, Swartz, Crump, & Kimmel, 2006; Marcucio, Cordero, Hu, & Helms, 2005). However, other approaches such as retroviral infection ectopically express *Shh* in the neural crest itself (Ahlgren et al., 2002). Given the pro-survival and growth functions of Shh signaling, such ectopic expression may not accurately model the underlying biology.

Direct genetic evidence for an interaction between Shh and alcohol focused on the genes *Shh* and *Gli2* and comes from analyses in mouse and zebrafish. Heterozygous *Shh* and *Gli2* mice are phenotypically indistinguishable from their wildtype littermates. However, ethanol exposure of *Shh* and *Gli2* heterozygous mice revealed a latent haploinsufficiency, wherein *Shh* and *Gli2* heterozygotes displayed more frequent malformations than siblings exposed *in utero* to the same ethanol dose (Kietzman, Everson, Sulik, & Lipinski, 2014). Similarly, zebrafish heterozygous for *shha* mutations (a zebrafish paralog of *Shh*) are sensitized to ethanol-induced midfacial defects (Everson, Batchu, & Eberhart, 2020).

While there is strong genetic evidence for Shh-alcohol interactions, whether these interactions are direct or not, is unclear. Cdon is a known co-receptor for the Shh pathway. Homozygous *Cdon* mutations are typically well tolerated and cause no overt malformations, however, Hong et al. discovered *Cdon* mutants were sensitized to alcohol induced malformations (Hong & Krauss, 2012). Loss of the negative regulator of Shh signaling *Patch1* (*Ptch1*), elevating Shh pathway activity, rescues ethanol-induced defects in *Cdon* mutants (Hong & Krauss, 2013). More recently, this group has determined that the Cdon-ethanol interaction is through disrupted Nodal signaling, operating upstream of Shh signaling (Hong, Christ, Christa, Willnow, & Krauss, 2020). Among the evidence for an indirect interaction, they found that heterozygosity for *Lefty2*, a nodal pathway inhibitor, and *Tdgf1*, the Nodal co-receptor, suppress and enhance the effects of ethanol on *Cdon* mutants, respectively. Together these series of experiments nicely demonstrate the role of Shh signaling in altering an embryo's susceptibility to ethanol-induced teratogenesis.

4.2.3.3 Ciliary genes

Phenotypes in FASD can also mirror those in ciliopathies (Schock & Brugmann, 2017), providing additional candidate genes. The ciliary genes *Kif3a* and *Mns1* have each been independently shown to modify alcohol teratogenesis (Boschen, Gong, Murdaugh, & Parnell, 2018). *Mns1* is required in motile cilia (Zhou, Yang, Leu, & Wang, 2012). *Mns1,* heterozygous mice were inbred to produce offspring of wildtype, heterozygous, and null embryos, which were exposed *in utero* to two doses of 2.9 g/kg ethanol at GD7 (gastrulation, 2–3 weeks in humans), the critical period for severe fetal alcohol syndrome (FAS) (Lipinski et al., 2012; Sulik, 1984; Sulik, Johnston, & Webb, 1981). While alcohol-exposed wildtypes had an incidence of eye defects of 41%, *Mns1* heterozygotes and homozygous nulls had incidences of 64% and 92%, respectively. In addition, the most severe defects (scores 5–7 on the authors' 1–7 severity scale) were only observed in heterozygous or null fetuses (Boschen et al., 2018). These heterozygous and null fetuses with the most severe eye defects also had brain and face malformations linked to FAS, including holoprosencephaly and a long upper lip with diminished upper lip notch. These phenotypes are also remarkably reminiscent of those caused by transient Hedgehog pathway inhibition at a close critical period of development (GD7.75) (Everson et al., 2019; Heyne et al., 2016).

The mechanism of these interactions is hypothesized to be *via* disrupted cell signaling of pathways like sonic hedgehog or retinoic acid. In the primitive node, motile cilia movements distribute morphogens for these pathways (reviewed in (Ryan & Izpisúa Belmonte, 2000)). Thus, disruption of motile cilia may reduce pathway activity because the morphogens do not reach their proper destinations. Evidence of this mechanism comes from the observation of a high incidence of *situs inversus* in *Mns1* mutants. *Situs inversus* is a left-right asymmetry defect that can result from disrupted nodal flow (reviewed in (Ryan & Izpisúa Belmonte, 2000)), and has previously been observed in *Mns1* mutants (Zhou et al., 2012). However, an alternative or parallel mechanism for the *Mns1*-alcohol interaction is *Mns1* may have a role in primary cilia, which are non-motile cilia that act as hubs of cell signaling (Anvarian, Mykytyn, Mukhopadhyay, Pedersen, & Christensen, 2019). For example, the sonic hedgehog pathway requires the primary cilia for signal transduction (Bangs & Anderson, 2017; Huangfu et al., 2003).

Currently, the contribution of each of these mechanisms has not been parsed. In support of a primary cilia model, the same authors more recently showed that mutations in the intra-flagellar transport primary cilia gene *Kif3a* sensitized embryos to alcohol-induced microcephaly (small head), which was associated with atypical neurobehavior (Boschen, Fish, & Parnell, 2021). The authors concluded that alcohol exposure may cause a "transient ciliopathy," resulting in abnormal cell signaling and abnormal development.

4.2.3.4 Nitric oxide pathway

Neuronal defects in FASD have led to the Nitric Oxide pathway being explored as a candidate for gene-ethanol interactions. Nitric Oxide (NO) is free radical with many known functions in normal physiology and disease pathology (reviewed in (Bruhwyler, Chleide, Liégeois, & Carreer, 1993)). In mammals, NO is produced from L-arginine by one of three isozymes of nitric oxide synthase (*NOS*), eNOS, nNOS, and iNOS (reviewed in (Bryan, Bian, & Murad, 2009; Davis & Syapin, 2005)). NO interactions have been described between alcohol and NOS in drinking adults (reviewed in (Davis & Syapin, 2005)). Less is known about the role of NO and NOS during development. However, Karacay et al. demonstrate *nNOS* null mice exposed to 0, 2.2, or 4.4 g/kg alcohol from postnatal days 4–9 (targeting the "brain growth spurt" which occurs during third trimester in humans (Dobbing & Sands, 1979)) display behavioral deficits that are not observed in alcohol-exposed wildtypes (Karacay, Bonthius, Plume, & Bonthius, 2015). Importantly, these behavioral deficits were coincident with neuronal cell death and structural brain malformations (microencephaly, small brain) (Karacay, Mahoney, Plume, & Bonthius, 2015), as had been observed previously (de Licona et al., 2009). Specifically, *nNOS* null male and female mice exposed to alcohol displayed significant reductions in brain weight at both the 2.2 and 4.4 g/kg dose levels, while wildtype animals showed only a non-significant reduction in brain weight. At the cellular level, combined loss of nNOS and exposure to alcohol resulted in reduced cell numbers in several important brain regions, the cerebral cortex, hippocampus, and dentate gyrus, which was not observed in alcohol-exposed wildtypes (Karacay, Mahoney, et al., 2015).

4.2.3.5 Growth factor and planar cell polarity pathways

The small size and high fecundity of zebrafish allow for a kind of hybrid between forward and reverse genetic approaches, the "shelf screen" or "phenotyping screen." In this approach, characterized mutant lines are tested for novel phenotypes. In an early shelf screen, five craniofacial mutants were tested for interactions with ethanol that exacerbated the craniofacial defects. Mutations in *platelet derived growth factor receptor alpha* (*pdgfra*) were found to sensitize embryos to ethanol-induced malformations (McCarthy et al., 2013). Homozygous *pdgfra* mutants displayed loss of the ethmoid plate, an analogous structure to the mammalian palate (Swartz, Sheehan-Rooney, Dixon, & Eberhart, 2011). However, *pdgfra* mutants exposed to a normally sub-teratogenic 1% ethanol dose displayed near complete loss of their facial skeletons (McCarthy et al., 2013). Interestingly, while unexposed *pdgfra* heterozygotes were indistinguishable from their wildtype siblings, ethanol exposure resulted in 62% of heterozygotes displaying ethmoid plate malformations, demonstrating a clear gene-alcohol interaction. Furthermore, while the defects in unexposed mutants are due to disrupted neural crest cell migration, apoptosis underlies the ethanol-treated phenotype. This highlights the strength of screening approaches, in their capacity to identify strongly synergistic interactions.

A follow-up shelf screen of 20 characterized mutants disrupting normal development was performed to identify gene-ethanol interactions that disrupted facial and/or neural development. Five of these mutations interacted with ethanol: *foxi1*, *hinfp*, *mars*, *plk1* and *vangl2* (Swartz et al., 2014). Of these, *vangl2* displayed midfacial defects ranging from hypotelorism (closely set eyes) to cyclopia (single medial eye). The planar cell polarity gene *vangl2* is required for the convergence and extension movements that drive embryogenesis (Yelin et al., 2005) and ethanol has been shown to disrupt convergent extension (Blader & Strahle, 1998). Bioinformatic analysis lead to the finding that this phenotype was due to the mispositioning and downstream perturbation of the Sonic Hedgehog pathway (Sidik et al., 2021).

4.2.4 Non-genetic factors in FASD

A growing body of evidence suggests environmental co-exposures can interact with alcohol to cause malformations at doses that are not hazardous on their own. Many teratogens cause malformations by perturbing important signaling pathways. Given this and the already defined gene-alcohol interactions, it follows that chemicals that perturb the activity of critical pathways or biological processes could also interact with ethanol to cause malformations.

4.2.4.1 Environmental toxicants

Few studies have examined interactions between alcohol and other environmental toxicants. However, the association between alcohol and the Shh pathway has predicted interactions with environmental inhibitors of the pathway. Piperonyl butoxide (PBO) is an EPA-registered pesticide synergist developed in the 1940s that is found in thousands of commercial and household pesticide products, making exposure to PBO nearly ubiquitous (Daiss & Edwards, 2006). Consistent with this, PBO was discovered to be a top 10 contaminant found in indoor dust (Rudel, Camann, Spengler, Korn, & Brody, 2003). PBO was discovered to block hedgehog signaling *via* smoothened (SMO) antagonism in 2012 (Wang et al., 2012). PBO exposure during early brain and face development dose-dependently caused HPE and associated midfacial defects in mice (Everson et al., 2019). In zebrafish, combined PBO and ethanol co-exposure synergistically caused craniofacial malformations at doses in which either chemical alone did not cause significant malformations (Everson et al., 2020). When this co-exposure was combined with heterozygosity for *shha*, the interaction was even more dramatic. This directly demonstrates the hazard potential of chemical mixtures, especially in the context of genetic predisposition, and raises the need for examination of chemical mixtures and gene-environment interactions in toxicant safety testing.

4.2.4.2 Cannabinoids

Another co-environmental interaction that unexpectedly appears to center upon the Shh signaling pathway is the interaction between alcohol and cannabinoids. The first direct evidence for an ethanol-cannabinoid interaction came from zebrafish (Boa-Amponsem, Zhang, Burton, Williams, & Cole, 2020). Zebrafish embryos were exposed to alcohol, a cannabinoid 1 receptor (Cb1r) agonist, or both. While single exposures did not alter tank-diving behavior (a fear/anxiety-associated behavior), co-exposure reduced tank diving, indicating increased risk-taking behavior (Boa-Amponsem et al., 2020). Phenotypes were rescued by addition of a Cb1r antagonist.

Around the same time, four cannabinoids, including Δ9-THC and cannabidiol (CBD), were tested and found to cause FASD-like phenotypes which also mimicked Shh mutants in mouse (Fish et al., 2019). Co-exposure of cannabinoids with alcohol caused more frequent malformations than exposure to cannabinoids or alcohol alone. The authors demonstrated that each cannabinoid was capable of inhibiting Shh pathway activity *via* direct antagonism of Smo (Fish et al., 2019). Cannabinoid-induced defects could be rescued by either addition of a Cb1r antagonist, consistent with the previous study (Boa-Amponsem et al., 2020), or restoration of Shh signaling by injection of *Shh* mRNA.

Together these studies provide evidence for alcohol interactions, wherein embryos can be chemically predisposed to alcohol-induced defects *via* co-exposure to Shh disruptors in the form of either cannabinoids or the pesticide synergist PBO.

5. Concluding remarks

What is a safe dose of alcohol? As discussed herein, FASD outcomes can be influenced by each individual's genetic susceptibility and co-exposures. This susceptibility is determined by a large number of known factors, and probably even more factors that have yet to be discovered. From this, it is clear that it is impossible at this time to define a dose of alcohol that is not potentially hazardous during pregnancy.

Acknowledgment

Funded by NIH F32DE030349 and T32AA007471 to JLE and R01AA023426, R35DE029086 and U01AA021651 to JKE.

References

Abecasis, G. R., et al. (2010). A map of human genome variation from population-scale sequencing. *Nature*, *467*, 1061–1073.

Abel, E. L. (1997). Was the fetal alcohol syndrome recognized in the ancient near east? *Alcohol and Alcoholism*, *32*, 3–7.

Abel, E. L. (1999). Was the fetal alcohol syndrome recognized by the Greeks and romans? *Alcohol and Alcoholism*, *34*, 868–872.

Addissie, Y. A., et al. (2020). Prenatal exposure to pesticides and risk for holoprosencephaly: A case-control study. *Environmental Health*, *19*, 65.

Ahlgren, S. C., Thakur, V., & Bronner-Fraser, M. (2002). Sonic hedgehog rescues cranial neural crest from cell death induced by ethanol exposure. *Proceedings of the National Academy of Sciences of the United States of America*, *99*, 10476–10481.

Anvarian, Z., Mykytyn, K., Mukhopadhyay, S., Pedersen, L. B., & Christensen, S. T. (2019). Cellular signalling by primary cilia in development, organ function and disease. *Nature Reviews. Nephrology*, *15*, 199–219.

Badger, T. M., et al. (2000). Cyclic expression of class I alcohol dehydrogenase in male rats treated with ethanol. *Biochemical and Biophysical Research Communications*, *274*, 684–688.

Bangs, F., & Anderson, K. V. (2017). Primary cilia and mammalian hedgehog signaling. *Cold Spring Harbor Perspectives in Biology*, *9*, a028175.

Beames, T. G., & Lipinski, R. J. (2020). Gene-environment interactions: Aligning birth defects research with complex etiology. *Development*, *147*, dev191064.

Beaty, T. H., et al. (2011). Evidence for gene-environment interaction in a genome wide study of nonsyndromic cleft palate. *Genetic Epidemiology*, *35*, 469–478.

Berres, M. E., Garic, A., Flentke, G. R., & Smith, S. M. (2017). Transcriptome profiling identifies ribosome biogenesis as a target of alcohol teratogenicity and vulnerability during early embryogenesis. *PLoS One*, *12*, e0169351.

Blader, P., & Strahle, U. (1998). Ethanol impairs migration of the prechordal plate in the zebrafish embryo. *Developmental Biology*, *201*, 185–201.

Boa-Amponsem, O., Zhang, C., Burton, D., Williams, K. P., & Cole, G. J. (2020). Ethanol and cannabinoids regulate zebrafish GABAergic neuron development and behavior in a sonic hedgehog and fibroblast growth factor-dependent mechanism. *Alcoholism, Clinical and Experimental Research*, *44*, 1366–1377.

Boschen, K. E., Fish, E. W., & Parnell, S. E. (2021). Prenatal alcohol exposure disrupts sonic hedgehog pathway and primary cilia genes in the mouse neural tube. *Reproductive Toxicology*, *105*, 136–147.

Boschen, K. E., Gong, H., Murdaugh, L. B., & Parnell, S. E. (2018). Knockdown of Mns1 increases susceptibility to craniofacial defects following gastrulation-stage alcohol exposure in mice. *Alcoholism, Clinical and Experimental Research*, *42*, 2136–2143.

Boschen, K. E., Ptacek, T. S., Berginski, M. E., Simon, J. M., & Parnell, S. E. (2021). Transcriptomic analyses of gastrulation-stage mouse embryos with differential susceptibility to alcohol. *Disease Models & Mechanisms*, *14*, dmm049012.

Boyles, A. L., et al. (2010). Maternal alcohol consumption, alcohol metabolism genes, and the risk of oral clefts: A population-based case-control study in Norway, 1996-2001. *American Journal of Epidemiology*, *172*, 924–931.

Branum, A. M., & Ahrens, K. A. (2017). Trends in timing of pregnancy awareness among US women. *Maternal and Child Health Journal*, *21*, 715–726.

Brent, R. L. (2004). Environmental causes of human congenital malformations: The pediatrician's role in dealing with these complex clinical problems caused by a multiplicity of environmental and genetic factors. *Pediatrics*, *113*, 957–968.

Brocardo, P. S., Gil-Mohapel, J., & Christie, B. R. (2011). The role of oxidative stress in fetal alcohol spectrum disorders. *Brain Research Reviews*, *67*, 209–225.

Brown, J. M., Bland, R., Jonsson, E., & Greenshaw, A. J. (2019). A brief history of awareness of the link between alcohol and fetal alcohol Spectrum disorder. *Canadian Journal of Psychiatry*, *64*, 164–168.

Brown, K. H., et al. (2012). Extensive genetic diversity and substructuring among zebrafish strains revealed through copy number variant analysis. *Proceedings of the National Academy of Sciences of the United States of America*, *109*, 529–534.

Bruhwyler, J., Chleide, E., Liégeois, J. F., & Carreer, F. (1993). Nitric oxide: A new messenger in the brain. *Neuroscience and Biobehavioral Reviews*, *17*, 373–384.

Bryan, N. S., Bian, K., & Murad, F. (2009). Discovery of the nitric oxide signaling pathway and targets for drug development. *Frontiers in Bioscience (Landmark Edition)*, *14*, 1–18.

Burton, D. F., et al. (2022). Pharmacological activation of the sonic hedgehog pathway with a smoothened small molecule agonist ameliorates the severity of alcohol-induced morphological and behavioral birth defects in a zebrafish model of fetal alcohol spectrum disorder. *Journal of Neuroscience Research*, *100*, 1585–1601.

Caputo, C., Wood, E., & Jabbour, L. (2016). Impact of fetal alcohol exposure on body systems: A systematic review. *Birth Defects Research. Part C, Embryo Today*, *108*, 174–180.

Charlap, J. H., Donahue, R. J., & Knudsen, T. B. (2003). Exposure-disease continuum for 2-chloro-2′-deoxyadenosine, a prototype ocular teratogen. 3. Intervention with PK11195. *Birth Defects Research. Part A, Clinical and Molecular Teratology*, *67*, 108–115.

Chen, Y., Ozturk, N. C., Ni, L., Goodlett, C., & Zhou, F. C. (2011). Strain differences in developmental vulnerability to alcohol exposure via embryo culture in mice. *Alcoholism, Clinical and Experimental Research*, *35*, 1293–1304.

Chevrier, C., et al. (2005). Interaction between the ADH1C polymorphism and maternal alcohol intake in the risk of nonsyndromic oral clefts: An evaluation of the contribution of child and maternal genotypes. *Birth Defects Research. Part A, Clinical and Molecular Teratology*, *73*, 114–122.

Clark, K. J., et al. (2011). In vivo protein trapping produces a functional expression codex of the vertebrate proteome. *Nature Methods*, *8*, 506–515.

Cordero, D., et al. (2004). Temporal perturbations in sonic hedgehog signaling elicit the spectrum of holoprosencephaly phenotypes. *The Journal of Clinical Investigation*, *114*, 485–494.

Coriale, G., et al. (2013). Fetal alcohol Spectrum disorder (FASD): Neurobehavioral profile, indications for diagnosis and treatment. *Rivista di Psichiatria*, *48*, 359–369.

Crabb, D. W., Matsumoto, M., Chang, D., & You, M. (2004). Overview of the role of alcohol dehydrogenase and aldehyde dehydrogenase and their variants in the genesis of alcohol-related pathology. *The Proceedings of the Nutrition Society*, *63*, 49–63.

Daiss, R., & Edwards, D. (2006). *Office of pesticide programs*. Washington, D.C.: United States Environment Protection Agency.

Das, U. G., Cronk, C. E., Martier, S. S., Simpson, P. M., & McCarver, D. G. (2004). Alcohol dehydrogenase 2*3 affects alterations in offspring facial morphology associated with maternal ethanol intake in pregnancy. *Alcoholism, Clinical and Experimental Research*, *28*, 1598–1606.

Davis, R. L., & Syapin, P. J. (2005). Interactions of alcohol and nitric-oxide synthase in the brain. *Brain Research. Brain Research Reviews*, *49*, 494–504.

de Licona, H. K., et al. (2009). A single exposure to alcohol during brain development induces microencephaly and neuronal losses in genetically susceptible mice, but not in wild type mice. *Neurotoxicology*, *30*, 459–470.

Debelak, K. A., & Smith, S. M. (2000). Avian genetic background modulates the neural crest apoptosis induced by ethanol exposure. *Alcoholism, Clinical and Experimental Research*, *24*, 307–314.

Denny, L., Coles, S., & Blitz, R. (2017). Fetal alcohol syndrome and fetal alcohol Spectrum disorders. *American Family Physician*, *96*, 515–522.

DeRoo, L. A., Wilcox, A. J., Drevon, C. A., & Lie, R. T. (2008). First-trimester maternal alcohol consumption and the risk of infant oral clefts in Norway: A population-based case-control study. *American Journal of Epidemiology*, *168*, 638–646.

Dobbing, J., & Sands, J. (1979). Comparative aspects of the brain growth spurt. *Early Human Development*, *3*, 79–83.

Downing, C., Balderrama-Durbin, C., Broncucia, H., Gilliam, D., & Johnson, T. E. (2009). Ethanol teratogenesis in five inbred strains of mice. *Alcoholism, Clinical and Experimental Research*, *33*, 1238–1245.

Duncan, J. M. (1888). On alcoholism in Gynaecology and obstetrics. *The Transactions of the Edinburgh Obstetrical Society*, *13*, 105–120.

Eberhart, J. K., Swartz, M. E., Crump, J. G., & Kimmel, C. B. (2006). Early hedgehog signaling from neural to oral epithelium organizes anterior craniofacial development. *Development*, *133*, 1069–1077.

Everson, J. L., Batchu, R., & Eberhart, J. K. (2020). Multifactorial genetic and environmental hedgehog pathway disruption sensitizes embryos to alcohol-induced craniofacial defects. *Alcoholism, Clinical and Experimental Research*, *44*, 1988–1996.

Everson, J. L., et al. (2019). Developmental toxicity assessment of Piperonyl Butoxide exposure targeting sonic hedgehog signaling and forebrain and face morphogenesis in the mouse: An in vitro and in vivo study. *Environmental Health Perspectives*, *127*, 107006.

Fainsod, A., Bendelac-Kapon, L., & Shabtai, Y. (2020). Fetal alcohol Spectrum disorder: Embryogenesis under reduced retinoic acid signaling conditions. *Sub-Cellular Biochemistry, 95*, 197–225.

Feldkamp, M. L., Carey, J. C., Byrne, J. L. B., Krikov, S., & Botto, L. D. (2017). Etiology and clinical presentation of birth defects: Population based study. *BMJ, 357*, j2249.

Finer, L. B., & Zolna, M. R. (2016). Declines in unintended pregnancy in the United States, 2008-2011. *The New England Journal of Medicine, 374*, 843–852.

Fish, E. W., et al. (2019). Cannabinoids exacerbate alcohol Teratogenesis by a CB1-hedgehog interaction. *Scientific Reports, 9*, 16057.

Garic, A., Berres, M. E., & Smith, S. M. (2014). High-throughput transcriptome sequencing identifies candidate genetic modifiers of vulnerability to fetal alcohol spectrum disorders. *Alcoholism, Clinical and Experimental Research, 38*, 1874–1882.

Godin, E. A., Dehart, D. B., Parnell, S. E., O'Leary-Moore, S. K., & Sulik, K. K. (2011). Ventromedian forebrain dysgenesis follows early prenatal ethanol exposure in mice. *Neurotoxicology and Teratology, 33*, 231–239.

Great Britain. Parliament. House of commons. Select committee on inquiry into drunkenness, & Buckingham, J. S. (1834). *Report from the Select committee on inquiry into drunkenness, with Minutes of evidence, and Appendix. (London, 1834)* (p. xii). Irish University Press. 445 p. [from old catalog].

Green, M. L., et al. (2007). Reprogramming of genetic networks during initiation of the fetal alcohol syndrome. *Developmental Dynamics, 236*, 613–631.

Grinblat, Y., & Lipinski, R. J. (2019). A forebrain undivided: Unleashing model organisms to solve the mysteries of holoprosencephaly. *Developmental Dynamics, 248*, 626–633.

Haddon, J. (1876). On intemperance in women, with special reference to its effects on the reproductive system. *British Medical Journal, 1*, 748–750.

Hammer, R., & Rapp, E. (2022). Women's views and experiences of occasional alcohol consumption during pregnancy: A systematic review of qualitative studies and their recommendations. *Midwifery, 111*, 103357.

Hanson, J. W., Streissguth, A. P., & Smith, D. W. (1978). The effects of moderate alcohol consumption during pregnancy on fetal growth and morphogenesis. *The Journal of Pediatrics, 92*, 457–460.

Harlap, S., & Shiono, P. H. (1980). Alcohol, smoking, and incidence of spontaneous abortions in the first and second trimester. *Lancet, 2*, 173–176.

Hemingway, S. J. A., et al. (2018). Twin study confirms virtually identical prenatal alcohol exposures can lead to markedly different fetal alcohol spectrum disorder outcomes-fetal genetics influences fetal vulnerability. *Advances in Pediatric Research, 5*, 23. https://doi.org/10.24105/apr.2019.5.23.

Heyne, G. W., et al. (2016). Gli2 gene-environment interactions contribute to the etiological complexity of holoprosencephaly: Evidence from a mouse model. *Disease Models & Mechanisms, 9*, 1307–1315.

Hingson, R., et al. (1982). Effects of maternal drinking and marijuana use on fetal growth and development. *Pediatrics, 70*, 539–546.

Hong, M., Christ, A., Christa, A., Willnow, T. E., & Krauss, R. S. (2020). Mutation and fetal alcohol converge on nodal signaling in a mouse model of holoprosencephaly. *eLife, 9*.

Hong, M., & Krauss, R. S. (2012). Cdon mutation and fetal ethanol exposure synergize to produce midline signaling defects and holoprosencephaly spectrum disorders in mice. *PLoS Genetics, 8*, e1002999.

Hong, M., & Krauss, R. S. (2013). Rescue of holoprosencephaly in fetal alcohol-exposed Cdon mutant mice by reduced gene dosage of Ptch1. *PLoS One, 8*, e79269.

Hong, M., & Krauss, R. S. (2017). Ethanol itself is a holoprosencephaly-inducing teratogen. *PLoS One, 12*, e0176440.

Hong, M., & Krauss, R. S. (2018). Modeling the complex etiology of holoprosencephaly in mice. *American Journal of Medical Genetics. Part C, Seminars in Medical Genetics*, *178*, 140–150.

Howard, J. T., et al. (2022). Trends in binge drinking and heavy alcohol consumption among pregnant women in the US, 2011 to 2020. *JAMA Network Open*, *5*, e2224846.

Huangfu, D., et al. (2003). Hedgehog signalling in the mouse requires intraflagellar transport proteins. *Nature*, *426*, 83–87.

Jacobson, S. W., et al. (2006). Protective effects of the alcohol dehydrogenase-ADH1B allele in children exposed to alcohol during pregnancy. *The Journal of Pediatrics*, *148*, 30–37.

Jones, K. L., & Smith, D. W. (1973). Recognition of the fetal alcohol syndrome in early infancy. *Lancet*, *302*, 999–1001.

Jones, K. L., Smith, D. W., Ulleland, C. N., & Streissguth, P. (1973). Pattern of malformation in offspring of chronic alcoholic mothers. *Lancet*, *1*, 1267–1271.

Kaminski, M., Rumeau-Rouquette, C., & Schwartz, D. (1976). Alcohol consumption among pregnant women and outcome of pregnancy (author's transl). *Revue d'Épidémiologie, Médecine Sociale et Santé Publique*, *24*, 27–40.

Karacay, B., Bonthius, N. E., Plume, J., & Bonthius, D. J. (2015). Genetic absence of nNOS worsens fetal alcohol effects in mice. I: Behavioral deficits. *Alcoholism, Clinical and Experimental Research*, *39*, 212–220.

Karacay, B., Mahoney, J., Plume, J., & Bonthius, D. J. (2015). Genetic absence of nNOS worsens fetal alcohol effects in mice. II: Microencephaly and neuronal losses. *Alcoholism, Clinical and Experimental Research*, *39*, 221–231.

Khokha, M. K., Mitchell, L. E., & Wallingford, J. B. (2017). White paper on the study of birth defects. *Birth Defects Research*, *109*, 180–185.

Kietzman, H. W., Everson, J. L., Sulik, K. K., & Lipinski, R. J. (2014). The teratogenic effects of prenatal ethanol exposure are exacerbated by sonic hedgehog or GLI2 haploinsufficiency in the mouse. *PLoS One*, *9*, e89448.

Kline, J., Shrout, P., Stein, Z., Susser, M., & Warburton, D. (1980). Drinking during pregnancy and spontaneous abortion. *Lancet*, *2*, 176–180.

Koppe, J. G., et al. (2006). Exposure to multiple environmental agents and their effect. *Acta Paediatrica. Supplement*, *95*, 106–113.

Kot-Leibovich, H., & Fainsod, A. (2009). Ethanol induces embryonic malformations by competing for retinaldehyde dehydrogenase activity during vertebrate gastrulation. *Disease Models & Mechanisms*, *2*, 295–305.

Krauss, R. S., & Hong, M. (2016). Gene–Environment Interactions and the Etiology of Birth Defects. *Current Topics in Developmental Biology*, *116*, 569–580.

Krewski, D., et al. (2010). Toxicity testing in the 21st century: A vision and a strategy. *Journal of Toxicology and Environmental Health. Part B, Critical Reviews*, *13*, 51–138.

Kruszka, P., et al. (2019). Cohesin complex-associated holoprosencephaly. *Brain: A Journal of Neurology*, *142*, 2631–2643.

Langevin, F., Crossan, G. P., Rosado, I. V., Arends, M. J., & Patel, K. J. (2011). Fancd2 counteracts the toxic effects of naturally produced aldehydes in mice. *Nature*, *475*, 53–58.

Lemoine, P. (1997). The history of alcoholic fetopathies. *Journal of Population Therapeutics and Clinical Pharmacology*, *19*, e224-226 (2012).

Lemoine, P., Harousseau, H., Borteyru, J. P., & Menuet, J. C. (2003). Children of alcoholic parents—Observed anomalies: discussion of 127 cases. *Therapeutic Drug Monitoring*, *25*, 132–136.

Li, Y. X., et al. (2007). Fetal alcohol exposure impairs hedgehog cholesterol modification and signaling. *Laboratory Investigation*, *87*, 231–240.

Lipinski, R. J., et al. (2012). Ethanol-induced face-brain Dysmorphology patterns are correlative and exposure-stage dependent. *PLoS One*, 7, e43067.

Loucks, E., & Carvan, M. J. (2004). Strain-dependent effects of developmental ethanol exposure in zebrafish. *Neurotoxicology and Teratology*, *26*, 745–755.

Lovely, C. B., Nobles, R. D., & Eberhart, J. K. (2014). Developmental age strengthens barriers to ethanol accumulation in zebrafish. *Alcohol*, *48*, 595–602.

Marcucio, R. S., Cordero, D. R., Hu, D., & Helms, J. A. (2005). Molecular interactions coordinating the development of the forebrain and face. *Developmental Biology*, *284*, 48–61.

Marrs, J. A., et al. (2010). Zebrafish fetal alcohol syndrome model: Effects of ethanol are rescued by retinoic acid supplement. *Alcohol*, *44*, 707–715.

Mattapallil, M. J., et al. (2012). The Rd8 mutation of the Crb1 gene is present in vendor lines of C57BL/6N mice and embryonic stem cells, and confounds ocular induced mutant phenotypes. *Investigative Ophthalmology & Visual Science*, *53*, 2921–2927.

Matthews, T. J., MacDorman, M. F., & Thoma, M. E. (2015). Infant mortality statistics from the 2013 period linked birth/infant death data set. *National Vital Statistics Reports*, *64*, 1–30.

May, P. A., et al. (2009). Prevalence and epidemiologic characteristics of FASD from various research methods with an emphasis on recent in-school studies. *Developmental Disabilities Research Reviews*, *15*, 176–192.

May, P. A., et al. (2014). Prevalence and characteristics of fetal alcohol spectrum disorders. *Pediatrics*, *134*, 855–866.

May, P. A., et al. (2020a). Fetal alcohol Spectrum disorders in a Midwestern City: Child characteristics, maternal risk traits, and prevalence. *Alcoholism, Clinical and Experimental Research*, *44*, 919–938.

May, P. A., et al. (2020b). Fetal alcohol Spectrum disorders in a Rocky Mountain Region City: Child characteristics, maternal risk traits, and prevalence. *Alcoholism, Clinical and Experimental Research*, *44*, 900–918.

May, P. A., et al. (2020c). Fetal alcohol Spectrum disorders in a Southeastern County of the United States: Child characteristics and maternal risk traits. *Alcoholism, Clinical and Experimental Research*, *44*, 939–959.

McCarthy, N., et al. (2013). Pdgfra protects against ethanol-induced craniofacial defects in a zebrafish model of FASD. *Development*, *140*, 3254–3265.

McCarver, D. G., Thomasson, H. R., Martier, S. S., Sokol, R. J., & Li, T. (1997). Alcohol dehydrogenase-2*3 allele protects against alcohol-related birth defects among African Americans. *The Journal of Pharmacology and Experimental Therapeutics*, *283*, 1095–1101.

McGovern, P. E., et al. (2004). Fermented beverages of pre- and proto-historic China. *Proceedings of the National Academy of Sciences of the United States of America*, *101*, 17593–17598.

Mitro, S. D., Johnson, T., & Zota, A. R. (2015). Cumulative chemical exposures during pregnancy and early development. *Current Environmental Health Reports*, *2*, 367–378.

National Conference on Infantile Mortality. (1906). *Report of the proceedings of the National Conference on Infantile Mortality: held in the Caxton Hall, Westminster, on the 13th and 14th June. Vol. 1906* (p. viii). P.S. King & son, London. 314 p: folded chart.

Obholzer, N., et al. (2012). Rapid positional cloning of zebrafish mutations by linkage and homozygosity mapping using whole-genome sequencing. *Development*, *139*, 4280–4290.

Obladen, M. (2021). Ignored papers, invented quotations: A history of fetal alcohol syndrome. *Neonatology*, *118*, 647–653.

O'Connor, S., Morzorati, S., Christian, J., & Li, T. K. (1998). Clamping breath alcohol concentration reduces experimental variance: Application to the study of acute tolerance to alcohol and alcohol elimination rate. *Alcoholism, Clinical and Experimental Research*, *22*, 202–210.

Ogawa, T., Kuwagata, M., Ruiz, J., & Zhou, F. C. (2005). Differential teratogenic effect of alcohol on embryonic development between C57BL/6 and DBA/2 mice: A new view. *Alcoholism, Clinical and Experimental Research, 29*, 855–863.
O'Hara, M. F., Charlap, J. H., Craig, R. C., & Knudsen, T. B. (2002). Mitochondrial transduction of ocular teratogenesis during methylmercury exposure. *Teratology, 65*, 131–144.
O'Hara, M. F., et al. (2003). Mitochondrial benzodiazepine receptors regulate oxygen homeostasis in the early mouse embryo. *Reproductive Toxicology, 17*, 365–375.
O'Leary-Moore, S. K., Parnell, S. E., Lipinski, R. J., & Sulik, K. K. (2011). Magnetic resonance-based imaging in animal models of fetal alcohol spectrum disorder. *Neuropsychology Review, 21*, 167–185.
Parnell, S. E., Sulik, K. K., Dehart, D. B., & Chen, S. Y. (2010). Reduction of ethanol-induced ocular abnormalities in mice through dietary administration of N-acetylcysteine. *Alcohol, 44*, 699–705.
Riley, E. P., Infante, M. A., & Warren, K. R. (2011). Fetal alcohol spectrum disorders: An overview. *Neuropsychology Review, 21*, 73–80.
Riley, E. P., et al. (1995). Abnormalities of the corpus callosum in children prenatally exposed to alcohol. *Alcoholism, Clinical and Experimental Research, 19*, 1198–1202.
Roessler, E., & Muenke, M. (2010). The molecular genetics of holoprosencephaly. *American Journal of Medical Genetics. Part C, Seminars in Medical Genetics, 154C*, 52–61.
Roessler, E., et al. (2003). Loss-of-function mutations in the human GLI2 gene are associated with pituitary anomalies and holoprosencephaly-like features. *Proceedings of the National Academy of Sciences of the United States of America, 100*, 13424–13429.
Ronchi, J. A., et al. (2013). A spontaneous mutation in the nicotinamide nucleotide transhydrogenase gene of C57BL/6J mice results in mitochondrial redox abnormalities. *Free Radical Biology & Medicine, 63*, 446–456.
Rouquette, J. (1957). *Influence de la toxicomanie alcoolique parentale sur le developpement physique & psychique des jeunes enfants*. Paris: University De Paris.
Rudel, R. A., Camann, D. E., Spengler, J. D., Korn, L. R., & Brody, J. G. (2003). Phthalates, alkylphenols, pesticides, polybrominated diphenyl ethers, and other endocrine-disrupting compounds in indoor air and dust. *Environmental Science & Technology, 37*, 4543–4553.
Ryan, A. K., & Izpisúa Belmonte, J. C. (2000). Establishing a left-right axis in the embryo. *IUBMB Life, 50*, 1–11.
Schock, E. N., & Brugmann, S. A. (2017). Discovery, diagnosis, and etiology of craniofacial ciliopathies. *Cold Spring Harbor Perspectives in Biology, 9*, a028258.
Shabtai, Y., & Fainsod, A. (2018). Competition between ethanol clearance and retinoic acid biosynthesis in the induction of fetal alcohol syndrome. *Biochemistry and Cell Biology, 96*, 148–160.
Sidik, A., et al. (2021). Exposure to ethanol leads to midfacial hypoplasia in a zebrafish model of FASD via indirect interactions with the Shh pathway. *BMC Biology, 19*, 134.
Stoler, J. M., Ryan, L. M., & Holmes, L. B. (2002). Alcohol dehydrogenase 2 genotypes, maternal alcohol use, and infant outcome. *The Journal of Pediatrics, 141*, 780–785.
Streissguth, A. P., & Dehaene, P. (1993). Fetal alcohol syndrome in twins of alcoholic mothers: Concordance of diagnosis and IQ. *American Journal of Medical Genetics, 47*, 857–861.
Su, B., Debelak, K. A., Tessmer, L. L., Cartwright, M. M., & Smith, S. M. (2001). Genetic influences on craniofacial outcome in an avian model of prenatal alcohol exposure. *Alcoholism, Clinical and Experimental Research, 25*, 60–69.
Sulik, K. K. (1984). Critical periods for alcohol teratogenesis in mice, with special reference to the gastrulation stage of embryogenesis. *Ciba Foundation Symposium, 105*, 124–141.

Sulik, K. K. (2005). Genesis of alcohol-induced craniofacial dysmorphism. *Experimental Biology and Medicine, 230*, 366–375.

Sulik, K. K., Johnston, M. C., & Webb, M. A. (1981). Fetal alcohol syndrome: Embryogenesis in a mouse model. *Science, 214*, 936–938.

Sullivan, W. C. (2011). A note on the influence of maternal inebriety on the offspring. 1899. *International Journal of Epidemiology, 40*, 278–282.

Swartz, M. E., Lovely, C. B., McCarthy, N., Kuka, T., & Eberhart, J. K. (2020). Novel ethanol-sensitive mutants identified in an F3 forward genetic screen. *Alcoholism, Clinical and Experimental Research, 44*, 56–65.

Swartz, M. E., Sheehan-Rooney, K., Dixon, M. J., & Eberhart, J. K. (2011). Examination of a palatogenic gene program in zebrafish. *Developmental Dynamics, 240*, 2204–2220.

Swartz, M. E., et al. (2014). A screen of zebrafish mutants identifies ethanol-sensitive genetic loci. *Alcoholism, Clinical and Experimental Research, 38*, 694–703.

Tennes, K., & Blackard, C. (1980). Maternal alcohol consumption, birth weight, and minor physical anomalies. *American Journal of Obstetrics and Gynecology, 138*, 774–780.

Thomasson, H. R., Beard, J. D., & Li, T. K. (1995). ADH2 gene polymorphisms are determinants of alcohol pharmacokinetics. *Alcoholism, Clinical and Experimental Research, 19*, 1494–1499.

Tice, R. R., Austin, C. P., Kavlock, R. J., & Bucher, J. R. (2013). Improving the human hazard characterization of chemicals: A Tox21 update. *Environmental Health Perspectives, 121*, 756–765.

Ulleland, C. N. (1972). The offspring of alcoholic mothers. *Annals of the New York Academy of Sciences, 197*, 167–169.

United States. Congress. Senate. Committee on Labor and Human Resources. Subcommittee on Alcoholism and Drug Abuse. (1982). *Effects of alcohol consumption during pregnancy: Hearing before the Subcommittee on Alcoholism and Drug Abuse of the Committee on Labor and Human Resources, United States Senate, Ninety-seventh Congress, second session, on to examine the effects of maternal alcohol consumption during pregnancy, September 21. Vol. 148* (p. dev192062). Washington: U.S. G.P.O.

Viljoen, D. L., et al. (2001). Alcohol dehydrogenase-2*2 allele is associated with decreased prevalence of fetal alcohol syndrome in the mixed-ancestry population of the Western Cape Province, South Africa. *Alcoholism, Clinical and Experimental Research, 25*, 1719–1722.

Wallingford, J. B. (2021). Aristotle, Buddhist scripture and embryology in ancient Mexico: Building inclusion by re-thinking what counts as the history of developmental biology. *Development, 148*.

Wang, J., et al. (2012). The insecticide synergist piperonyl butoxide inhibits hedgehog signaling: Assessing chemical risks. *Toxicological Sciences, 128*, 517–523.

Warner, R. H., & Rosett, H. L. (1975). The effects of drinking on offspring: An historical survey of the American and British literature. *Journal of Studies on Alcohol, 36*, 1395–1420.

Warren, K. R. (2015). A review of the history of attitudes toward drinking in pregnancy. *Alcoholism, Clinical and Experimental Research, 39*, 1110–1117.

Warren, K. R., & Li, T. K. (2005). Genetic polymorphisms: Impact on the risk of fetal alcohol spectrum disorders. *Birth Defects Research. Part A, Clinical and Molecular Teratology, 73*, 195–203.

Wentzel, P., & Eriksson, U. J. (2006). Ethanol-induced fetal dysmorphogenesis in the mouse is diminished by high antioxidative capacity of the mother. *Toxicological Sciences, 92*, 416–422.

Wilhoit, L. F., Scott, D. A., & Simecka, B. A. (2017). Fetal alcohol Spectrum disorders: Characteristics, complications, and treatment. *Community Mental Health Journal, 53*, 711–718.

Wozniak, J. R., Riley, E. P., & Charness, M. E. (2019). Clinical presentation, diagnosis, and management of fetal alcohol spectrum disorder. *Lancet Neurology*, *18*, 760–770.
Wozniak, J. R., et al. (2009). Microstructural corpus callosum anomalies in children with prenatal alcohol exposure: An extension of previous diffusion tensor imaging findings. *Alcoholism, Clinical and Experimental Research*, *33*, 1825–1835.
Yang, Y., et al. (2011). Abnormal cortical thickness alterations in fetal alcohol Spectrum disorders and their relationships with facial Dysmorphology. *Cerebral Cortex*, *22*, 1170–1179.
Yang, Y., et al. (2012). Callosal thickness reductions relate to facial dysmorphology in fetal alcohol spectrum disorders. *Alcoholism, Clinical and Experimental Research*, *36*, 798–806.
Yelin, R., et al. (2005). Ethanol exposure affects gene expression in the embryonic organizer and reduces retinoic acid levels. *Developmental Biology*, *279*, 193–204.
Yin, X., Li, J., Li, Y., & Zou, S. (2019). Maternal alcohol consumption and oral clefts: A meta-analysis. *The British Journal of Oral & Maxillofacial Surgery*, *57*, 839–846.
Zaso, M. J., et al. (2018). Interaction between the ADH1B*3 allele and drinking motives on alcohol use among black college students. *The American Journal of Drug and Alcohol Abuse*, *44*, 329–338.
Zhang, Y., Wang, H., Li, Y., & Peng, Y. (2018). A review of interventions against fetal alcohol spectrum disorder targeting oxidative stress. *International Journal of Developmental Neuroscience*, *71*, 140–145.
Zhou, J., Yang, F., Leu, N. A., & Wang, P. J. (2012). MNS1 is essential for spermiogenesis and motile ciliary functions in mice. *PLoS Genetics*, *8*, e1002516.

CHAPTER FOUR

The role of genes and environment in the etiology of congenital diaphragmatic hernias

Nathan G. Burns and Gabrielle Kardon*
Department of Human Genetics, University of Utah, Salt Lake City, UT, United States
*Corresponding author: e-mail address: gkardon@genetics.utah.edu

Contents

Abstract

Structural birth defects are a common cause of abnormalities in newborns. While there are cases of structural birth defects arising due to monogenic defects or environmental exposures, many birth defects are likely caused by a complex interaction between genes and the environment. A structural birth defect with complex etiology is congenital diaphragmatic hernias (CDH), a common and often lethal disruption in diaphragm development. Mutations in more than 150 genes have been implicated in CDH pathogenesis. Although there is generally less evidence for a role for environmental factors in the etiology of CDH, deficiencies in maternal vitamin A and its derivative embryonic retinoic acid are strongly associated with CDH. However, the incomplete penetrance of CDH-implicated genes and environmental factors such as vitamin A deficiency suggest that interactions between genes and environment may be necessary to cause CDH. In this review, we examine the genetic and environmental factors implicated in diaphragm and CDH development. In addition, we evaluate the potential for gene-environment

Current Topics in Developmental Biology, Volume 152
ISSN 0070-2153
https://doi.org/10.1016/bs.ctdb.2022.10.004

interactions in CDH etiology, focusing on the potential interactions between the CDH-implicated gene, *Gata4*, and maternal vitamin A deficiency.

1. Introduction

One of every 33 babies suffers from serious structural birth defects such as congenital heart defects, cleft palate, urogenital and limb malformations, and congenital diaphragmatic hernias (Centers for Disease Control and Prevention, 2008). Birth defects can be caused by both genetic and environmental factors. Some structural birth defects, such as achondroplasia, craniofrontonasal syndrome, and focal dermal hypoplasia, are monogenic defects (Brady et al., 2015; Colvin, Bohne, Harding, McEwen, & Ornitz, 1996; Harmsen et al., 2009; Niethamer et al., 2020; Twigg et al., 2006). Other defects arise from environmental insults, such as teratogen exposure. For example, defects in limb, heart, and vertebrae development are associated with exposure to thalidomide, alcohol, tobacco, and antiepileptic drugs (Alexander, Clark, & Tuan, 2016; Caputo, Wood, & Jabbour, 2016; Rogers, 2009). Overall, chromosomal and genetic defects are estimated to account for 20–25% of birth defects, while environmental insults cause 10–12% birth defects; thus, the remaining 63–70% are presumed to result from a combination of genetic mutation(s) and environmental exposure(s) (Alexander et al., 2016).

A relatively small number of structural birth defects have been formally demonstrated to be caused by the interaction of genetic and environmental factors. Craniofacial defects are an excellent example where such interactions have been experimentally tested and shown using zebrafish and mouse models (Hong, Christ, Christa, Willnow, & Krauss, 2020; Hong & Krauss, 2012; Kietzman, Everson, Sulik, & Lipinski, 2014; McCarthy et al., 2013). Mutations in *Cdon*, *Gli2*, or *Shh* coupled with maternal alcohol exposure during organogenesis lead to a higher incidence of holoprosencephaly (a defect in midline pattering of the forebrain and midface) than those with only alcohol exposure (Hong et al., 2020; Hong & Krauss, 2012; Kietzman et al., 2014), showing that alcohol exacerbates the effect of genetically sensitized individuals. Another example is congenital scoliosis, a lateral curvature of the spine, caused by defects in vertebral development. Based on human genetic studies, deleterious mutations in *HES7* were found to be a dominant, but weakly penetrant cause of scoliosis (Sparrow et al., 2012). Using mice that were *Hes7*+/−, the authors found that short-term gestational hypoxia increased the severity and penetrance of vertebral defects and thus demonstrated a synergistic interaction between genes and environment. Finally, congenital heart defects are the most common type of structural birth defect, and at least 100 genes have been implicated as well as a multitude of environmental factors, including environmental teratogens and maternal alcohol exposure (Moreau et al., 2019). Recently, short-term gestational hypoxia has been shown to increase the incidence and severity of heart defects of mouse embryos harboring heterozygous null mutations for several genes (*Fgfr1*, *Fgfr2*, and *Tbx1*) implicated in heart development (Moreau et al., 2019).

Another structural birth defect in which both genetic and environmental factors have been implicated in its etiology is Congenital Diaphragmatic Hernia (CDH). The diaphragm is a skeletal muscle that is essential for respiration and acts as a barrier separating the abdominal and thoracic cavities. CDH is a common birth defect (1/3000 births; Torfs, Curry, Bateson, & Honoré, 1992) in which the diaphragm does not develop completely, its barrier function is lost, and abdominal contents herniate into the thoracic cavity (Pober, 2007). The resulting lung hypoplasia is the cause of the 50% mortality rate in CDH patients (Pober, 2007). The etiology of CDH is complex with over 100 genes implicated in its pathogenesis (Kardon et al., 2017). Correlative studies suggest that environmental factors may also contribute to the etiology of CDH, although there is currently no direct evidence linking an environmental insult with development of CDH in humans. The prevalence of CDH and the lack of common, highly penetrant genetic or environmental causes of CDH suggest that interactions between genetic variants and environmental factors may contribute to CDH etiology (Clugston, Zhang, Lvarez, De Lera, & Greer, 2010; Rocke et al., 2021; Warkany & Roth, 1948; Wilson, Roth, & Warkany, 1953). In this review, we will explore the role of genetic, environmental, and gene-environment interactions in the development of CDH.

2. Diaphragm development

The diaphragm is composed of costal and crural domains (Fig. 1A; Merrell & Kardon, 2013). The crural muscle surrounds the esophagus and aorta, while the costal diaphragm is composed of a radial array of muscle fibers surrounded by muscle connective tissue that connect medially to a central tendon and laterally to the ribs. It is the costal diaphragm that primarily acts as a barrier between the abdominal and thoracic cavities and is breached in CDH. The muscle, muscle connective tissue, and central

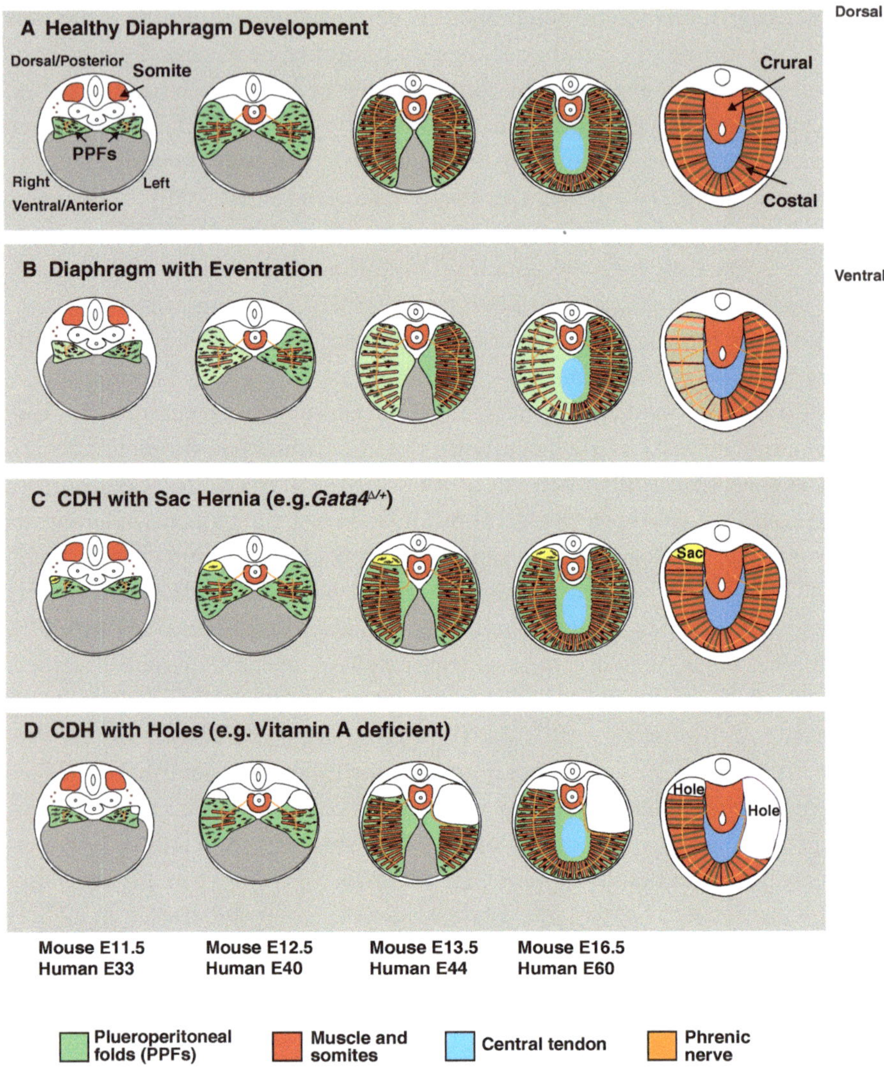

Fig. 1 Diaphragm and CDH development. (A) In a healthy diaphragm, muscle progenitors (red) migrate from the cranial somites toward the pleuroperitoneal folds (PPFs, green). The PPFs, along with the myogenic progenitors then expand dorsally and ventrally along the surface of the liver and differentiate into connective tissue and central tendon or myofibers. (B) Eventration arises when the diaphragm is poorly muscularized. The eventrated region bulges into the thoracic cavity, however, there is no overt sac or hole. (C) Sac hernias (yellow), such as in *Gata4*$^{\Delta/+}$ diaphragms, arise early due to a localized loss of muscle tissue. The loss of muscle tissue provides a weak point through which the liver herniates into the thoracic cavity and is surrounded by a connective tissue sac. (D) Hernias with holes (white), such as in vitamin A deficient diaphragms, arise due to a total loss of both connective tissue and muscle in a localized region of the diaphragm, which allows the liver to herniate into the thoracic cavity.

tendon are derived from two embryonic sources (Babiuk, Zhang, Clugston, Allan, & Greer, 2003; Merrell et al., 2015). The connective tissue and central tendon are derived from the pleuroperitoneal folds (PPFs), two pyramidal structures lying on either side of the esophagus (Merrell et al., 2015; Sefton, Gallardo, & Kardon, 2018). The PPFs arise from the lateral plate mesoderm and begin to thicken by embryonic day (E) 10.5 of mouse development (human E28) and are distinct structures by E11.5 (human E33; Sefton et al., 2018). The PPFs expand both dorsally and ventrally to cover the surface of the liver and differentiate into the muscle connective tissue and central tendon by E16.5 (human E60; Merrell et al., 2015; Sefton et al., 2018). The muscle progenitors, which ultimately give rise to the diaphragm's myofibers, are derived from a different embryonic structure, the cranial somites (Fig. 1A; Merrell et al., 2015; Sefton et al., 2018). Beginning at E9.5, the muscle progenitors delaminate from the cranial somites and migrate toward the PPFs at E9.5 and then reside within the core of the PPFs by E11.5 (Allan & Greer, 1997; Babiuk et al., 2003; Sefton et al., 2018). As the PPFs expand, the muscle precursors migrate with them and differentiate into a radial array of muscle fibers by E16.5 (Fig. 1A; Merrell et al., 2015; Sefton et al., 2018). Thus diaphragm development is an early embryonic event, largely complete during the first 3 months of human development.

3. Congenital diaphragmatic hernia (CDH)

Defects in the development of the costal diaphragm lead to CDH (Torfs et al., 1992). In the majority (80–86%) of cases the defect is a hole in the diaphragm, allowing abdominal contents to directly herniate into the thoracic cavity (Fig. 1D; Heiwegen et al., 2020). In other cases, there is no overt hole in the diaphragm. Instead the diaphragm may be poorly muscularized and this weaker area bulges into the thoracic cavity (termed an eventration; Fig. 1B; Pober, 2007). Alternatively, the diaphragm may contain localized connective tissue regions that completely lack muscle, and these weaker regions balloon into the thoracic cavity, leading to hernias covered with a connective tissue sac (Fig. 1C; Pober, 2007). Hernias in the dorsal (referred to as posterior in medical parlance) region, called Bochdalek hernias, comprise the vast majority (90%) of diagnosed and symptomatic CDH cases (Pober, 2007; Torfs et al., 1992). Hernias within the ventral (anterior) region of the diaphragm, termed Morgagni hernias, are less often diagnosed; however, this may be due to an ascertainment bias, as hernias in this region generally do not lead to lung defects. Interestingly, CDH more commonly (~80%) occurs on the left side, as opposed to the right side, of the diaphragm (Pober, 2007; Schulz et al., 2021; Yang et al., 2008).

4. Development of CDH

The phenotypic heterogeneity of diaphragmatic hernias suggests that CDH develops by different mechanisms depending on the type of hernia. The formation of a hole in the diaphragm, allowing communication between the abdominal and thoracic cavities, has generally been attributed to defects in the development of the PPFs (Fig. 1D; Babiuk & Greer, 2002; Clugston et al., 2006). Decreased proliferation, increased apoptosis, defective migration, and altered differentiation of PPF fibroblasts have all been implicated in the development of holes in the diaphragm (Cleal et al., 2021; Clugston, Zhang, & Greer, 2009; Coles & Ackerman, 2013; Paris, Coles, & Ackerman, 2015). Analysis of gene expression, mouse genetic mutants, and embryonic mice subject to pharmacological inhibitors have shown that defects in several molecular pathways and transcription factors cause PPF defects and lead to hernias with holes: signaling pathways include retinoic acid (Clugston et al., 2006, 2009, 2010), Wnt/β-catenin (Paris et al., 2015), Hedgehog (Coles & Ackerman, 2013), and PDGF (Bleyl et al., 2007) and transcription factors include WT1 (Carmona et al., 2016; Cleal et al., 2021; Clugston et al., 2006; Paris et al., 2015). Defects in other embryonic tissues, such as the post-hepatic mesenchymal plate and the septum transversum, have also been suggested to cause diaphragmatic hernias with holes (Carmona et al., 2016; Cleal et al., 2021; Iritani, 1984).

The development of hernias covered with sacs is also closely tied to defects in the PPF fibroblasts. The most detailed analysis of the etiology of this type of hernia comes from studies from our lab analyzing the role of the transcription factor *Gata4* (Fig. 1C; Merrell et al., 2015). GATA4 is expressed in the PPF fibroblasts and conditional deletion of *Gata4* in these cells in mice leads to diaphragmatic hernias covered by a connective tissue sac. Unlike hernias with a hole, these hernias with a sac have no defect in the migration or differentiation of the PPF fibroblasts. Instead, loss of *Gata4* leads to localized muscleless regions that are biomechanically weaker and allow abdominal contents to herniate. Although we originally thought that the development of such muscleless regions was sufficient to initiate herniation, our recent analysis of *Hgf* mutants now shows that additional defects in the connective tissue extracellular matrix or vascularization are also necessary (Sefton et al., 2022).

Finally, there is the least amount of data concerning the origin of poorly muscularized, eventrated diaphragms (Fig. 1B). Defects in the muscularization of the diaphragm may arise from loss of PPF-derived signals necessary to recruit and support muscle progenitors in the nascent diaphragm (Sefton et al., 2022). Alternatively, there may be cell-autonomous defects in muscle cells that prevent full muscularization of the diaphragm.

Overall, the preponderance of evidence suggests that the PPFs are critical for normal morphogenesis of the diaphragm and defects in the PPFs are central to the formation of diaphragmatic hernias (Kardon et al., 2017). Therefore, the PPFs are likely the focal region in which genetic and/or environmental factors act and interact to produce CDH.

5. Genetic etiology of CDH

CDH is generally thought to be caused by genetic factors, although its genetic etiology is highly heterogeneous (Kardon et al., 2017; Pober, 2007). Currently, genetic causes have been identified in about 30% of cases (Kardon et al., 2017; Russell et al., 2012; Yu et al., 2015) and these causes include aneuploidies, cytogenetic rearrangements, copy number variants, and single-gene mutations (see references in Kardon et al., 2017). While there are familial cases of CDH (e.g., Longoni et al., 2015; Yu et al., 2013), the sibling recurrence rate of CDH is rare, at 0.7% (Pober et al., 2005). Instead, most CDH cases arise sporadically without a family history of CDH, and so *de novo* mutations are expected to contribute significantly to the etiology of these cases. A focus of current research is the identification of gene-disrupting variants in trios of CDH-affected child and unaffected parents (Longoni et al., 2017, 2014; Qi et al., 2018; Yu et al., 2015, 2012).

More than 150 genes have been implicated in CDH (Bogenschutz, Fox, et al., 2020), many of which have been identified by trio studies. The CDH genes with the strongest level of support are those genes in which gene-disrupting variants have been identified in CDH individuals and functional studies of these genes have been conducted in human fibroblast cultures or *in vivo* in mice. Three genes with particularly high levels of support are the zinc finger transcription factor *GATA4*, the zinc finger transcriptional co-factor *ZFPM2*, and the membrane-associated transcription factor *MYRF*.

The transcription factor *GATA4* is one of the genes most strongly implicated in CDH. *GATA4* is located in a chromosomal region, 8p23.1, that is associated with recurrent CDH-associated copy number variants (Holder et al., 2007; Longoni et al., 2012), and *GATA4* variants are found in two cases of familial CDH (Arrington et al., 2012; Yu et al., 2013). While mice null for *Gata4* die prior to diaphragm development (Kuo et al., 1997; Molkentin, Lin, Duncan, & Olson, 1997), mice heterozygous for *Gata4* exhibit CDH (Jay et al., 2007). Furthermore, as discussed above, conditional null deletion of *Gata4* specifically in PPF fibroblasts leads to CDH with complete penetrance and shows that *Gata4* in these fibroblasts is required to support the survival and proliferation of neighboring myogenic cells (Merrell et al., 2015). Thus these experiments demonstrate conclusively that loss of *Gata4* in the PPFs leads to CDH. Yet patients with *GATA4*-associated CDH are, at most, missing one allele of *GATA4* (as *GATA4* null mutations would lead to early embryonic lethality). In humans, *GATA4* haploinsufficiency is incompletely penetrant for CDH (Longoni et al., 2012; Wat et al., 2009), and similarly mice with heterozygous null mutations in *Gata4* cause CDH in 29–50% of pups (our unpublished data and Jay et al., 2007). In addition, individuals with shared missense (Yu et al., 2013) or intronic (Arrington et al., 2012) *GATA4* variants exhibit hernias that differ in location and severity. The incomplete penetrance and variable expressivity of *GATA4* suggests that *GATA4* haploinsufficiency is not sufficient, but rather sensitizes individuals to develop CDH. Given the noted dosage sensitivity of GATA4 protein in regulating morphogenesis (Pu, Ishiwata, Juraszek, Ma, & Izumo, 2004), other genetic mutations or environmental modifiers may be necessary to lower GATA4 protein expression or function below a critical threshold that then induces CDH.

ZFPM2 is another gene strongly implicated in CDH. Like *GATA4*, *ZFPM2* is located in a chromosomal region, 8q22-23, in which copy number variants are associated with CDH (Holder et al., 2007; Temple, Barber, James, & Burge, 1994; Wat et al., 2011). In addition, CDH patients have been identified with gene-disrupting mutations in *ZFPM2* (Ackerman et al., 2005; Bleyl et al., 2007; Brady et al., 2014; Longoni et al., 2015). Null mutations in *Zfpm2* in mice lead to defective diaphragms in which regions develop that contain connective tissue, but no muscle (Ackerman et al., 2005). The ZFPM2 protein is a zinc finger protein that does not directly interact with DNA, but rather functionally interacts with GATA proteins to regulate GATA4-mediated transcription (Chlon & Crispino, 2012). Given the similarity in diaphragm defects in mice with mutations in

Gata4 or *Zfpm2*, it is likely that ZFPM2 synergistically regulates GATA4-mediated transcription in PPF fibroblasts. Also, like CDH patients with *GATA4* mutations, individuals harboring pathogenic mutations in *ZFPM2* dominantly exhibit CDH with a penetrance estimated at only 37.5% (Longoni et al., 2015; Wat et al., 2011).

Finally, an analysis of 362 proband-parent trios identified damaging *de novo* variants in *MYRF* in 4 unrelated individuals with CDH (Qi et al., 2018). A subsequent review of a large database of exome sequencing studies identified three more unrelated CDH patients harboring deleterious *MYRF* variants (Rossetti et al., 2019). Although there is less experimental support for the functional significance of *MYRF* in CDH, RNA-seq on diaphragm fibroblast cell cultures derived from neonatal CDH patients revealed that a large number of genes, including *GATA4*, are differentially expressed in *MYRF* mutant fibroblasts (Qi et al., 2018). Although these data strongly link *MYRF* to CDH, nevertheless like *GATA4* and *ZFPM2*, mutations in *MYRF* are incompletely penetrant for CDH (Qi et al., 2018).

Altogether the literature indicates that mutations in particular genes or chromosomal regions are critical for the development of CDH. However, the etiology of CDH is complex. Even for genes (such as *GATA4*, *ZFPM2*, and *MYRF*) strongly implicated in CDH, nevertheless mutations in these genes exhibit incomplete penetrance and variable expressivity for CDH. Thus, while mutations in these genes are critical for the development of CDH, they are not by themselves sufficient. Instead, mutations in other genes, epigenetic modifications, or environmental factors must also be required for CDH to actually arise.

6. Environmental factors contributing to CDH

Environmental factors have been postulated to also contribute to CDH etiology (Kardon et al., 2017; Pober, 2007; Torfs et al., 1992). Most data on the role of environmental factors comes from analyses of CDH patient records coupled with parent questionnaires. Some studies have shown that nicotine may be associated with developing CDH (Balayla & Abenhaim, 2014; Finn et al., 2022), although other studies have either found the opposite effect or no association at all (Caspers et al., 2010; Felix et al., 2008; McAteer, Hecht, De Roos, & Goldin, 2014; Perry, Mulcahy, & DeFranco, 2019). Similar conflicting results have been found regarding maternal alcohol intake, with some studies finding that it is correlated with CDH (Balayla & Abenhaim, 2014; Felix et al., 2008; McAteer et al., 2014) and others finding that it is not (Caspers et al., 2010; Finn et al., 2022). Thus there appears to be no consensus on whether nicotine or alcohol contribute to CDH etiology. However, it should be noted that differences in study design, including time during which nicotine or alcohol intake was investigated (such as periconception, 1st trimester, or anytime throughout pregnancy) may lead to these inconclusive results. An environmental factor for which there is stronger support for its role in CDH is maternal vitamin A deficiency (Greer, Babiuk, & Thebaud, 2003). Vitamin A is processed into retinoic acid (RA), which is essential for diaphragm development. Below, we will discuss in more detail the evidence that vitamin A and the retinoic acid signaling pathway are critical for development of the diaphragm and CDH. Because Vitamin A and retinoic acid signaling also interact with CDH-implicated genes, we also explore the potential for gene-environment interactions in the etiology of CDH.

7. Vitamin A in CDH development

Vitamin A (retinol) is an essential nutrient obtained through the diet and converted, through a series of oxidative steps, to its active metabolite retinoic acid, which in turn regulates gene transcription (see Duester, 2008; Rhinn & Dollé, 2012; Shannon, Moise, & Trainor, 2017). Retinol can be obtained in two forms: preformed vitamin A, primarily coming from animal sources such as meats and dairy; or as provitamin A carotenoids, primarily derived from plants including carrots, sweet potatoes, and leafy greens (Harrison, 2012). During pregnancy, retinol must be obtained through the maternal diet and passed to the embryo through the placenta. Cellular retinol uptake is mediated by the STRA6 transmembrane receptor and oxidized to retinal, primarily by retinal dehydrogenase 10 (RDH10) in the embryo (Fig. 2; Duester, 2008; Rhinn & Dollé, 2012; Sandell, Lynn, Inman, McDowell, & Trainor, 2012; Sandell et al., 2007). Retinal is then irreversibly oxidized to all-*trans*-retinoic acid (ATRA) by the ALDH1A (formerly called RALDH) family of enzymes and translocates to the nucleus. ALDH1A2 is the most broadly expressed ALDH1A enzyme in the embryo, suggesting it is the critical ALDH1A enzyme needed for RA synthesis (Duester, 2008; Niederreither, McCaffery, Drager, Chambon, & Dolle, 1997; Rhinn & Dollé, 2012; Shannon et al., 2017). Once inside the nucleus, ATRA binds to the retinoic acid receptor (RAR) on the retinoic acid response element (RARE) in the promoter region of a target gene to initiate transcription (Fig. 2; Al Tanoury, Piskunov, & Rochette-Egly, 2013; Delva et al., 1999; Mic, Molotkov, Benbrook, & Duester, 2003).

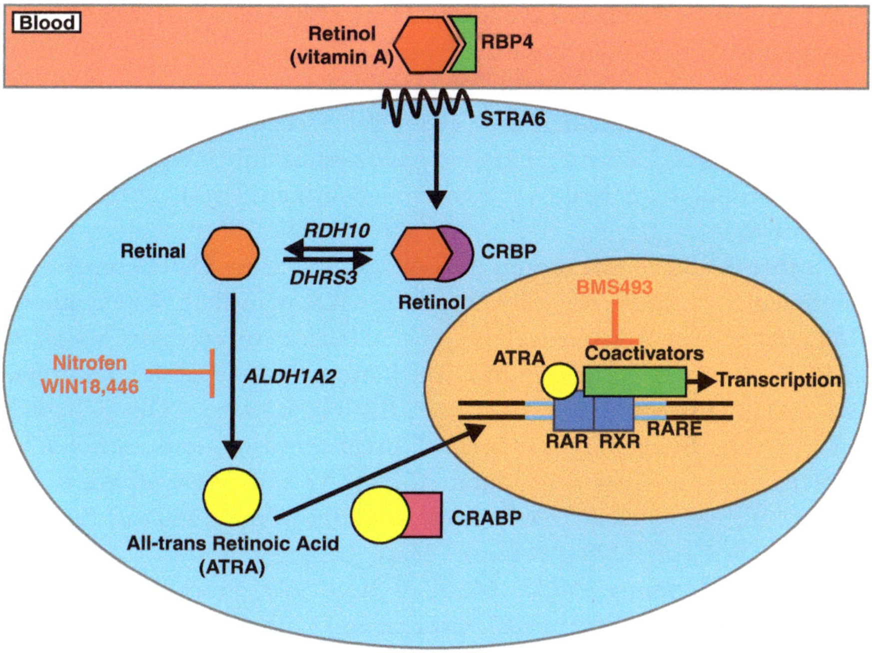

Fig. 2 The RA biosynthesis pathway. Circulating vitamin A (retinol, red) is brought into the cell by STRA6 and is bound by the cellular retinoic acid-binding protein (CRBP, purple) until it is oxidized to retinal (orange) by RDH10. Retinal is further irreversibly oxidized to all-*trans*-retinoic acid (ATRA, yellow) by ALDH1A2 and translocates to the nucleus, mediated by the cellular retinoic acid-binding protein (CRABP, pink). Once inside the nucleus, ATRA binds to the retinoic acid receptor (RAR, blue) bound to a retinoic acid response element (RARE), leading to recruitment of a transcription co-activator complex and subsequent transcription of a gene. ALDH1A2, can be pharmacologically inhibited with nitrofen and WIN18,446 and the transcriptional activity of ATRA can be inhibited *via* BMS493.

Vitamin A deficiency during pregnancy causes multiple birth defects, including CDH. The importance of vitamin A consumption during pregnancy was first demonstrated in animal studies as early as the 1930s. When kept on a vitamin A deficient (VAD) diet, female pigs gave birth to piglets with a variety of developmental anomalies including craniofacial defects and missing or malformed eyes (Hale, 1935). Subsequent studies in rodents found that females on a VAD diet gave birth to pups with similar malformations as well as defects in heart and lung development, vertebral defects, and CDH (Kostetskii et al., 1999; See, Kaiser, White, & Clagett-Dame, 2008; Warkany & Roth, 1948; Wilson et al., 1953), which can be rescued by exogenous vitamin A supplementation (See et al., 2008; Warkany & Roth, 1948; Wilson et al., 1953). As a consequence of these animal studies, researchers have analyzed whether vitamin A deficiency is associated with babies born with CDH. Reduced maternal dietary intake of vitamin A is associated with an increased risk of babies with CDH (Beurskens et al., 2013; Michikawa et al., 2019; Yang et al., 2008). In addition, CDH infants have been found to have significantly lower levels of circulating retinol than healthy newborns (Beurskens et al., 2010; Major et al., 1998).

Further evidence that defects in RA signaling cause CDH comes from experiments pharmacologically blocking RA signaling or synthesis. BMS493 is a pan-RAR antagonist that stabilizes co-repressors bound to RAR (Germain et al., 2009) and thus inhibits RA signaling. Treatment with BMS493 of rat embryos between E8-E11 leads to CDH (Clugston et al., 2010). Similarly, treatment of pregnant dams with nitrofen or WIN18,446, teratogens that inhibit ALDH1A2 and RA synthesis (Chen et al., 2018; Noble et al., 2007), leads to CDH in mice and rats (Babiuk, Thébaud, & Greer, 2004; Kilburn, Hess, Lesser, & Oster, 1982; Kluth et al., 1990; Momma, Ando, Mori, & Ito, 1992; Rocke et al., 2021; Taleporos, Salgo, & Oster, 1978). Experiments with these pharmacological inhibitors have also provided important mechanistic insights into how defects in RA signaling lead to CDH. Analysis of nitrofen-treated rat embryos or mouse PPF explants shows that hernias are caused by a failure of the PPFs to expand, due to inhibition of proliferation of PPF fibroblasts (Bogenschutz, Sefton, & Kardon, 2020; Clugston et al., 2006, 2009).

Altogether these studies strongly support a role for vitamin A and RA signaling deficiencies in the etiology of CDH. However, maternal vitamin A deficiency in rodents generally does not lead to CDH with complete penetrance (See et al., 2008; Wilson et al., 1953). Similarly, pharmacological perturbation of RA signaling pathway only leads to CDH with 50–59% penetrance (Babiuk et al., 2004; Clugston et al., 2010; Rocke et al., 2021). Therefore, it is likely that low maternal levels of vitamin A and deficient embryonic RA signaling are generally not completely penetrant and suggests that other collaborating factors to may be necessary to cause diaphragmatic hernias.

8. Gene–environment interactions in etiology of CDH

While mutations in particular genes, such as *Gata4*, and environmental perturbations, such as vitamin A deficiency, cause CDH, all are notably incompletely penetrant for development of diaphragmatic hernias.

This suggests that the etiology of CDH is more complex and involves a combination of different genetic mutations, several environmental perturbations, or CDH-inducing gene-environment interactions. While gene-environment interactions are likely to play a role in CDH, establishing that such generally complex interactions are indeed taking place and the molecular and cellular mechanisms underlying these interactions is challenging.

To test whether gene-environment interactions are important in CDH etiology, it is important to define what is meant by gene-environment interactions and establish the criteria that must be met to demonstrate that such interactions are functioning. A gene-environment interaction is defined as "a different effect of an environmental exposure on disease risks in persons with different genotypes" or conversely as "a different effect of a genotype on disease risk in persons with different environmental exposures" (Ottman, 1996). A useful concept for designing studies to test gene-interactions is the epidemiological analysis of risk, *via* the calculation of an odds ratio (OR). An OR is the odds of an outcome, e.g., CDH, occurring after an exposure (genetic or environmental) compared to the odds of the same outcome occurring in the absence of the exposure (Szumilas, 2010). For example, if an OR is calculated to be 1, this indicates that the exposure does not affect the odds of CDH. If the OR is greater or less than 1, however, it indicates that exposure is associated with higher or lower odds of CDH, respectively (Szumilas, 2010).

To test, using the concept of odds ratios, whether gene-environment interactions contribute to CDH it is necessary to determine the OR of CDH with a genetic mutation (OR_G), the OR of CDH with an environmental perturbation (OR_E), and then the OR of CDH with both a genetic mutation and an environmental perturbation (OR_{GE}). To illustrate such a series of experiments in mice, we will use the example of the effects of a genetic mutation in *Gata4* and maternal vitamin A deficiency (Fig. 3). In our lab, we found that mouse embryos heterozygous for a null mutation in *Gata4* (*Gata4Δ*/+ on a C57BL/6J background) exhibit small dorsal sac hernias generally on the right side with an $OR_G = 4.0$ (see calculation in Fig. 3B). Several labs have found that embryos developing in mothers on a VAD diet develop hernias with holes in left and right dorsal regions of the diaphragm (Fig. 3C; See et al., 2008; Wilson et al., 1953). The frequency of CDH depends on the level of vitamin A deficiency, but one epidemiological study estimated an adjusted OR_E of 1.8 for risk of CDH in babies born to mothers with low vitamin A (Yang et al., 2008). To test in mice whether *Gata4* mutations genetically interact with low maternal vitamin

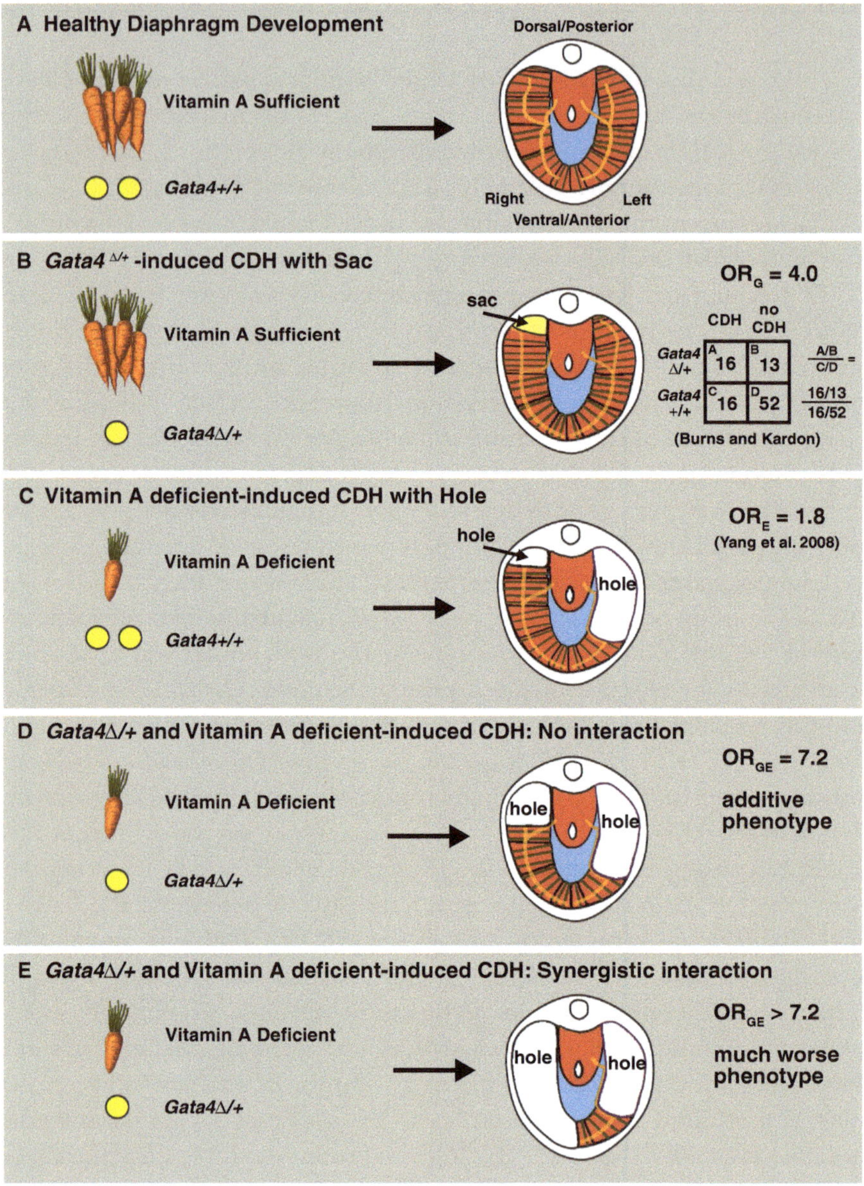

Fig. 3 Gene–environment interactions in diaphragm development. (A) A normal level of vitamin A (full bunch of carrots) coupled with two alleles of *Gata4* (yellow circles) leads to healthy diaphragm development. (B) Loss of one copy of *Gata4* under normal levels of vitamin A induces a sac hernia (yellow). Sac hernias arise in $Gata4^{\Delta/+}$ mice with an OR = 4. (C) Vitamin A deficiency (one carrot) in an individual with both copies of *Gata4* leads to hole hernias (white) with an OR = 1.8. (D) If there is no interaction between vitamin A deficiency and loss of one copy of *Gata4*, the diaphragm will develop a slightly larger hole hernia on the right due to the addition of the sac and hole hernias with an OR = 7.2. (E) If there is a synergistic interaction between vitamin A deficiency and loss of one copy of *Gata4*, a large hole hernia on the right will develop on both sides of the diaphragm with an OR > 7.2.

A, *Gata4Δ*/+ pups would be generated in VAD dams. If the OR_{GE} of CDH in these pups is 7.2 (Fig. 3D), this would indicate that there is no gene-environment interaction because the OR_{GE} is simply equivalent to OR_G X OR_E (assuming a multiplicative model; see discussion in Ottman, 1996). However, if $OR_{GE} > 7.2$ (Fig. 3E), this would indicate that there is a greater than expected odds of developing CDH when *Gata4Δ*/+ pups develop with low vitamin A and thus there is a synergistic gene-environment interaction. Likewise, if $OR_{GE} < 7.2$, this would indicate there is an antagonistic gene-environment interaction. In addition to the frequency of CDH, gene-environment interaction can also be revealed by the nature of the hernias that develop. If there is no interaction between *Gata4* and vitamin A, then *Gata4Δ*/+ pups born to dams on a VAD diet would be expected to show the additive phenotype of moderate-sized hernias with holes on right and left sides of the diaphragm (Fig. 3D). However, if there is a gene-environment synergistic interaction, the pups might be expected to have much larger hernias on the right side than expected with a simple additive phenotype (Fig. 3E).

Previous studies in rodents suggest mechanistically how genetic mutations and environmental factors may act synergistically to increase the incidence and severity of CDH. To date, the preponderance of evidence points to the PPFs as a major target of CDH. In *Gata4* mutants (Bogenschutz, Sefton, et al., 2020; Merrell et al., 2015) and embryos in which RA signaling is pharmacologically inhibited (Bogenschutz, Sefton, et al., 2020; Clugston et al., 2006, 2009), PPF fibroblast proliferation is decreased and overall PPF morphogenic expansion is defective in mice; thus, it is plausible that *Gata4* mutants developing on a vitamin A deficient diet could synergistically interact to inhibit PPF development and lead to a higher incidence of CDH with more severe hernias. The timing of the effects of genetic and environmental factors may also be critical. Rodent studies indicate that CDH initiates early in development, likely prior to E11.5 in mouse. Mutations in *Gata4* cause visible right-sided defects in diaphragm development by E12.5 (Merrell et al., 2015). VAD diet or teratogen inhibition of RA experiments indicate that development of CDH is extraordinarily sensitive to timing of reduced RA signaling (Clugston et al., 2010; See et al., 2008); RA is required before E11.5 of mouse development, with RA inhibition prior to E9.5 leading preferentially to left hernias and inhibition E9.5–11.5 leading to right hernias (Clugston et al., 2010). The likely temporal overlap of GATA4 and RA function suggests that they may act synergistically to regulate the formation of right-sided hernias. On a molecular level, RA may either directly or indirectly regulate *Gata4* expression in the diaphragm similar to results in heart development, in which a VAD diet (Kostetskii et al., 1999) or nitrofen exposure (Takayasu, Sato, Sugimoto, & Puri, 2008) decreases *Gata4* expression.

Other scenarios in which genetic mutations and environmental factors synergistically interact are also possible. For instance, deficiencies in vitamin A may interact with mutations in several genes. In addition to *Gata4*, RA signaling also regulates the CDH-implicated receptors *COUP-TFII* (Doi, Sugimoto, & Puri, 2009; Jonk et al., 1994; Kruse et al., 2008) and *Stra6* (Zheng et al., 2020). The effects of genetic mutations and environmental perturbations may be temporally separated. Genetic mutations may act early in development to produce muscleless regions, but these regions may only herniate in the presence of later environmental perturbations, such as transient hypoxia that could reduce vascularization, biomechanically weaken the diaphragm, and permit the liver to herniate. Finally, the clinical severity of CDH is largely determined by the magnitude of the effects on lung development, and a "dual-hit" hypothesis has been proposed (Keijzer, Liu, Deimling, Tibboel, & Post, 2000), whereby both the developing diaphragm and lungs are affected by genetic mutations and/or environmental perturbations. Therefore, a complex scenario may be at work, in which genetic mutations disrupt diaphragm development, environmental perturbations affect lung development, and these effects interact synergistically to produce clinically severe CDH.

While gene–environment interactions are likely critical to the etiology of CDH, it is important to acknowledge the challenges of designing experiments to test such interactions. As the diaphragm is only present in mammals (Perry, Similowski, Klein, & Codd, 2010) and currently mice are the most tractable mammal for genetic manipulation, this limits such gene-environment experiments to mice. Engineering mutations in CDH candidate genes and characterizing the prevalence and phenotype of hernias that arise is relatively straightforward in mice. However, designing and testing environmental perturbations is more difficult. For example, testing the effects of vitamin A deficiency has been a central focus of CDH research, but designing appropriate VAD diets for mice that do not cause embryonic lethality but lead to reproducible levels of CDH is complex (see discussion in Rocke et al., 2021). The challenges are further compounded when trying to test for gene–environment interactions. Detecting a statistically significant increase between observed OR_{GE} and OR_{GE} expected from $OR_G \times$ w OR_E will generally require hundreds of embryos and so is often prohibitively time-consuming and expensive. An important strategy to maximize the chance to detect gene–environment interactions is to design alleles of CDH-implicated genes or environmental perturbations (such as VAD diets; Rocke et al., 2021) that sensitize embryos to develop CDH, but by themselves do not cause hernias.

9. Concluding remarks

CDH is a common and often devastating birth defect associated with high rates of morbidity and mortality. Its etiology is notably complex, involving genetic and environmental factors. Recent research has largely concentrated on the role of genetic mutations and has revealed more than 150 candidate CDH-implicated genes. Yet, even genes for which there is strong experimental evidence in mice for their role in CDH are incompletely penetrant for CDH. This suggests that single genetic mutations are rarely sufficient to induce CDH. At present, there is limited evidence for the role of environmental factors, but maternal vitamin A insufficiency is one environmental perturbation for which there is strong support for its role as a sensitizing factor for development of CDH. Altogether, the lack of common, highly penetrant genetic or environmental causes of CDH suggests that gene–environment interactions may also contribute to CDH etiology. A future challenge is designing sensitive and cost-effective experiments to test such gene–environment interactions.

Are there opportunities to reduce the risk of CDH? Because CDH most often arises spontaneously by *de novo* mutations and CDH-implicated genes are incompletely penetrant for CDH, prenatal screening for potential genetic risk factors is less likely to be beneficial. However, vitamin A insufficiency is an environmental risk factor that is preventable. Vitamin A deficiency is a major global health problem, particularly in low- and middle-income countries, most notably in sub-Saharan Africa and southern Asia (Stevens et al., 2015). Even in the United States, as many as 15% of pregnant women in the United States consume insufficient vitamin A (Bailey, Pac, Fulgoni, Reidy, & Catalano, 2019). Efforts to reduce maternal vitamin A insufficiency, particularly during the first trimester of pregnancy, would lower the risk of CDH and other birth defects associated with low vitamin A intake.

Acknowledgments

We thank C. Goclowski, E.M. Sefton, and J. Shapiro for critical reading of the manuscript. Research in the Kardon lab on diaphragm development and CDH is supported by NIH R01HD104317, and the Wheeler Foundation. N.G.B. is supported by NIH T32HD007491.

References

Ackerman, K. G., Herron, B. J., Vargas, S. O., Huang, H., Tevosian, S. G., Kochilas, L., et al. (2005). Fog2 is required for normal diaphragm and lung development in mice and humans. *PLoS Genetics*, *1*, 58–65.

Al Tanoury, Z., Piskunov, A., & Rochette-Egly, C. (2013). Vitamin A and retinoid signaling: Genomic and nongenomic effects. *Journal of Lipid Research*, *54*, 1761–1775.

Alexander, P. G., Clark, K. L., & Tuan, R. S. (2016). Prenatal exposure to environmental factors and congenital limb defects. *Birth Defects Research. Part C, Embryo Today*, *108*, 243–273.

Allan, D. W., & Greer, J. J. (1997). Embryogenesis of the phrenic nerve and diaphragm in the fetal rat. *The Journal of Comparative Neurology*, *382*, 459–468.

Arrington, C. B., Bleyl, S. B., Matsunami, N., Bowles, N. E., Leppert, T. I., Demarest, B. L., et al. (2012). A family-based paradigm to identify candidate chromosomal regions for isolated congenital diaphragmatic hernia. *American Journal of Medical Genetics. Part A*, *158A*, 3137–3147.

Babiuk, R. P., & Greer, J. J. (2002). Diaphragm defects occur in a CDH hernia model independently of myogenesis and lung formation. *American Journal of Physiology. Lung Cellular and Molecular Physiology*, *283*, L1310–L1314.

Babiuk, R. P., Thébaud, B., & Greer, J. J. (2004). Reductions in the incidence of nitrofen-induced diaphragmatic hernia by vitamin A and retinoic acid. *American Journal of Physiology—Lung Cellular and Molecular Physiology*, *286*, 970–973.

Babiuk, R. P., Zhang, W., Clugston, R., Allan, D. W., & Greer, J. J. (2003). Embryological origins and development of the rat diaphragm. *The Journal of Comparative Neurology*, *455*, 477–487.

Bailey, R. L., Pac, S. G., Fulgoni, V. L., 3rd, Reidy, K. C., & Catalano, P. M. (2019). Estimation of total usual dietary intakes of pregnant women in the United States. *JAMA Network Open*, *2*, e195967.

Balayla, J., & Abenhaim, H. A. (2014). Incidence, predictors and outcomes of congenital diaphragmatic hernia: A population-based study of 32 million births in the United States. *The Journal of Maternal-Fetal & Neonatal Medicine*, *27*, 1438–1444.

Beurskens, L. W., Schrijver, L. H., Tibboel, D., Wildhagen, M. F., Knapen, M. F., Lindemans, J., et al. (2013). Dietary vitamin a intake below the recommended daily intake during pregnancy and the risk of congenital diaphragmatic hernia in the offspring. *Birth Defects Research. Part A, Clinical and Molecular Teratology*, *97*, 60–66.

Beurskens, L. W. J. E., Tibboel, D., Lindemans, J., Duvekot, J. J., Cohen-Overbeek, T. E., Veenma, D. C. M., et al. (2010). Retinol status of newborn infants is associated with congenital diaphragmatic hernia. *Pediatrics*, *126*, 712.

Bleyl, S. B., Moshrefi, A., Shaw, G. M., Saijoh, Y., Schoenwolf, G. C., Pennacchio, L. A., et al. (2007). Candidate genes for congenital diaphragmatic hernia from animal models: Sequencing of FOG2 and PDGFRalpha reveals rare variants in diaphragmatic hernia patients. *European Journal of Human Genetics*, *15*, 950–958.

Bogenschutz, E. L., Fox, Z. D., Farrell, A., Wynn, J., Moore, B., Yu, L., et al. (2020). Deep whole-genome sequencing of multiple proband tissues and parental blood reveals the complex genetic etiology of congenital diaphragmatic hernias. *HGG Advances*, *1*.

Bogenschutz, E. L., Sefton, E. M., & Kardon, G. (2020). Cell culture system to assay candidate genes and molecular pathways implicated in congenital diaphragmatic hernias. *Developmental Biology*, *467*, 30–38.

Brady, P. D., Van Esch, H., Fieremans, N., Froyen, G., Slavotinek, A., Deprest, J., et al. (2015). Expanding the phenotypic spectrum of PORCN variants in two males with syndromic microphthalmia. *European Journal of Human Genetics*, *23*, 551–554.

Brady, P. D., Van Houdt, J., Callewaert, B., Deprest, J., Devriendt, K., & Vermeesch, J. R. (2014). Exome sequencing identifies ZFPM2 as a cause of familial isolated congenital

diaphragmatic hernia and possibly cardiovascular malformations. *European Journal of Medical Genetics*, *57*, 247–252.
Caputo, C., Wood, E., & Jabbour, L. (2016). Impact of fetal alcohol exposure on body systems: A systematic review. *Birth Defects Research. Part C, Embryo Today*, *108*, 174–180.
Carmona, R., Cañete, A., Cano, E., Ariza, L., Rojas, A., & Muñoz-Chápuli, R. (2016). Conditional deletion of WT1 in the septum transversum mesenchyme causes congenital diaphragmatic hernia in mice. *eLife*, *5*, 1–17.
Caspers, K. M., Oltean, C., Romitti, P. A., Sun, L., Pober, B. R., Rasmussen, S. A., et al. (2010). Maternal periconceptional exposure to cigarette smoking and alcohol consumption and congenital diaphragmatic hernia. *Birth Defects Research. Part A, Clinical and Molecular Teratology*, *88*, 1040–1049.
Centers for Disease Control and Prevention. (2008). Update on overall prevalence of major birth defects—Atlanta, Georgia, 1978–2005. *Morbidity and Mortality Weekly Report*, *57*, 1–5.
Chen, Y., Zhu, J.-Y., Ho Hong, K., Mikles, D. C., Georg, G. I., Goldstein, A. S., et al. (2018). Structural basis of ALDH1A2 inhibition by irreversible and reversible small molecule inhibitors HHS public access. *ACS Chemical Biology*, *13*, 582–590.
Chlon, T. M., & Crispino, J. D. (2012). Combinatorial regulation of tissue specification by GATA and FOG factors. *Development*, *139*, 3905–3916.
Cleal, L., McHaffie, S. L., Lee, M., Hastie, N., Martinez-Estrada, O. M., & Chau, Y. Y. (2021). Resolving the heterogeneity of diaphragmatic mesenchyme: A novel mouse model of congenital diaphragmatic hernia. *Disease Models & Mechanisms*, *14*.
Clugston, R. D., Klattig, J., Englert, C., Clagett-Dame, M., Martinovic, J., Benachi, A., et al. (2006). Teratogen-induced, dietary and genetic models of congenital diaphragmatic hernia share a common mechanism of pathogenesis. *The American Journal of Pathology*, *169*, 1541–1549.
Clugston, R. D., Zhang, W., & Greer, J. J. (2009). Early development of the primordial mammalian diaphragm and cellular mechanisms of nitrofen-induced congenital diaphragmatic hernia. *Birth Defects Research Part A: Clinical and Molecular Teratology*, *88*.
Clugston, R. D., Zhang, W., Lvarez, S. A., De Lera, A. R., & Greer, J. J. (2010). Understanding abnormal retinoid signaling as a causative mechanism in congenital diaphragmatic hernia. *American Journal of Respiratory Cell and Molecular Biology*, *42*, 276–285.
Coles, G. L., & Ackerman, K. G. (2013). Kif7 is required for the patterning and differentiation of the diaphragm in a model of syndromic congenital diaphragmatic hernia. *Proceedings of the National Academy of Sciences of the United States of America*, *110*, E1898.
Colvin, J. S., Bohne, B. A., Harding, G. W., McEwen, D. G., & Ornitz, D. M. (1996). Skeletal overgrowth and deafness in mice lacking fibroblast growth factor receptor 3. *Nature Genetics*, *12*, 390–397.
Delva, L., Bastie, J. N., Rochette-Egly, C., Kraiba, R., Balitrand, N., Despouy, G., et al. (1999). Physical and functional interactions between cellular retinoic acid binding protein II and the retinoic acid-dependent nuclear complex. *Molecular and Cellular Biology*, *19*, 7158–7167.
Doi, T., Sugimoto, K., & Puri, P. (2009). Prenatal retinoic acid up-regulates pulmonary gene expression of COUP-TFII, FOG2, and GATA4 in pulmonary hypoplasia. *Journal of Pediatric Surgery*, *44*, 1933–1937.
Duester, G. (2008). Retinoic acid synthesis and signaling during early organogenesis. *Cell*, *6*, 921–931.
Felix, J. F., van Dooren, M. F., Klaassens, M., Hop, W. C., Torfs, C. P., & Tibboel, D. (2008). Environmental factors in the etiology of esophageal atresia and congenital diaphragmatic hernia: Results of a case-control study. *Birth Defects Research. Part A, Clinical and Molecular Teratology*, *82*, 98–105.

Finn, J., Suhl, J., Kancherla, V., Conway, K. M., Oleson, J., Sidhu, A., et al. (2022). Maternal cigarette smoking and alcohol consumption and congenital diaphragmatic hernia. *Birth Defects Research*, *114*(13), 746–758.

Germain, P., Gaudon, C., Pogenberg, V., Sanglier, S., Van Dorsselaer, A., Royer, C. A., et al. (2009). Differential action on coregulator interaction defines inverse retinoid agonists and neutral antagonists. *Chemistry & Biology*, *16*, 479–489.

Greer, J. J., Babiuk, R. P., & Thebaud, B. (2003). Etiology of congenital diaphragmatic hernia: The retinoid hypothesis. *Pediatric Research*, *53*, 726–730.

Hale, F. (1935). The relation of vitamin A to anophthalmos in pigs. *American Journal of Ophthalmology*, *18*, 1087–1093.

Harmsen, M. B., Azzarello-Burri, S., Garcia Gonzalez, M. M., Gillessen-Kaesbach, G., Meinecke, P., Muller, D., et al. (2009). Goltz-Gorlin (focal dermal hypoplasia) and the microphthalmia with linear skin defects (MLS) syndrome: No evidence of genetic overlap. *European Journal of Human Genetics*, *17*, 1207–1215.

Harrison, E. H. (2012). Mechanisms involved in the intestinal absorption of dietary vitamin A and provitamin A carotenoids. *Biochimica et Biophysica Acta*, *1821*, 70–77.

Heiwegen, K., van Heijst, A. F., Daniels-Scharbatke, H., van Peperstraten, M. C., de Blaauw, I., & Botden, S. M. (2020). Congenital diaphragmatic eventration and hernia sac compared to CDH with true defects: A retrospective cohort study. *European Journal of Pediatrics*, *179*, 855–863.

Holder, A. M., Klaassens, M., Tibboel, D., de Klein, A., Lee, B., & Scott, D. A. (2007). Genetic factors in congenital diaphragmatic hernia. *American Journal of Human Genetics*, *80*, 825–845.

Hong, M., Christ, A., Christa, A., Willnow, T. E., & Krauss, R. S. (2020). Cdon mutation and fetal alcohol converge on Nodal signaling in a mouse model of holoprosencephaly. *eLife*, *9*.

Hong, M., & Krauss, R. S. (2012). Cdon mutation and fetal ethanol exposure synergize to produce midline signaling defects and holoprosencephaly spectrum disorders in mice. *PLoS Genetics*, *8*, e1002999.

Iritani, I. (1984). Experimental study on embryogenesis of congenital diaphragmatic hernia. *Anatomy and Embryology*, *169*, 33–139.

Jay, P. Y., Bielinska, M., Erlich, J. M., Mannisto, S., Pu, W. T., Heikinheimo, M., et al. (2007). Impaired mesenchymal cell function in Gata4 mutant mice leads to diaphragmatic hernias and primary lung defects. *Developmental Biology*, *301*, 602–614.

Jonk, L. J., de Jonge, M. E., Pals, C. E., Wissink, S., Vervaart, J. M., Schoorlemmer, J., et al. (1994). Cloning and expression during development of three murine members of the COUP family of nuclear orphan receptors. *Mechanisms of Development*, *47*, 81–97.

Kardon, G., Ackerman, K. G., McCulley, D. J., Shen, Y., Wynn, J., Shang, L., et al. (2017). Congenital diaphragmatic hernias: From genes to mechanisms to therapies. *Disease Models and Mechanisms*, *10*, 955–970.

Keijzer, R., Liu, J., Deimling, J., Tibboel, D., & Post, M. (2000). Dual-hit hypothesis explains pulmonary hypoplasia in the nitrofen model of congenital diaphragmatic hernia. *The American Journal of Pathology*, *156*, 1299–1306.

Kietzman, H. W., Everson, J. L., Sulik, K. K., & Lipinski, R. J. (2014). The teratogenic effects of prenatal ethanol exposure are exacerbated by Sonic Hedgehog or GLI2 haploinsufficiency in the mouse. *PLoS One*, *9*, e89448.

Kilburn, K. H., Hess, R. A., Lesser, M., & Oster, G. (1982). Perinatal death and respiratory apparatus dysgenesis due to a bis (dichloroacetyl) diamine. *Teratology*, *26*, 155–162.

Kluth, D., Kangah, R., Reich, P., Tenbrinck, R., Tibboel, D., & Lambrecht, W. (1990). Nitrofen-induced diaphragmatic hernias in rats: An animal model. *Journal of Pediatric Surgery*, *25*, 850–854.

Kostetskii, I., Jiang, Y., Kostetskaia, E., Yuan, S., Evans, T., & Zile, M. (1999). Retinoid signaling required for normal heart development regulates GATA-4 in a pathway distinct from cardiomyocyte differentiation. *Developmental Biology, 206*, 206–218.

Kruse, S. W., Suino-Powell, K., Zhou, X. E., Kretschman, J. E., Reynolds, R., Vonrhein, C., et al. (2008). Identification of COUP-TFII orphan nuclear receptor as a retinoic acid-activated receptor. *PLoS Biology, 6*, e227.

Kuo, C. T., Morrisey, E. E., Anandappa, R., Sigrist, K., Lu, M. M., Parmacek, M. S., et al. (1997). GATA4 transcription factor is required for ventral morphogenesis and heart tube formation. *Genes & Development, 11*, 1048–1060.

Longoni, M., High, F. A., Qi, H., Joy, M. P., Hila, R., Coletti, C. M., et al. (2017). Genome-wide enrichment of damaging de novo variants in patients with isolated and complex congenital diaphragmatic hernia. *Human Genetics, 136*, 679–691.

Longoni, M., High, F. A., Russell, M. K., Kashani, A., Tracy, A. A., Coletti, C. M., et al. (2014). Molecular pathogenesis of congenital diaphragmatic hernia revealed by exome sequencing, developmental data, and bioinformatics. *Proceedings of the National Academy of Sciences of the United States of America, 111*, 12450–12455.

Longoni, M., Lage, K., Russell, M. K., Loscertales, M., Abdul-Rahman, O. A., Baynam, G., et al. (2012). Congenital diaphragmatic hernia interval on chromosome 8p23.1 characterized by genetics and protein interaction networks. *American Journal of Medical Genetics. Part A, 158A*, 3148–3158.

Longoni, M., Russell, M. K., High, F. A., Darvishi, K., Maalouf, F. I., Kashani, A., et al. (2015). Prevalence and penetrance of ZFPM2 mutations and deletions causing congenital diaphragmatic hernia. *Clinical Genetics, 87*, 362–367.

Major, D., Cadenas, M., Fournier, L., Leclerc, S., Lefebvre, M., & Cloutier, R. (1998). Retinol status of newborn infants with congenital diaphragmatic hernia. *Pediatric Surgery International, 13*, 547–549.

McAteer, J. P., Hecht, A., De Roos, A. J., & Goldin, A. B. (2014). Maternal medical and behavioral risk factors for congenital diaphragmatic hernia. *Journal of Pediatric Surgery, 49*, 34–38, discussion 38.

McCarthy, N., Wetherill, L., Lovely, C. B., Swartz, M. E., Foroud, T. M., & Eberhart, J. K. (2013). Pdgfra protects against ethanol-induced craniofacial defects in a zebrafish model of FASD. *Development, 140*, 3254–3265.

Merrell, A. J., Ellis, B. J., Fox, Z. D., Lawson, J. A., Weiss, J. A., & Kardon, G. (2015). Muscle connective tissue controls development of the diaphragm and is a source of congenital diaphragmatic hernias. *Nature Genetics, 47*.

Merrell, A. J., & Kardon, G. (2013). Development of the diaphragm—A skeletal muscle essential for mammalian respiration. *The FEBS Journal, 280*, 4026–4035.

Mic, F. A., Molotkov, A., Benbrook, D. M., & Duester, G. (2003). Retinoid activation of retinoic acid receptor but not retinoid X receptor is sufficient to rescue lethal defect in retinoic acid synthesis. *Proceedings of the National Academy of Sciences of the United States of America, 100*, 7135–7140.

Michikawa, T., Yamazaki, S., Sekiyama, M., Kuroda, T., Nakayama, S. F., Isobe, T., et al. (2019). Maternal dietary intake of vitamin A during pregnancy was inversely associated with congenital diaphragmatic hernia: The Japan Environment and Children's Study. *The British Journal of Nutrition, 122*, 1295–1302.

Molkentin, J. D., Lin, Q., Duncan, S. A., & Olson, E. N. (1997). Requirement of the transcription factor GATA4 for heart tube formation and ventral morphogenesis. *Genes & Development, 11*, 1061–1072.

Momma, K., Ando, M., Mori, Y., & Ito, T. (1992). Hypoplasia of the lung and heart in fetal rats with diaphragmatic hernia. *Fetal Diagnosis and Therapy, 7*, 46–52.

Moreau, J. L. M., Kesteven, S., Martin, E., Lau, K. S., Yam, M. X., O'Reilly, V. C., et al. (2019). Gene-environment interaction impacts on heart development and embryo survival. *Development, 146*.

Niederreither, K., McCaffery, P., Drager, U. C., Chambon, P., & Dolle, P. (1997). Restricted expression and retinoic acid-induced downregulation of the retinaldehyde dehydrogenase type 2 (RALDH-2) gene during mouse development. *Mechanisms of Development, 62*, 67–78.

Niethamer, T. K., Teng, T., Franco, M., Du, Y. X., Percival, C. J., & Bush, J. O. (2020). Aberrant cell segregation in the craniofacial primordium and the emergence of facial dysmorphology in craniofrontonasal syndrome. *PLoS Genetics, 16*, e1008300.

Noble, B. R., Babiuk, R. P., Clugston, R. D., Underhill, T. M., Sun, H., Kawaguchi, R., et al. (2007). Mechanisms of action of the congenital diaphragmatic hernia-inducing teratogen nitrofen. *American Journal of Physiology. Lung Cellular and Molecular Physiology, 293*, L1079–L1087.

Ottman, R. (1996). Gene-environment interaction: Definitions and study designs. *Preventive Medicine, 25*, 764–770.

Paris, N. D., Coles, G. L., & Ackerman, K. G. (2015). Wt1 and beta-catenin cooperatively regulate diaphragm development in the mouse. *Developmental Biology, 407*, 40–56.

Perry, M. F., Mulcahy, H., & DeFranco, E. A. (2019). Influence of periconception smoking behavior on birth defect risk. *American Journal of Obstetrics and Gynecology, 220*. 588 e581-588 e587.

Perry, S. F., Similowski, T., Klein, W., & Codd, J. R. (2010). The evolutionary origin of the mammalian diaphragm. *Respiratory Physiology & Neurobiology, 171*, 1–16.

Pober, B. R. (2007). Overview of epidemiology, genetics, birth defects, and chromosome abnormalities associated with CDH. *American Journal of Medical Genetics, 145C*, 158–171.

Pober, B. R., Lin, A., Russell, M., Ackerman, K. G., Chakravorty, S., Strauss, B., et al. (2005). Infants with Bochdalek diaphragmatic hernia: Sibling precurrence and monozygotic twin discordance in a hospital-based malformation surveillance program. *American Journal of Medical Genetics. Part A, 138A*, 81–88.

Pu, W. T., Ishiwata, T., Juraszek, A. L., Ma, Q., & Izumo, S. (2004). GATA4 is a dosage-sensitive regulator of cardiac morphogenesis. *Developmental Biology, 275*, 235–244.

Qi, H., Yu, L., Zhou, X., Wynn, J., Zhao, H., Guo, Y., et al. (2018). De novo variants in congenital diaphragmatic hernia identify MYRF as a new syndrome and reveal genetic overlaps with other developmental disorders. *PLoS Genetics, 14*, 1–26.

Rhinn, M., & Dollé, P. (2012). Retinoic acid signalling during development. *Development, 139*, 843–858.

Rocke, A. W., Clarke, T. G., Dalmer, T. R. A., McCluskey, S. A., Rivas, J. F. G., & Clugston, R. D. (2021). Low maternal vitamin A intake increases the incidence of teratogen induced congenital diaphragmatic hernia in mice. *Pediatric Research, 91*, 83–91.

Rogers, J. M. (2009). Tobacco and pregnancy. *Reproductive Toxicology, 28*, 152–160.

Rossetti, L. Z., Glinton, K., Yuan, B., Liu, P., Pillai, N., Mizerik, E., et al. (2019). Review of the phenotypic spectrum associated with haploinsufficiency of MYRF. *American Journal of Medical Genetics. Part A, 179*, 1376–1382.

Russell, M. K., Longoni, M., Wells, J., Maalouf, F. I., Tracy, A. A., Loscertales, M., et al. (2012). Congenital diaphragmatic hernia candidate genes derived from embryonic transcriptomes. *Proceedings of the National Academy of Sciences of the United States of America, 109*, 2978–2983.

Sandell, L. L., Lynn, M. L., Inman, K. E., McDowell, W., & Trainor, P. A. (2012). RDH10 oxidation of vitamin A is a critical control step in synthesis of retinoic acid during mouse embryogenesis. *PLoS One, 7*.

Sandell, L. L., Sanderson, B. W., Moiseyev, G., Johnson, T., Mushegian, A., Young, K., et al. (2007). RDH10 is essential for synthesis of embryonic retinoic acid and is required for limb, craniofacial, and organ development. *Genes & Development*, *21*, 1113–1124.

Schulz, F., Jenetzky, E., Zwink, N., Bendixen, C., Kipfmueller, F., Rafat, N., et al. (2021). Parental risk factors for congenital diaphragmatic hernia—A large German case-control study. *BMC Pediatrics*, *21*, 278.

See, A. W. M., Kaiser, M. E., White, J. C., & Clagett-Dame, M. (2008). A nutritional model of late embryonic vitamin A deficiency produces defects in organogenesis at a high penetrance and reveals new roles for the vitamin in skeletal development. *Developmental Biology*, *316*, 171–190.

Sefton, E. M., Gallardo, M., & Kardon, G. (2018). Developmental origin and morphogenesis of the diaphragm, an essential mammalian muscle. *Developmental Biology*, *440*, 64–73.

Sefton, E. M., Gallardo, M., Tobin, C. E., Collins, B. C., Colasanto, M. P., & Kardon, G. (2022). Fibroblast-derived Hgf controls recruitment and expansion of muscle during morphogenesis of the mammalian diaphragm. *eLife*, *11*, e74592. https://doi.org/10.7554/eLife.74592.

Shannon, S. R., Moise, A. R., & Trainor, P. A. (2017). New insights and changing paradigms in the regulation of vitamin A metabolism in development. *Wiley Interdisciplinary Reviews: Developmental Biology*, *6*.

Sparrow, D. B., Chapman, G., Smith, A. J., Mattar, M. Z., Major, J. A., O'Reilly, V. C., et al. (2012). A mechanism for gene-environment interaction in the etiology of congenital scoliosis. *Cell*, *149*, 295–306.

Stevens, G. A., Bennett, J. E., Hennocq, Q., Lu, Y., De-Regil, L. M., Rogers, L., et al. (2015). Trends and mortality effects of vitamin A deficiency in children in 138 low-income and middle-income countries between 1991 and 2013: A pooled analysis of population-based surveys. *The Lancet Global Health*, *3*, e528–e536.

Szumilas, M. (2010). Explaining odds ratios. *Journal of Canadian Academy of Child and Adolescent Psychiatry*, *19*, 227–229.

Takayasu, H., Sato, H., Sugimoto, K., & Puri, P. (2008). Downregulation of GATA4 and GATA6 in the heart of rats with nitrofen-induced diaphragmatic hernia. *Journal of Pediatric Surgery*, *43*, 362–366.

Taleporos, P., Salgo, M. P., & Oster, G. (1978). Teratogenic action of bis(dichloroacetyl) diamine on rats: Patterns of malformations produced in high incidence at time-limited periods of development. *Teratology*, *18*, 5–15.

Temple, I. K., Barber, J. C., James, R. S., & Burge, D. (1994). Diaphragmatic herniae and translocations involving 8q22 in two patients. *Journal of Medical Genetics*, *31*, 735–737.

Torfs, C. P., Curry, C. J. R., Bateson, T. F., & Honoré, L. H. (1992). A population-based study of congenital diaphragmatic hernia. *Teratology*, *46*, 555–565.

Twigg, S. R., Matsumoto, K., Kidd, A. M., Goriely, A., Taylor, I. B., Fisher, R. B., et al. (2006). The origin of EFNB1 mutations in craniofrontonasal syndrome: Frequent somatic mosaicism and explanation of the paucity of carrier males. *American Journal of Human Genetics*, *78*, 999–1010.

Warkany, J., & Roth, C. B. (1948). Congenital malformations induced in rats by maternal vitamin A deficiency. *The Journal of Nutrition*, *35*, 1–11.

Wat, M. J., Shchelochkov, O. A., Holder, A. M., Breman, A. M., Dagli, A., Bacino, C., et al. (2009). Chromosome 8p23.1 deletions as a cause of complex congenital heart defects and diaphragmatic hernia. *American Journal of Medical Genetics. Part A*, *149A*, 1661–1677.

Wat, M. J., Veenma, D., Hogue, J., Holder, A. M., Yu, Z., Wat, J. J., et al. (2011). Genomic alterations that contribute to the development of isolated and non-isolated congenital diaphragmatic hernia. *Journal of Medical Genetics*, *48*, 299–307.

Wilson, J. G., Roth, C. B., & Warkany, J. (1953). An analysis of the syndrome of malformations induced by maternal vitamin a deficiency. Effects of restoration of vitamin a at various times during gestation. *American Journal of Anatomy*, *92*, 189–217.

Yang, W., Shaw, G. M., Carmichael, S. L., Rasmussen, S. A., Waller, D. K., Pober, B. R., et al. (2008). Nutrient intakes in women and congenital diaphragmatic hernia in their offspring. *Birth Defects Research. Part A, Clinical and Molecular Teratology*, *82*, 131–138.

Yu, L., Sawle, A. D., Wynn, J., Aspelund, G., Stolar, C. J., Arkovitz, M. S., et al. (2015). Increased burden of de novo predicted deleterious variants in complex congenital diaphragmatic hernia. *Human Molecular Genetics*, *24*, 4764–4773.

Yu, L., Wynn, J., Cheung, Y. H., Shen, Y., Mychaliska, G. B., Crombleholme, T. M., et al. (2013). Variants in GATA4 are a rare cause of familial and sporadic congenital diaphragmatic hernia. *Human Genetics*, *132*, 285–292.

Yu, L., Wynn, J., Ma, L., Guha, S., Mychaliska, G. B., Crombleholme, T. M., et al. (2012). De novo copy number variants are associated with congenital diaphragmatic hernia. *Journal of Medical Genetics*, *49*, 650–659.

Zheng, J., He, Q., Tang, H., Li, J., Xu, H., Mao, X., et al. (2020). Overexpression of miR-455-5p affects retinol (vitamin A) absorption by downregulating STRA6 in a nitrofen-induced CDH with lung hypoplasia rat model. *Pediatric Pulmonology*, *55*, 1433–1439.

CHAPTER FIVE

Gene–environment interactions in the pathogenesis of common craniofacial anomalies

Sharien Fitriasari[a] and Paul A. Trainor[a,b,*]
[a]Stowers Institute for Medical Research, Kansas City, MO, United States
[b]Department of Anatomy and Cell Biology, University of Kansas Medical Center, Kansas City, KS, United States
*Corresponding author: e-mail address: pat@stowers.org

Contents

Abstract

Craniofacial anomalies often exhibit phenotype variability and non-mendelian inheritance due to their multifactorial origin, involving both genetic and environmental factors. A combination of epidemiologic studies, genome-wide association, and analysis of animal models have provided insight into the effects of gene–environment interactions on craniofacial and brain development and the pathogenesis of congenital disorders. In this chapter, we briefly summarize the etiology and pathogenesis of common craniofacial anomalies, focusing on orofacial clefts, hemifacial microsomia, and microcephaly. We then discuss how environmental risk factors interact with genes to modulate the incidence and phenotype severity of craniofacial anomalies. Identifying environmental risk factors and dissecting their interaction with different genes and modifiers is central to improved strategies for preventing craniofacial anomalies.

Current Topics in Developmental Biology, Volume 152
ISSN 0070-2153
https://doi.org/10.1016/bs.ctdb.2022.10.005

1. Introduction

Craniofacial anomalies are a diverse group of congenital malformations that affect the head and face, and individual malformations may occur in isolation (non-syndromic) or in combination (syndromic). Craniofacial anomalies, which constitute one-third of all birth defects in the US, are major contributors to infant mortality (Mai et al., 2019) due to the impairment of essential processes such as breathing and feeding. However, craniofacial anomalies also affect speech and social integration later in life. The high incidence of craniofacial anomalies highlights the need for and importance of identifying the factors that modulate normal craniofacial morphogenesis, which is an intricate developmental process involving complex tissue interactions and integrated three-dimensional cell movement, growth, proliferation, survival, and differentiation. Although much progress has been made in identifying the genetic etiology underpinning many craniofacial anomalies, the mechanisms by which exogenous environmental and other factors exert their influence on the etiology and pathogenesis of craniofacial malformations are less well understood.

Clinically, craniofacial disorders often lack genotype-phenotype correlation and present with a wide spectrum of phenotype variability, ranging from normal or subclinical phenotypes to extreme structural malformations. Craniofacial malformations can also occur spontaneously without prior history of causative mutations within the family, suggesting non-genetic factors may also be involved (Trainor, Dixon, & Dixon, 2009). Craniofacial anomalies are often attributed to perturbation of the development of neural crest cells (NCC), which comprise migratory stem and progenitor cells that give rise to most of the cartilage, bone, and connective tissues of the head and face. Although the discordance in genotype-phenotype correlation is largely attributed to genetic variants and variable penetrance, considerable evidence points to the role of environmental factors and gene–environment interactions in modulating the severity of craniofacial malformations (Fitriasari & Trainor, 2021). For example, NCC are susceptible to exogenous stress such as hyperglycemia and oxidative stress (Sakai, Dixon, Achilleos, Dixon, & Trainor, 2016; Sakai & Trainor, 2016). Furthermore, defects in global processes required for the proliferation and survival of all cell types, such as ribosome biogenesis and DNA damage repair, often result in tissue-specific craniofacial and brain malformation phenotypes (Watt, Macintosh, Bernard, & Trainor, 2022; Yamaguchi et al., 2021). Thus, extrinsic factors can exacerbate the effects of a genetic mutation on NCC development and increase the severity of a craniofacial phenotype. To further understand the mechanisms and contributions of gene–environment interactions in craniofacial disorders, here we summarize the cellular and molecular basis of common craniofacial abnormalities, outline evidence of gene–environment interactions, and discuss exogenous contributions in the pathogenesis of craniofacial malformations.

2. Overview of craniofacial morphogenesis

In humans, the most critical and foundational steps of craniofacial morphogenesis occur within the first trimester of gestation. Around the third week of gestation, gastrulation occurs during which the three germ layers—ectoderm, endoderm, and mesoderm are delineated (Müller & O'Rahilly, 2004). As the embryo develops, the ectoderm is further subspecialized into surface ectoderm (non-neural ectoderm), which will eventually become skin, nails, and hair, and the neural ectoderm, which gives rise to the future central nervous system and components of the peripheral nervous system. Collective cell movement and signaling from a primary organizer induce neural plate formation from the neural ectoderm, after which the neural plate on both sides of the midline elevates and fuses to form a neural tube, the process of which is called neurulation (Keller, Shih, Sater, & Moreno, 1992; Rubenstein, Shimamura, Martinez, & Puelles, 1998), During neurulation, NCC are specified in the dorsal neural ectoderm at the interface between the neural plate and the surface ectoderm, a region termed the neural plate border (Selleck & Bronner-Fraser, 1995). NCC are formed via epithelial to mesenchymal transition (EMT) and subsequently delaminate and migrate away from the neural plate or neural tube, depending on the species, progressively along the neural axis of the embryo (Fig. 1A) (Zhao & Trainor, 2022). Depending on the axial level from which they originate and migrate, NCC can be classified as cranial, cardiac, trunk, vagal, or sacral NCC (Kirby, Gale, & Stewart, 1983; Le Douarin & Teillet, 1974; LeDouarin & Kalcheim, 1999). Different axial populations of NCC exhibit distinct differentiation capacities. For example, cranial NCC have the potential to form bone and cartilage, whereas trunk NCC generally do not (Le Douarin, Creuzet, Couly, & Dupin, 2004), although there are exceptions (Cebra-Thomas et al., 2007, 2013; Clark et al., 2001; Dash & Trainor, 2020; Gilbert, Bender, Betters, Yin, & Cebra-Thomas, 2007;

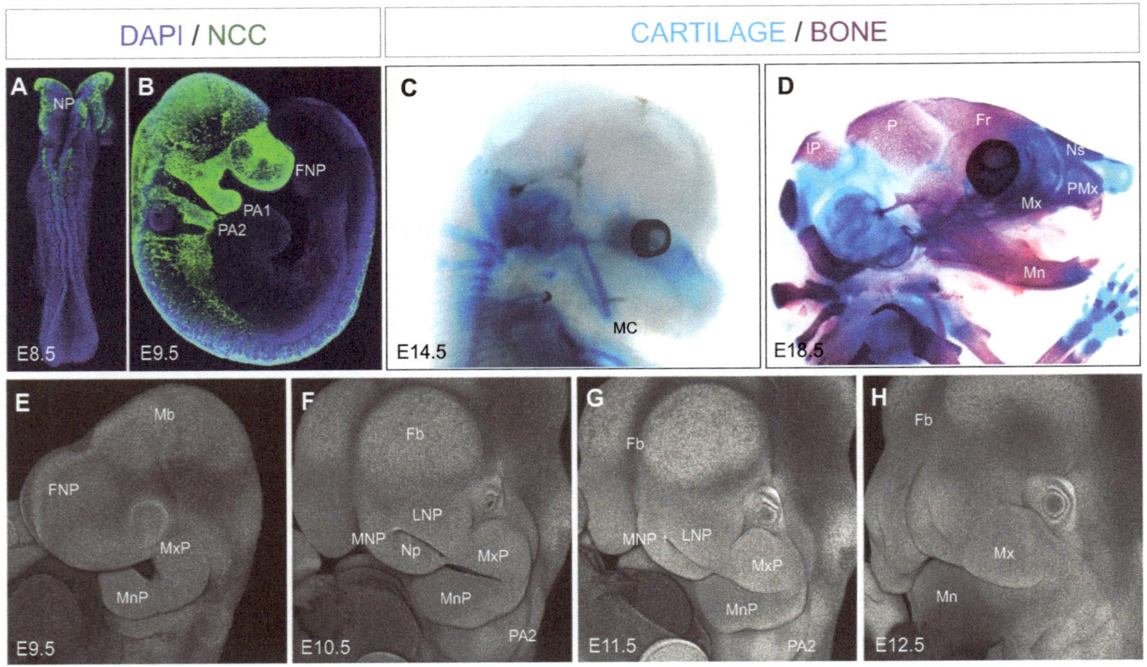

Fig. 1 Craniofacial morphogenesis is largely dependent on NCC development. (A) Dorsal view of E8.5 *Wnt1-Cre; YFP* control mouse embryo illustrating NCC migration from the neural plate (NP). (B) Lateral view of E9.5 *Wnt1-Cre; YFP* control mouse embryo showing distinct streams of NCC migration into the frontonasal prominence (FNP) and pharyngeal arches (PA). (C) Wildtype E14.5 mouse embryo stained with alizarin red and alcian blue (MC, Meckel's cartilage). (D) Wildtype E18.5 mouse embryo stained with alizarin red and alcian blue (IP, interparietal bone; P, parietal bone; Fr, frontal bone; Ns, nasal bone; PMx, premaxilla; Mx, maxilla; Mn, mandible). (E) Pseudo-SEM image of wildtype E9.5 mouse embryo (Mb, Midbrain; MxP, maxillary prominence; MnP, mandibular prominence). (F–H) Pseudo-SEM images of wildtype E10.5—E12.5 mouse embryos (Fb, forebrain; MNP, medial nasal processes; LNP, lateral nasal processes; Np, nasal pit).

Graham & McGonnell, 1999; Rice et al., 2017; Rice, Kallonen, Cebra-Thomas, & Gilbert, 2016; Rice, Riccio, Gilbert, & Cebra-Thomas, 2015; Trainor, 2005, 2016).

During the early stages of craniofacial development, cranial NCC migrate in distinct, segregated streams into the frontonasal prominence and pharyngeal arches in response to microenvironmental signals from the neural tube and surrounding ectoderm, paraxial mesoderm and endoderm (Fig. 1B) (Farlie et al., 1999; Kulesa & Fraser, 1998; Morrison et al., 2017; Noden & Trainor, 2005; Trainor, Sobieszczuk, Wilkinson, & Krumlauf, 2002; Trainor & Tam, 1995). A combination of intrinsic transcription factors and signaling cues from surrounding tissues subsequently regulate cranial NCC fate specification and the patterning of its derivatives (Bhatt, Diaz, & Trainor, 2013; Simoes-Costa & Bronner, 2016; Trainor & Krumlauf, 2001). For example, NCC destined for neuronal and glial fates express Foxd3 and Sox10 during their migration and differentiation, while NCC fated for osteochondral differentiation switch off these factors and instead express Sox9 (Akiyama et al., 2005; Britsch et al., 2001; Dottori, Gross, Labosky, & Goulding, 2001). Signaling from the surrounding epithelia, through FGFs, Wnts, TGFβ/BMPs and others, then induces cranial NCC aggregation and mature differentiation into the chondrocytes and osteoblasts that become craniofacial cartilage and bone respectively (Fig. 1C and D) (Dash & Trainor, 2020).

Due to the major contribution NCC make to the formation of the craniofacial complex, it is not surprising many craniofacial anomalies stem from perturbations in cranial NCC development. In fact, distinct phenotypes occur depending on which phase of NCC development is affected (Trainor, 2010; Watt & Trainor, 2014). Although proper cranial NCC development is central to craniofacial morphogenesis, reciprocal tissue interactions between NCC, mesoderm, ectoderm, and endoderm are required for generating a fully functional head. Patterning of the pharyngeal arches, for example, is achieved through a combination of Bmp/Fgf signaling from the underlying pouch endoderm and cleft ectoderm, in concert with cell–cell interactions between migrating NCC and mesodermal cells (McKinney et al., 2020; Trainor & Krumlauf, 2000; Trainor & Tam, 1995; Veitch, Begbie, Schilling, Smith, & Graham, 1999). Perturbation of the formation, migration, and/or differentiation of NCC or in signaling from the surrounding tissues can impair tissue interactions resulting in abnormal craniofacial morphogenesis. This suggests that intrinsic and extrinsic factors impact cranial NCC development and contribute significantly to the pathogenesis of craniofacial anomalies.

3. Developmental basis of common craniofacial anomalies

3.1 Orofacial clefts

Orofacial clefts (OFC) are the most common craniofacial anomaly and occur with an estimated frequency of 1 in every 700 live births worldwide, however the prevalence of OFC varies substantially across racial groups, lifestyles, and socioeconomic status (Mai et al., 2019; Mossey & Castilla, 2003). The term OFC refers to anatomical defects involving the lip and/or roof of the oral cavity (palate), which can be classified further into cleft lip alone (CL), cleft lip with cleft palate (CLP), and cleft palate only (CP) (Huang et al., 2019; Reynolds et al., 2020). A cleft lip is a condition where a gap occurs between the philtrum (cupid's bow) and the upper lip, whereas a cleft palate is characterized by an opening in the hard and/or soft palate (Gorlin, Cohen, & Hennekam, 2001). A wide continuum of phenotype severity can be observed within each of these anatomical subdivisions. For example, CL can manifest as a microform type, where the lip and the muscles around the orifice of the mouth (orbicularis oris) are slightly disoriented but otherwise perfectly intact, or a complete unilateral or bilateral CL in which the cleft involves the full vertical height of the lip, starting from the lateral part of the upper lip and extending up to the nostril (Bernheim, Georges, Malevez, De Mey, & Mansbach, 2006; Shkoukani, Chen, & Vong, 2013). Similarly, CP can be complete or incomplete, and depending on its severity, CP may affect the primary palate, the soft and hard portions of the secondary palate, the uvula, and the bony sockets of the teeth (alveolus) unilaterally or bilaterally (Burg, Chai, Yao, Magee, & Figueiredo, 2016). In the absence of a cleft, the palate may be high arched in association with a small oral cavity which can still compromise the breathing and feeding of affected individuals (Conley et al., 2016). The development of the upper lip and primary palate is distinct from the secondary palate. Epigenome-wide association studies have identified distinct DNA methylation profiles associated with CL, CLP, and CP, suggesting that the three OFC subtypes may be etiologically distinct (Sharp et al., 2017).

In human embryos, the development of the upper lip and primary palate begins during the fourth week of gestation (approximately E9.0 in mice) with the establishment of five facial primordia around a primitive oral cavity: a rostral median frontonasal prominence, and caudal lateral pairs of maxillary prominences and mandibular prominences (Fig. 1E) (Burdi, 2006; Frisdal & Trainor, 2014; Hinrichsen, 1985). The frontonasal prominence (FNP) consists of NCC-derived mesenchyme originating from the forebrain and anterior midbrain. As the forebrain increases in size, the FNP elongates while the mandibular prominences enlarge and begin to merge in the midline (Yoon, Chung, Seol, Park, & Park, 2000). By the beginning of the fifth week (approximately E10.0 in mice), the surface ectoderm gives rise to the nasal placodes in the ventrolateral region of the FNP and as the surrounding tissues grow and expand, the lateral and medial nasal processes (LNP and MNP) develop along the margins of the nasal placodes, transforming the placodes into surface depressions called nasal pits (Fig. 1F) (Hinrichsen, 1985; Yoon et al., 2000). Mesenchymal cells within the maxillary prominences (MxP) then undergo selective differentiation and proliferation, causing the MxP to enlarge and push the nasal pits toward the center of the face (Fig. 1G). The continuous growth of the MxP pushes the MNP toward each other and facilitates the fusion of the MNP with the LNP (Fig. 1H). By approximately 38 days of gestation, rapid proliferation within the MxP and the MNP brings both structures together into direct contact (Fig. 1H), a process that is independent of MNP-LNP fusion (Burdi, 2006; Jiang, Bush, & Lidral, 2006; Ohbayashi & Eto, 1986; Warbrick, 1960). Fusion between LNP, MNP, and MxP during the seventh week of gestation in humans, results in a contiguous primitive upper lip (Yoon et al., 2000). These nasal pits are transformed into nasal cavities via programmed cell death, and this induces the formation of an epithelial seam between the MNP-derived intermaxillary segment and the MxP, called the nasal fin (Gaare & Langman, 1977; Warbrick, 1960). The nasal fin then degenerates as a mesenchymal bridge form between the intermaxillary segment and the MxP, allowing for continuity of cell movement between the two structures and proper choanae formation (Kurosaka et al., 2021; Kurosaka, Wang, Sandell, Yamashiro, & Trainor, 2017). By the end of the 7th week the intermaxillary prominences grow into the oral cavity to form the primary palate and the philtrum, concluding upper lip and primary palate formation (Hinrichsen, 1985). Interestingly, amniote faces share a period of reduced shape variance and convergent growth trajectories from prominence formation through fusion. This is thought to select against the pathogenesis of maladaptive facial clefts and illustrates how epigenetic factors such as embryo geometry and shape impact facial morphogenesis and variation (Young et al., 2014).

The secondary palate begins to develop at the end of the sixth week when bilateral outgrowth of the maxillary processes, termed the palatal shelves, grow vertically downward beside the developing tongue (Sperber, 2002). The palatal shelves consist of NCC-derived mesenchyme and mesoderm-derived endothelial and muscle cells, surrounded by pharyngeal ectoderm-derived epithelial cells (Chai et al., 2000; Yoshida, Vivatbutsiri, Morriss-Kay, Saga, & Iseki, 2008). During the 7–8 weeks of gestation, the palatal shelves grow horizontally above the tongue, until the medial edge epithelia come into contact with each other, after which contact, adhesion and fusion continues bidirectionally (Burdi & Faist, 1967; Dixon, Marazita, Beaty, & Murray, 2011; Hammond & Dixon, 2022). The fusion process involves degeneration of the epithelial seam and is thought to depend upon a combination of programmed cell death, EMT and cell extrusion (Cuervo, Valencia, Chandraratna, & Covarrubias, 2002; Gritli-Linde, 2008, 2012; Kim et al., 2015; Teng, Teng, Kaartinen, & Bush, 2022; Vaziri Sani et al., 2005). The pace of midline epithelial seam degradation appears to be different in the anterior portion of the palate, which gives rise to the hard palate, compared to the posterior palate which will become the soft palate and uvula, suggesting that distinct anterior-posterior signaling pathways or mechanisms may regulate palate closure (Danescu, Mattson, Dool, Diewert, & Richman, 2015). In addition to fusing with the opposing palatal shelf, the shelves of the secondary palate also fuse with the primary palate and the nasal septum to separate the oral and nasal cavities (Sperber, 2002). After fusion is complete, the NCC-derived mesenchymal cells condense and differentiate into osteoblasts via intramembranous ossification to form the bony structures of the hard palate (Chai et al., 2000; Dash & Trainor, 2020).

Proper morphogenesis of the lip and palate requires integrated cell proliferation, differentiation, apoptosis, adhesion, EMT and cell extrusion. Defective fusion between the MNP, LNP, and MxP due to disruption in NCC proliferation, lack of cell adhesion, epithelial seam persistence, and aberrant signaling can result in cleft lip or cleft primary palate (Everson et al., 2017; Jiang et al., 2006; Sperber & Sperber, 2013). Similarly, insufficient or incomplete merging of the facial processes can lead to microforms of CL (Suzuki et al., 2009). It is therefore critical that the facial prominences come into contact with each other for proper fusion to occur, however each epithelial seam where facial prominences fuse is a developmental weak point and potential site for a cleft. Uneven tissue mass or delayed cell migration on one side of the facial primordia may create asymmetric pulling forces between the MNP and MxP, leading to a unilateral CL, while independent growths of the facial primordia on either side of the face can result in bilateral CL (Shkoukani et al., 2013). Mechanistically, insufficient fusion of the facial prominences has also been linked to reduced or absent epithelial filopodia at the free ends of each tissue projection, leading to higher susceptibility to clefts in the primary palate (Sun, Baur, & Hay, 2000). Similarly, clefts of the secondary palate can be caused by failure of palatal shelf fusion, which may be preceded by insufficient growth of the palatal shelves, failure of shelf elevation and contact, lack of cell adhesion between the shelves, or epithelial seams rupturing upon fusion (Baek et al., 2011; Ji et al., 2020; Kim et al., 2015). Furthermore, in the case of Pierre Robin Sequence, a cleft palate phenotype is associated with mandibular retrognathia, which may prevent the tongue from descending such that it physically obstructs reorientation and fusion of the palatal shelves (Dash, Bhatt, Falcon, Sandell, & Trainor, 2021; Huang et al., 2016). Interestingly, parental philtrum width has been implicated in non-syndromic OFC inheritance, suggesting facial geometry may modulate susceptibility to OFC due to differential distance between each facial prominences (Howe et al., 2018; Weinberg et al., 2009; Young et al., 2014).

3.2 Hemifacial microsomia

Hemifacial microsomia (HFM), also known as craniofacial microsomia (CFM), lateral facial dysplasia, oculoauriculovertebral spectrum (OAVS), first and second branchial arch anomaly, or facioauriculovertebral syndrome (FAVS), is the second most common congenital craniofacial anomaly after OFC, and affects approximately 1 in 3500–6500 newborns in the United States (Akram, McKnight, Bellardie, Beale, & Evans, 2015; Hartsfield, 2007; Kelberman et al., 2001). The clinical features of HFM are heterogeneous and involve malformations of the first and second pharyngeal arches and their derivatives on one or both sides of the face. Some of the more common phenotypes characteristic of HFM include hypoplasia of the mandible and the zygomatic complex, hypodontia, and auricular deformities affecting the incus, malleus, and tympanic bone (Chen, Zhao, Shen, & Dai, 2018; Gorlin et al., 2001). HFM also affects the NCC-derived cranial nerve and connective tissues and the mesoderm-derived craniofacial muscles (Heude, Rivals, Couly, & Levi, 2011; Vento, Labrie, & Mulliken, 1991). Defective trigeminal ganglia can result in unilateral hearing loss, whereas underdeveloped masticatory muscles may contribute to feeding and speech impediment (Monahan et al., 2001). Additionally, severe cases of HFM are associated with concurrent OFC on the affected side, suggesting an etiological link between HFM and OFC (Fan, Mulliken, & Padwa, 2005).

The majority of HFM patients exhibit significant facial asymmetry due to underdevelopment of craniofacial bones, cartilage, and muscles on one side of the face (Beleza-Meireles, Clayton-Smith, Saraiva, & Tassabehji, 2014; Maruko, Hayes, Evans, Padwa, & Mulliken, 2001; Rollnick et al., 1987). Most HFM cases are sporadic, but a segregation analysis has provided evidence for familial cases where HFM is inherited in autosomal dominant manner (Kaye et al., 1992). HFM can also be observed as part of other syndromes, such as Treacher Collins syndrome and Goldenhar syndrome, implying potentially genetic causes of HFM. Although several whole exome sequencing studies have identified candidate genes associated with HFM, the cellular and molecular basis of its pathogenesis remain to be elucidated (Chen, Liu, Mar Aung, Zhang, & Chai, 2021; Güleray et al., 2021; Liu et al., 2021; Lopez et al., 2016; Rengasamy Venugopalan et al., 2019; Timberlake et al., 2021). Currently there are two prevailing theories for the pathogenic mechanisms underlying HFM: vascular hemorrhage and disruption of cranial NCC development.

The first theory is based on mid-gestational treatment of CS1 mice and *Macaca irus* monkeys, with triazene and thalidomide respectively, which induced hemorrhaging within the stapedial arteries and disrupted blood supply to the first and second pharyngeal arches during early embryo development (Poswillo, 1973). The resulting hematomas were thought to disrupt the development of the adjacent branchial arch mesenchyme, leading to defects in Meckel's cartilage formation, hypoplasia of the mandible, and malformations of the ear (Otani et al., 1991; Poswillo, 1973). While this model may explain the asymmetric defects in HFM, localized hemorrhage has yet to be observed in human HFM patients and the exact cause of arterial disruption is unknown. A subsequent triazene study in mice showed that ear malformations can occur without the presence of local vascular hemorrhaging, suggesting that the auricular and mandibular defects in HFM are independent of embryonic hematomas (Louryan, Heymans, & Goffard, 1995).

The second hypothesis involves disruption of cranial NCC development as a mechanism to account for the broad range of tissues affected in HFM. Cranial NCC differentiate into osteoblasts and chondrocytes, giving rise to most craniofacial bone and cartilages respectively, including Meckel's cartilage, the mandible, temporo-mandibular joint, maxilla, zygomatic bones, and components of the external and middle ear such as incus, malleus, and tympanic ring. Additionally, cranial NCC are important for providing spatial cues to ensure proper migration and segmentation of mesodermal cells, which gives rise to facial and masticatory muscles (Heude et al., 2010, 2011; Trainor & Tam, 1995). Perturbation of cranial NCC formation, migration, and differentiation can therefore lead to mandibular and auricular hypoplasia due to insufficient bone and cartilage progenitor cells, as well as underdevelopment of the craniofacial soft tissues (Dixon et al., 2006; Heude et al., 2011; Jones et al., 2008).

It is unclear why craniofacial malformations in HFM are frequently asymmetrical. Variation in the pace, timing, and pattern of NCC development on the right vs left side of the head may lead to different degrees of susceptibility to environmental insults, therefore altering the development only on one side of the face (Danescu et al., 2021; Inman, Purcell, Kume, & Trainor, 2013). Additionally, heart development, which occurs predominantly on the left side of the body axis has been proposed to influence facial asymmetry through variation in the level of *Fgf8*, an important factor for survival and patterning of NCC. Asymmetric expression of *Fgf8* in the developing heart, which comes into close contact with the pharyngeal endoderm, may lead to higher or sustained Fgf8 signaling on the right developing mandibular process, resulting in longer dentary bones and temporalis volumes compared to the left side of the face (Zbasnik et al., 2022). However, population studies indicate that the frequency of HFM occurrence is similar on the right and left sides of the face (Maruko et al., 2001; Vento et al., 1991), suggesting that other exogenous factors may influence the facial asymmetry characteristic of HFM.

3.3 Microcephaly

Microcephaly is clinically defined as having an occipitofrontal head circumference and brain weight of more than two standard deviations below the expected average for the patient's sex, age, and ethnicity (Passemard, Kaindl, & Verloes, 2013). The incidence of microcephaly varies widely according to population and methodology, ranging from 1.53 to 8.7 in 10,000 live births (Auger, Quach, Healy-Profitós, Lowe, & Arbour, 2018; Cragan et al., 2016; Hansen, Armstrong, Bower, & Baynam, 2017; Morris et al., 2016). The clinical features of microcephaly may encompass a severe reduction in brain volume, intellectual deficiency, cerebellar hypoplasia, as well as lissencephaly where the folds of the cerebral cortex (gyri and sulci) are absent (Abuelo, 2007; Boonsawat et al., 2019). Microcephaly may be evident at birth (primary or congenital microcephaly) or develop progressively postnatally (secondary microcephaly). Almost 800 genes have been linked to microcephaly, but the processes underlying its pathogenesis and phenotypic variability remain to be fully elucidated.

Congenital microcephaly is often associated with perturbation during early brain development (Alcantara & O'Driscoll, 2014). Mutations affecting centrosome cycle and DNA damage repair processes, which control cell division, cell cycle progression, and survival have been described in patients with congenital microcephaly (Boonsawat et al., 2019; Passemard et al., 2013; Rump et al., 2016). To understand how defective cell cycle progression can lead to microcephaly, it is important to understand the normal development of the central nervous system (CNS), which begins with the process of neurulation during the third week of gestation in humans and at E8.5 in mice. During neurulation, the neural ectoderm undergoes massive cellular rearrangement where each side elevates to fuse at the dorsal midline, forming a cylindrical neural tube. Three primary brain vesicles are formed from the lumen of the neural tube—the prosencephalon (forebrain), mesencephalon (midbrain), and rhombencephalon (hindbrain) (Altmann & Brivanlou, 2001). The prosencephalon subdivides into the telencephalon, which gives rise to the cerebral hemispheres, olfactory cortex, hippocampus, and basal ganglia, and the diencephalon, which forms the optic cup, thalamus, and hypothalamus. The mesencephalon does not subdivide, but generates the tectum, cerebral peduncle and substantia nigra. The rhombencephalon subdivides into the metencephalon, the precursor of the pons and cerebellum, and the myelencephalon which forms the medulla oblongata (Stiles & Jernigan, 2010).

The neural tube initially consists of uncommitted cells within a single-layer pseudostratified neuroepithelium (Ericson et al., 1997; Kicheva et al., 2014; Pierani, Brenner-Morton, Chiang, & Jessell, 1999), however its morphology rapidly changes in concert with cell proliferation. The neuroepithelial cells, which later gives rise to the cerebral cortex and the spinal cord, divide symmetrically such that both daughter cells re-enter the cell cycle after mitosis to expand the pool of neuronal progenitor cells (Smart, 1973). During cranial neurogenesis, which occurs between the 8th and 24th weeks of gestation in humans and begins at E11.5 in mice, the neuroepithelial cells express glial markers and transform into radial glial cells. Radial glial cells can divide asymmetrically to generate an undifferentiated neural stem cell capable of proliferation and a differentiation, and a non-dividing neuron, or divide symmetrically to generate two neurons

(Götz & Huttner, 2005; Mione, Cavanagh, Harris, & Parnavelas, 1997). Since neurons are post-mitotic and incapable of dividing to produce new cells, the balance between undifferentiated progenitor cells and neurons is an important determinant of eventual brain size and is precisely controlled by the number and duration of cell cycles (Caviness et al., 2003). Incorrectly positioned mitotic spindles due to mutations in centrosome-related genes are thought to promote asymmetric cell division, leading to premature differentiation and fewer proliferating progenitor cells. Therefore, disruption of centriole biogenesis, centrosome maturation, and mitotic spindle formation may result in a smaller pool of neurons overall at the end of neurogenesis and result in microcephaly (Caviness et al., 2003; Gilmore & Walsh, 2013; Jean, Stuart, & Tarailo-Graovac, 2020). In addition, mutations in DNA damage repair genes can lead to the accumulation of unrepaired DNA breaks that trigger cell cycle arrest. Prolonged cell cycle arrest due to severe DNA damage prevents cells from proliferating, eventually inducing cell death through the p53 pathway (Hartwell & Weinert, 1989). Indeed, increased cell death was observed in the cerebellum of mice deficient in the DNA damage repair gene *Nbs1*, indicating that a loss of neuronal cells results in a smaller cerebral cortex and microcephaly (Li, Yang, Gao, Wang, & Tong, 2012).

4. Common risk factors and gene–environment interaction in craniofacial anomalies

4.1 Smoking and environmental tobacco smoke

Genome-wide association studies and epidemiology data have provided insights into potential gene–environment interactions in the pathogenesis of common craniofacial anomalies. Maternal smoking or exposure to environmental tobacco smoke (ETS) are known to be major risk factors for OFC. A recent cohort study of more than 12 million births in the US found that low intensity maternal smoking (1–5 cigarettes per day) is enough to be significantly associated with OFC and several other congenital anomalies, regardless of the timing of smoking during pregnancy (Yang et al., 2022). On a global scale, nearly 53% of women who smoked daily continue to smoke during pregnancy. Therefore, maternal smoking remains a highly prevalent risk factor for craniofacial birth defects (Lange, Probst, Rehm, & Popova, 2018).

Smoke from tobacco products, including e-cigarettes, contains more than 7,000 compounds, but nicotine, polycyclic aromatic hydrocarbons, and carbon monoxide make up the majority (Guerin, 1980). Maternal smoking has been shown to alter the metabolomic profile of intrauterine blood supply from the umbilical cord, which correlates with reduced head circumference in human newborns (Yong-Ping et al., 2020). Furthermore, mouse embryos subjected to cigarette smoke inhalation during early neurulation exhibited changes in mitochondria morphology within the neural tube, which is suggestive of metabolic or oxidative stress (Bnait & Seller, 1995). In agreement with this data, exposure of *Xenopus* embryos to e-cigarette vapors also results in OFC, midface hypoplasia, and microcephaly. Interestingly, this occurs in association with upregulation of a human ortholog of the constitutive P450 metabolizing enzyme *CYP1A1* (Collier, Tingle, Paxton, Mitchell, & Keelan, 2002; Dickinson et al., 2022; Hessel-Pras, Ehlers, Braeuning, & Lampen, 2018; Kennedy, Kandalam, Olivares-Navarrete, & Dickinson, 2017). *Cyp1a1* has been shown to be tissue-specifically expressed during mouse embryogenesis with dynamic elevated levels in the dorsal hindbrain at E8.5, which coincides with the formation and migration of cranial neural crest cells (Campbell et al., 2005). Consistent with this observation, further analysis suggests that the craniofacial anomalies observed in smoke-exposed embryos occur as a result of NCC loss, triggered by an increase in oxidative stress and cell death (Kennedy et al., 2017; Zhao & Reece, 2005). More research however is required to determine whether the increase in oxidative stress is a consequence of higher enzymatic activity of CYP1A1, which is known to metabolize polycyclic aromatic hydrocarbons into epoxide intermediates. Nevertheless, epoxide intermediates need to be rapidly conjugated onto glutathione via glutathione transferase enzymes, such as GSTT1, otherwise the highly reactive molecules can react with DNA to form bulky DNA adducts (Henkler, Stolpmann, & Luch, 2012; Seidegård & Ekström, 1997). DNA adducts then trigger the nucleotide excision repair pathway, subsequently activating p53-dependent apoptosis in the case of severe DNA damage. Taken together, these data may explain the correlation between increased risk of non-syndromic OFC and human variants of *CYP1A1* and *GSTT1* (Shi, Wehby, & Murray, 2008), such that a genetic predisposition for OFC is amplified by the detrimental effect of smoking or ETS.

Interestingly, certain variants of genes known to cause OFC characteristic of different syndromes, such as *IRF6* (Van der Woude syndrome) and *RUNX2* (Cleidocranial Dysplasia), show no significant genome-wide association when considered alone, but become significantly associated with increased incidence of OFC once ETS exposure is taken into account (Wu et al., 2010, 2012). This further supports the notion that the severity of OFC and other craniofacial disorders are shaped by complex interactions between genetic and environmental influences throughout development. One mechanism by which environmental factors exert their influence on gene expression is via alteration of epigenetic landmarks. Several epigenome-wide association studies have shown that cigarette smoking alters DNA methylation patterns at multiple CpG sites, some of which are passed down to progeny (Everson et al., 2021; Howe et al., 2019; Joubert et al., 2012). A differential methylation pattern has been found at CpG sites in aryl hydrocarbon receptor genes as well as their downstream target *CYP1A1* (Joubert et al., 2012). Furthermore, nicotine from smoking and ETS can alter gene expression at the miRNA level, disrupting the molecular signaling required for normal facial morphogenesis such as is the case for PDGF which has critical roles in mesenchymal cell proliferation and migration (Li et al., 2011; Zhang, Pan, Zhong, & Cheng, 2014). Collectively, these studies provide evidence of gene—smoking/ETS interaction in the pathogenesis of craniofacial anomalies, particularly OFC.

4.2 Alcohol exposure

Despite the high number of studies linking alcohol to adverse pregnancy outcomes, alcohol exposure still remains a significant risk factor for craniofacial anomalies. Alcohol is an established teratogen, capable of inducing a broad range of congenital malformations collectively termed fetal alcohol spectrum disorder (FASD). FASD encompasses diverse clinical phenotypes such as microcephaly, short palpebral fissures, smooth philtrum, midface hypoplasia, and cognitive impairment (Hoyme et al., 2016). Approximately 10% of mothers worldwide consume alcohol during pregnancy, and are three times more likely to deliver a child with FASD compared to the general population (Popova, Lange, Probst, Gmel, & Rehm, 2017). Prevention efforts are complicated by the timing of an embryo's vulnerability to alcohol during craniofacial morphogenesis and neurogenesis early in the first trimester, typically before women are aware they are pregnant. However, a twin study found that identical prenatal alcohol exposure can lead to different FASD severity in dizygotic twins, whereas the outcome of alcohol exposure is concordant in monozygotic twins, suggesting that embryos can be more resistant or susceptible to alcohol due to the confounding effects of genetic background (Hemingway et al., 2018).

Similar effects are observed in mouse embryos of different genetic backgrounds, with C57BL/6J embryos being more susceptible to craniofacial malformations following ethanol exposure compared to DBA/2J embryos (Downing et al., 2012). This variation in alcohol susceptibility is attributed to changes in the expression of genes involved in ribosome biogenesis, oxidative stress, methylation, chromatin organization, and mRNA splicing (Berres, Garic, Flentke, & Smith, 2017; Downing et al., 2012). Studies in chicken embryos have corroborated this finding to the extent that partial knockdown of ribosomal genes led to heightened sensitivity to alcohol-induced apoptosis and craniofacial malformations (Berres et al., 2017). Interestingly, disruption of ribosome biogenesis is known to underlie the pathogenesis of Treacher Collins syndrome and Acrofacial Dysostosis, Cincinnati type, both of which are congenital disorders characterized by craniofacial malformations such as CP, HFM, hypoplasia of the zygomatic arches, and occasionally microcephaly. Insufficient ribosome production results in apoptosis of NCC and a reduction of osteochondral progenitor cells, causing hypoplasia of the craniofacial bone and cartilage as evidenced in mouse and zebrafish models of Treacher Collins syndrome and Acrofacial Dysostosis, Cincinnati type (Berres et al., 2017; Dixon et al., 2006; Falcon et al., 2022; Fitriasari & Trainor, 2021; Jones et al., 2008; Vlachos et al., 2008; Watt, Achilleos, Neben, Merrill, & Trainor, 2016; Watt, Neben, Hall, Merrill, & Trainor, 2018; Weaver et al., 2015). Alcohol exposure can also directly impair sonic hedgehog signaling (SHH), which induces apoptosis in osteochondral and neural progenitor cells thereby disrupting patterning of the face and brain (Aoto, Shikata, Higashiyama, Shiota, & Motoyama, 2008; Beames & Lipinski, 2020; Lipinski et al., 2012). This suggests there are threshold requirements for SHH signaling and ribosome biogenesis during craniofacial morphogenesis and neurogenesis, such that genetic predispositions can be exacerbated by alcohol exposure, which reduces ribosome biogenesis and SHH signaling below the threshold required for normal morphogenesis, leading to craniofacial malformations.

In addition to genes involved in ribosome biogenesis and SHH signaling, multiple other loci including *NOS1*, *VGLL2*, *SMC2*, *SLCO3A1*, *MSX1*, *ARHGEF10*, and *IGF2* have been associated with an increased risk of OFC and microcephaly in humans when combined with prenatal alcohol exposure (Beaty et al., 2011; Carlson et al., 2022; Haaland et al., 2017, 2019; Marjonen, Kahila, & Kaminen-Ahola, 2017; Romitti et al., 1999; Zhang et al., 2021). NOS1 produces nitric oxide (NO), a signaling molecule which can post-translationally modify proteins through S-nitrosylation of histone acetyltransferases and histone deacetylases. NO signaling is required for neurogenesis and neuronal differentiation, and a deficiency was shown to exacerbate microcephaly in alcohol-treated mouse embryos (Bonthius et al., 2002; Nott, Watson, Robinson, Crepaldi, & Riccio, 2008). Furthermore, in zebrafish, inhibition of NO signaling alters cranial NCC patterning and prevents their differentiation into chondrocyte progenitor cells (Kong et al., 2014). *MSX1* mutations are associated with CP and tooth agenesis in humans (Vastardis, Karimbux, Guthua, Seidman, & Seidman, 1996) and similar gene–environment interactions were observed in cranial NCC for Msx1, a transcription factor required for osteoprogenitor cell proliferation and bone growth, with ethanol (Nassif et al., 2014). Furthermore, in chicken embryos exposed to ethanol, *Msx1* expression was down-regulated in concert with increased cranial NCC apoptosis (Zhang et al., 2017). The remaining loci identified through epidemiology studies have not previously been affiliated with craniofacial anomalies, and therefore offer the potential to discover novel mechanisms by which alcohol exert its influence during craniofacial development. Recent screens in zebrafish have identified numerous additional novel alcohol responsive genes associated with the pathogenesis of craniofacial anomalies (Swartz et al., 2014; Swartz, Lovely, McCarthy, Kuka, & Eberhart, 2020). Interestingly, *pdgfrα* was found to protect against ethanol-induced craniofacial defects in a zebrafish model of FASD (McCarthy et al., 2013). Collectively, these studies contribute to the growing number of gene–alcohol interactions that might underlie susceptibility to craniofacial malformations, particularly those characteristic of FASD (Lovely, Fernandes, & Eberhart, 2016; Lovely, Rampersad, Fernandes, & Eberhart, 2017).

4.3 Vitamin deficiency and excess

Since the 1970s, prenatal vitamins containing folate or vitamin B9 have been routinely recommended or prescribed to pregnant women. Folate deficiency during neurulation is known to result in neural tube defects such as anencephaly and spina bifida (Greenberg, Bell, Guan, & Yu, 2011). In an effort to reduce the incidence of neural tube defects, many countries around the world adopted policies that fortified grain products with folate (Al-Gailani, 2014; Duerbeck, Dowling, & Duerbeck, 2014). In addition to a decrease in spina bifida and other open neural tube defects, the occurrence of non-syndromic OFC also decreased after folate fortification, suggesting direct involvement of folate metabolism in craniofacial morphogenesis (Millacura, Pardo, Cifuentes, & Suazo, 2017; Yazdy, Honein, & Xing, 2007). Consistent with this idea, polymorphisms in folate metabolism genes, such as *MTHFR* (5,10 methylenetetrahydrofolate reductase), *RFC1* (reduced folate carrier), *TCN1,* and *TCN2* (transcobalamin) have been associated with increased risk of OFC (Martinelli et al., 2001, 2006; Pezzetti et al., 2004; Vieira, Cooper, Marazita, Castilla, & Orioli, 2008). Folate is an important substrate for thymidine and purine biosynthesis, a process required for DNA/RNA synthesis and damage repair (Crider, Yang, Berry, & Bailey, 2012), and during craniofacial morphogenesis, folate promotes cell survival by facilitating nucleotide synthesis and DNA repair in Xenopus embryos. Faulty folate metabolism can therefore lead to increased apoptosis which results in median clefts (Wahl et al., 2015).

Other vitamin deficiencies that have been implicated in abnormal craniofacial bone and cartilage development include vitamin D and K, but most studies are performed in animal models and their relevance to human craniofacial anomalies have yet to be established (Howe & Webster, 1994; Kwon, 2019). Two recent studies have revealed the importance of Vitamin B12 or cobalamin in craniofacial and neural development (Chern et al., 2020, 2022). Methylmalonic Aciduria type C and Homocystinuria (*MMACHC*) encodes a protein thought to be a cobalamin-processing enzyme (Lerner-Ellis, Gradinger, et al., 2006; Lerner-Ellis, Tirone, et al., 2006) and that also intersects with folate metabolism. Patients with mutations in *MMACHC* often exhibit mild craniofacial dysmorphia and microcephaly among other anomalies. Interestingly, perturbation of cobalamin function was associated with altered ribosome biogenesis in concert with defects in NCC survival and differentiation, illustrating the importance of vitamin deficiencies and their extrinsic effects on NCC and craniofacial development (Chern et al., 2020, 2022).

Although the use of multivitamins periconceptionally and throughout pregnancy is considered beneficial, a couple of recent meta-analysis studies have flagged an association between vitamin consumption and increased risk for OFC (Haaland et al., 2019; Zhang et al., 2021). A significant gene–vitamin interaction was identified for three SNPs in *ESRRG*, which encodes the estrogen-related receptor gamma (Haaland et al., 2019). No specific association has been made for *ESRRG* variants alone with elevated risks for OFC in humans, but abnormal estrogen levels have been linked with perturbation in craniofacial cartilage development in zebrafish (He, Wang, Tang, Yang, & Xu, 2018; Huang et al., 2021; Kramer, Vuthiganon, & Lassiter, 2016). In addition, *ESRRG* overexpression results in craniofacial, axial and limb skeletal hypoplasia (Cardelli et al., 2013). Further studies are needed to verify the mechanism, but gene–vitamin interactions that result in hormone modulation have been proposed to explain sex-specific differences in OFC risk (Awotoye et al., 2021; Carlson et al., 2018).

Additionally, vitamin A (retinol) and its metabolite retinoic acid, are tightly regulated during craniofacial development. Retinoic acid, which is an active derivative of vitamin A, is a signaling molecule involved in the anterior-posterior patterning of the neural tube, and cranial NCC formation, migration and differentiation in large part via its regulation of *Hox* genes among many others (Williams & Bohnsack, 2019). Gene regulation by RA is direct, through the presence of retinoic acid response elements in the promotor regions of target genes which bind retinoic acid receptors (RAR) and retinoid receptors (RXR). Both insufficient and excess RA signaling can elicit craniofacial anomalies mediated by aberrant cranial NCC development (Shannon et al., 2020; Shannon, Moise, & Trainor, 2017) and it's remarkable how subtle temporal or spatial perturbations of vitamin A and retinoic acid activity can lead to craniofacial anomalies such as choanal atresia (Kurosaka et al., 2017, 2021), cleft palate (Wang et al., 2019), midfacial clefts (Wu et al., 2022), or abnormal branchiomeric muscle patterning (Wang et al., 2021). Interestingly, polymorphisms within genes involved in RA metabolism, such as *RBP4* (retinol-binding protein), which plays a critical role in transporting vitamin A from the liver and other stores to the tissues that need it, or signaling via *RARα,* are associated with increased risk for OFC in human populations (Xavier et al., 2013; Zhang et al., 2018). Furthermore, a derivative of vitamin A called isotretinoin is often used to treat dermatologic conditions such as severe cystic acne, but oral intake of isotretinoin during the first week of pregnancy can lead to morphological defects termed retinoic acid embryopathies which typically include craniofacial malformations (Forbat, Ali, & Al-Niaimi, 2018; Lammer et al., 1985). Thus gene–environment interactions involving vitamin A and retinoic acid signaling have a major impact on the pathogenesis of common craniofacial anomalies.

5. Conclusion

Gene–environment interactions in craniofacial anomalies can take many different forms (Khoury, Adams, & Flanders, 1988) (Fig. 2A). A genetic variant may be innocuous until an additive effect from an environmental

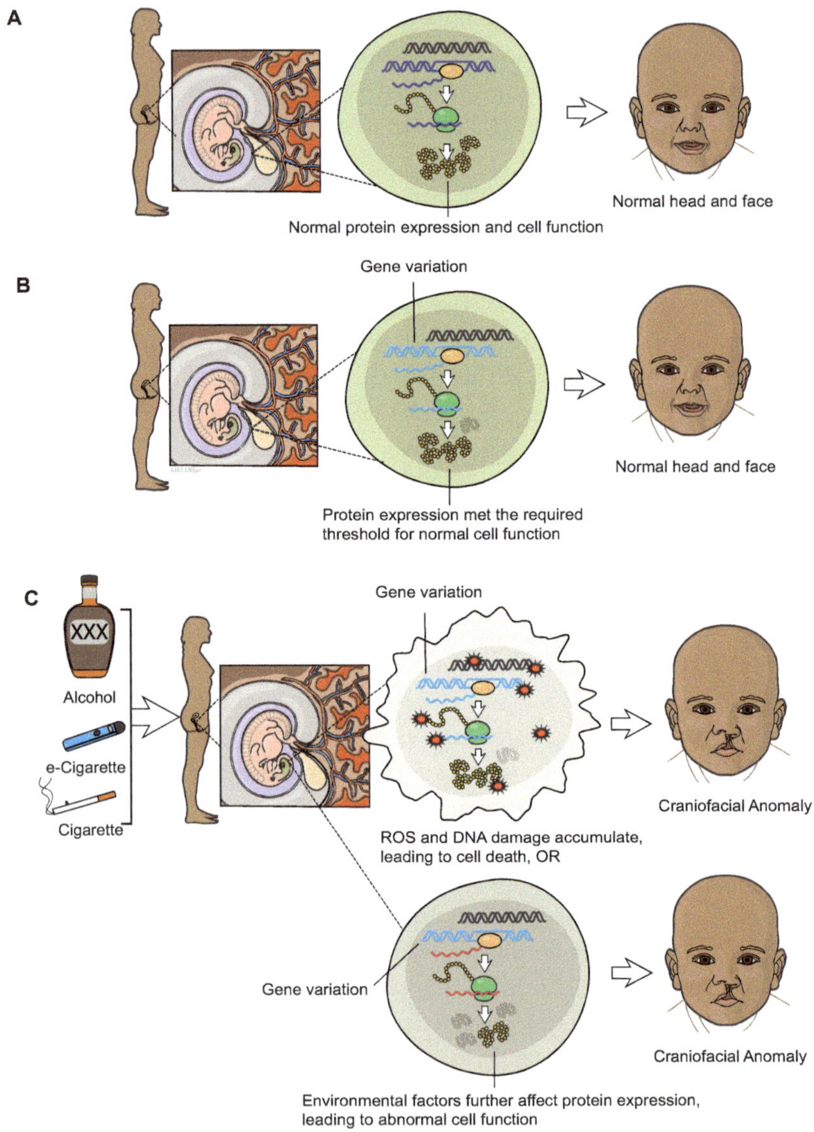

Fig. 2 Different mechanisms for gene–environment interaction in modulating craniofacial anomaly. (A) A schematic representation of a normal cell during a healthy pregnancy. (B) Gene variant alone may not be strong enough to induce craniofacial anomaly. (C) Environmental insults may disrupt important cell functions and induce cell death in addition to the effect from a detrimental gene variant, or the detrimental effect from a gene variant is further amplified by an environmental insult through post-transcriptional or post-translational modifications.

factor occurs (Fig. 2B). This concept potentially explains the intriguing relationship between *IRF6* polymorphisms and smoking/ETS, where the genotype and the environmental factor have seemingly independent effects on craniofacial morphogenesis. In this case, the variant alone is not enough to disrupt normal orofacial morphogenesis, but smoking/ETS exposure acting in concert with reduced IRF6 activity leads to increased cell death and OFCs (Fig. 2C). The opposite also applies in which an environmental insult may not cause an obvious phenotype unless a causative genetic covariate is present. For example, certain variants may sensitize C57BL/6J mice to exhibit more severe craniofacial anomalies compared to DBA/2J mice when exposed to alcohol. Alternatively, certain protective polymorphisms present in DBA/2J but absent in C57BL/6J may underpin the difference in alcohol resistance between these two strains. Gene–environment interaction studies should therefore screen for protective as well as detrimental loci and variants. Additionally, environmental factors can directly affect the activity of a specific gene or protein through alteration of their transcription or translation (Fig. 2C). For example, ethanol reduces the transcription of ribosome biogenesis genes, potentially resulting in ribosome deficiencies which affects cranial NCC survival and proliferation (Berres et al., 2017; Dixon et al., 2006; Haft et al., 2014). Gene–environment interaction can also be quite straightforward, such that variants in metabolic genes needed to process a teratogen result in increased severity of a craniofacial anomaly when embryos are exposed to said teratogen. For example, an individual possessing a gain of function variant in *RBP4* may develop more severe OFC when exposed to exogenous retinoic acid, compared to an individual on a vitamin A deficient diet.

Identification of gene–environment interactions in the human population remains challenging due to the statistical power required to make a significant genome-wide association. Furthermore, characterization of a genetic variant in relation to craniofacial anomalies are often dichotomous and qualitative, such that subclinical phenotypes are often mistakenly treated as "unaffected" thus masking the actual effect of said variant (Dixon et al., 2011). Epidemiology studies also rely heavily on self-reported exposure to environmental insults, which may not be completely accurate or even unknown to many individuals. Nonetheless, integrating metabolomics data in addition to genetic counseling during prenatal consultation may prove beneficial in estimating the risk for craniofacial anomalies. With more than 3,000 known chemicals circulating in our environment, high-throughput chemical screening in animal models will be beneficial to identify novel risk factors for craniofacial anomalies (Lovely et al., 2016, 2017). However, it is crucial for researchers to characterize the effects of genetic variants and environmental conditions using different animal models and genetic backgrounds to mimic the heterogeneity of the human population more accurately. Finally, careful phenotyping and attention to subclinical phenotypes is fundamental to improving our understanding of the impact of gene–environment interactions on the etiology and pathogenesis of craniofacial malformations (Trainor & Richtsmeier, 2015).

Acknowledgments

The authors thank Dr. Amanda Barlow and Dr. Hiroshi Kurosaka for the images used in Fig. 1, as well as Mark Miller for his artistic efforts in creating Fig. 2. This work was performed to fulfill, in part, requirements for S.F.'s thesis research in the Graduate School of the Stowers Institute for Medical Research. Research in the Trainor laboratory is supported by the Stowers Institute for Medical Research.

References

Abuelo, D. (2007). *Seminars in Pediatric Neurology, 14*, 118–127.

Akiyama, H., Kim, J.-E., Nakashima, K., Balmes, G., Iwai, N., Deng, J. M., et al. (2005). *Proceedings of the National Academy of Sciences of the United States of America, 102*, 14665–14670.

Akram, A., McKnight, M. M., Bellardie, H., Beale, V., & Evans, R. D. (2015). *British Dental Journal, 218*, 129–141.

Alcantara, D., & O'Driscoll, M. (2014). American Journal of Medical Genetics Part C: Seminars in Medical. *Genetics, 166*, 124–139.

Al-Gailani, S. (2014). *Studies in History and Philosophy of Biological and Biomedical Sciences, 47*(Pt. B), 278–289.

Altmann, C. R., & Brivanlou, A. H. (2001). Neural patterning in the vertebrate embryo. In *International review of cytology* (pp. 447–482). Academic Press.

Aoto, K., Shikata, Y., Higashiyama, D., Shiota, K., & Motoyama, J. (2008). *Birth Defects Research. Part A, Clinical and Molecular Teratology, 82*, 224–231.

Auger, N., Quach, C., Healy-Profitós, J., Lowe, A.-M., & Arbour, L. (2018). *Archives of Disease in Childhood - Fetal and Neonatal Edition, 103*, F167.

Awotoye, W., Comnick, C., Pendleton, C., Zeng, E., Alade, A., Mossey, P. A., et al. (2021). *Journal of Dental Research, 101*, 465–472.

Baek, J.-A., Lan, Y., Liu, H., Maltby, K. M., Mishina, Y., & Jiang, R. (2011). *Developmental Biology, 350*, 520–531.

Beames, T. G., & Lipinski, R. J. (2020). *Development, 147*.

Beaty, T. H., Ruczinski, I., Murray, J. C., Marazita, M. L., Munger, R. G., Hetmanski, J. B., et al. (2011). *Genetic Epidemiology, 35*, 469–478.

Beleza-Meireles, A., Clayton-Smith, J., Saraiva, J. M., & Tassabehji, M. (2014). *Journal of Medical Genetics, 51*, 635.

Bernheim, N., Georges, M., Malevez, C., De Mey, A., & Mansbach, A. (2006). *B-ENT, 2*(Suppl. 4), 11–19.

Berres, M. E., Garic, A., Flentke, G. R., & Smith, S. M. (2017). *PLoS One, 12*, e0169351.

Bhatt, S., Diaz, R., & Trainor, P. A. (2013). *Cold Spring Harbor Perspectives in Biology, 5*, a008326.

Bnait, K. S., & Seller, M. J. (1995). *Experimental and Toxicologic Pathology*, *47*, 453–461.
Bonthius, D. J., Tzouras, G., Karacay, B., Mahoney, J., Hutton, A., McKim, R., et al. (2002). *Brain Research. Developmental Brain Research*, *138*, 45–59.
Boonsawat, P., Joset, P., Steindl, K., Oneda, B., Gogoll, L., Azzarello-Burri, S., et al. (2019). *Genetics in Medicine*, *21*, 2043–2058.
Britsch, S., Goerich, D. E., Riethmacher, D., Peirano, R. I., Rossner, M., Nave, K. A., et al. (2001). *Genes & Development*, *15*, 66–78.
Burdi, A. R. (2006). Developmental biology and morphogenesis of the face, lip and palate. In S. Berkowitz (Ed.), *Cleft lip and palate* (pp. 3–12). Berlin, Heidelberg: Springer Berlin Heidelberg.
Burdi, A. R., & Faist, K. (1967). *American Journal of Anatomy*, *120*, 149–159.
Burg, M. L., Chai, Y., Yao, C. A., Magee, W., & Figueiredo, J. C. (2016). *Frontiers in Physiology*, 7.
Campbell, S. J., Henderson, C. J., Anthony, D. C., Davidson, D., Clark, A. J., & Wolf, C. R. (2005). *Journal of Biological Chemistry*, *280*, 5828–5835.
Cardelli, M., Zirngibl, R. A., Boetto, J. F., McKenzie, K. P., Troy, T. C., Turksen, K., et al. (2013). *PLoS One*, *8*, e81511.
Carlson, J. C., Nidey, N. L., Butali, A., Buxo, C. J., Christensen, K., Deleyiannis, F. W., et al. (2018). *Genetic Epidemiology*, *42*, 664–672.
Carlson, J. C., Shaffer, J. R., Deleyiannis, F., Hecht, J. T., Wehby, G. L., Christensen, K., et al. (2022). *Frontiers in Cell and Development Biology*, *10*.
Caviness, V. S., Jr., Goto, T., Tarui, T., Takahashi, T., Bhide, P. G., & Nowakowski, R. S. (2003). *Cerebral Cortex*, *13*, 592–598.
Cebra-Thomas, J. A., Betters, E., Yin, M., Plafkin, C., McDow, K., & Gilbert, S. F. (2007). *Evolution & Development*, *9*, 267–277.
Cebra-Thomas, J. A., Terrell, A., Branyan, K., Shah, S., Rice, R., Gyi, L., et al. (2013). *Developmental Dynamics*, *242*, 1223–1235.
Chai, Y., Jiang, X., Ito, Y., Bringas, P., Han, J., Rowitch, D. H., et al. (2000). *Development*, *127*, 1671–1679.
Chen, X., Liu, F., Mar Aung, Z., Zhang, Y., & Chai, G. (2021). *Frontiers in Genetics*, *12*, 580761.
Chen, Q., Zhao, Y., Shen, G., & Dai, J. (2018). *Journal of Dental Research*, *97*, 1297–1305.
Chern, T., Achilleos, A., Tong, X., Hill, M. C., Saltzman, A. B., Reineke, L. C., et al. (2022). *Nature Communications*, *13*, 134.
Chern, T., Achilleos, A., Tong, X., Hsu, C. W., Wong, L., & Poche, R. A. (2020). *Developmental Biology*, *468*, 1–13.
Clark, K., Bender, G., Murray, B. P., Panfilio, K., Cook, S., Davis, R., et al. (2001). *Genesis*, *31*, 111–117.
Collier, A. C., Tingle, M. D., Paxton, J. W., Mitchell, M. D., & Keelan, J. A. (2002). *Human Reproduction*, *17*, 2564–2572.
Conley, Z. R., Hague, M., Kurosaka, H., Dixon, J., Dixon, M. J., & Trainor, P. A. (2016). *Developmental Biology*, *415*, 296–305.
Cragan, J. D., Isenburg, J. L., Parker, S. E., Alverson, C. J., Meyer, R. E., Stallings, E. B., et al. (2016). *Birth Defects Research Part A: Clinical and Molecular Teratology*, *106*, 972–982.
Crider, K. S., Yang, T. P., Berry, R. J., & Bailey, L. B. (2012). *Advances in Nutrition*, *3*, 21–38.
Cuervo, R., Valencia, C., Chandraratna, R. A., & Covarrubias, L. (2002). *Developmental Biology*, *245*, 145–156.
Danescu, A., Mattson, M., Dool, C., Diewert, V. M., & Richman, J. M. (2015). *Journal of Anatomy*, *227*, 474–486.
Danescu, A., Rens, E. G., Rehki, J., Woo, J., Akazawa, T., Fu, K., et al. (2021). *Development*, *148*, dev193755.
Dash, S., Bhatt, S., Falcon, K. T., Sandell, L. L., & Trainor, P. A. (2021). *Journal of Dental Research*, *100*, 406–414.

Dash, S., & Trainor, P. A. (2020). *Bone, 137*, 115409.
Dickinson, A. J. G., Turner, S. D., Wahl, S., Kennedy, A. E., Wyatt, B. H., & Howton, D. A. (2022). *Developmental Biology, 481*, 14–29.
Dixon, J., Jones, N. C., Sandell, L. L., Jayasinghe, S. M., Crane, J., Rey, J.-P., et al. (2006). *Proceedings of the National Academy of Sciences of the United States of America, 103*, 13403–13408.
Dixon, M. J., Marazita, M. L., Beaty, T. H., & Murray, J. C. (2011). *Nature Reviews. Genetics, 12*, 167–178.
Dottori, M., Gross, M. K., Labosky, P., & Goulding, M. (2001). *Development, 128*, 4127–4138.
Downing, C., Flink, S., Florez-McClure, M. L., Johnson, T. E., Tabakoff, B., & Kechris, K. J. (2012). *Alcoholism, Clinical and Experimental Research, 36*, 1519–1529.
Duerbeck, N. B., Dowling, D. D., & Duerbeck, J. M. (2014). *Obstetrical & Gynecological Survey, 69*, 777–788.
Ericson, J., Rashbass, P., Schedl, A., Brenner-Morton, S., Kawakami, A., van Heyningen, V., et al. (1997). *Cell, 90*, 169–180.
Everson, J. L., Fink, D. M., Yoon, J. W., Leslie, E. J., Kietzman, H. W., Ansen-Wilson, L. J., et al. (2017). *Development, 144*, 2082–2091.
Everson, T. M., Vives-Usano, M., Seyve, E., Cardenas, A., Lacasaña, M., Craig, J. M., et al. (2021). *Nature Communications, 12*, 5095.
Falcon, K. T., Watt, K. E. N., Dash, S., Zhao, R., Sakai, D., Moore, E. L., et al. (2022). *Proceedings of the National Academy of Sciences of the United States of America, 119*, e2116974119.
Fan, W. S., Mulliken, J. B., & Padwa, B. L. (2005). *Journal of Oral and Maxillofacial Surgery, 63*, 330–334.
Farlie, P. G., Kerr, R., Thomas, P., Symes, T., Minichiello, J., Hearn, C. J., et al. (1999). *Developmental Biology, 213*, 70–84.
Fitriasari, S., & Trainor, P. A. (2021). *Frontiers in Cell and Development Biology, 9*, 644410.
Forbat, E., Ali, F. R., & Al-Niaimi, F. (2018). *The Journal of Dermatological Treatment, 29*, 698–705.
Frisdal, A., & Trainor, P. A. (2014). *Wiley Interdisciplinary Reviews: Developmental Biology, 3*, 403–418.
Gaare, J. D., & Langman, J. (1977). *American Journal of Anatomy, 150*, 477–499.
Gilbert, S. F., Bender, G., Betters, E., Yin, M., & Cebra-Thomas, J. A. (2007). *Integrative and Comparative Biology, 47*, 401–408.
Gilmore, E. C., & Walsh, C. A. (2013). *WIREs Developmental Biology, 2*, 461–478.
Gorlin, R. J., Cohen, M. M., & Hennekam, R. C. M. (2001). *Syndromes of the head and neck*. Oxford University Press.
Götz, M., & Huttner, W. B. (2005). *Nature Reviews. Molecular Cell Biology, 6*, 777–788.
Graham, A., & McGonnell, I. (1999). *Current Biology, 9*, R630–R632.
Greenberg, J. A., Bell, S. J., Guan, Y., & Yu, Y. H. (2011). *Reviews in Obstetrics & Gynecology, 4*, 52–59.
Gritli-Linde, A. (2008). *Current Topics in Developmental Biology, 84*, 37–138.
Gritli-Linde, A. (2012). *Frontiers in Oral Biology, 16*, 32–51.
Guerin, M. R. (1980). Chemical composition of cigarette smoke. In G. B. Gori, & F. G. Bock (Eds.), *Banbury Report No. 3: A Safe Cigarette?* (pp. 191–204). Cold Spring Harbor, New York: Cold Spring Harbor Laboratory.
Güleray, N., Koşukcu, C., Oğuz, S., Ürel Demir, G., Taşkıran, E. Z., Kiper, P.Ö.Ş., et al. (2021). *The Cleft Palate-Craniofacial Journal, 10556656211038115*.
Haaland, Ø. A., Jugessur, A., Gjerdevik, M., Romanowska, J., Shi, M., Beaty, T. H., et al. (2017). *PLoS One, 12*, e0184358.

Haaland, Ø. A., Romanowska, J., Gjerdevik, M., Lie, R. T., Gjessing, H. K., & Jugessur, A. (2019). *F1000Research, 8*, 960.
Haft, R. J., Keating, D. H., Schwaegler, T., Schwalbach, M. S., Vinokur, J., Tremaine, M., et al. (2014). *Proceedings of the National Academy of Sciences of the United States of America, 111*, E2576–E2585.
Hammond, N. L., & Dixon, M. J. (2022). *Oral Diseases, 28*, 1306–1326.
Hansen, M., Armstrong, P. K., Bower, C., & Baynam, G. S. (2017). *Medical Journal of Australia, 206*, 351–356.
Hartsfield, J. K. (2007). *Orthodontics & Craniofacial Research, 10*, 121–128.
Hartwell, L. H., & Weinert, T. A. (1989). *Science, 246*, 629–634.
He, H., Wang, C., Tang, Q., Yang, F., & Xu, Y. (2018). *Toxicological Letters, 289*, 22–27.
Hemingway, S. J. A., Bledsoe, J. M., Brooks, A., Davies, J. K., Jirikowic, T., Olson, E. M., et al. (2018). *Advances in Pediatric Research, 5*.
Henkler, F., Stolpmann, K., & Luch, A. (2012). *Experimentia Supplementum, 101*, 107–131.
Hessel-Pras, S., Ehlers, A., Braeuning, A., & Lampen, A. (2018). *EXCLI Journal, 17*, 246–256.
Heude, É., Bouhali, K., Kurihara, Y., Kurihara, H., Couly, G., Janvier, P., et al. (2010). *Proceedings of the National Academy of Sciences of the United States of America, 107*, 11441–11446.
Heude, É., Rivals, I., Couly, G., & Levi, G. (2011). *American Journal of Medical Genetics Part A, 155*, 1991–1995.
Hinrichsen, K. (1985). The early development of morphology and patterns of the face in the human embryo. In *Advances in anatomy, embryology and cell biology*. Berlin, Heidelberg: Springer.
Howe, L. J., Lee, M. K., Sharp, G. C., Davey Smith, G., St Pourcain, B., Shaffer, J. R., et al. (2018). *PLoS Genetics, 14*, e1007501.
Howe, A. M., & Webster, W. S. (1994). *Australian Dental Journal, 39*, 88–92.
Howe, C. G., Zhou, M., Wang, X., Pittman, G. S., Thompson, I. J., Campbell, M. R., et al. (2019). *Environmental Health Perspectives, 127*, 47009.
Hoyme, H. E., Kalberg, W. O., Elliott, A. J., Blankenship, J., Buckley, D., Marais, A. S., et al. (2016). *Pediatrics*, 138.
Huang, L., Jia, Z., Shi, Y., Du, Q., Shi, J., Wang, Z., et al. (2019). *PLoS Genetics, 15*, e1008357.
Huang, W., Wang, X., Zheng, S., Wu, R., Liu, C., & Wu, K. (2021). *Ecotoxicology and Environmental Safety, 212*, 111991.
Huang, H., Yang, X., Bao, M., Cao, H., Miao, X., Zhang, X., et al. (2016). *Journal of Biological Chemistry, 291*, 7107–7118.
Inman, K. E., Purcell, P., Kume, T., & Trainor, P. A. (2013). *PLoS Genetics, 9*, e1003949.
Jean, F., Stuart, A., & Tarailo-Graovac, M. (2020). *Frontiers in Neurology*, 11.
Ji, Y., Garland, M. A., Sun, B., Zhang, S., Reynolds, K., McMahon, M., et al. (2020). *Birth Defects Research, 112*, 1558–1587.
Jiang, R., Bush, J. O., & Lidral, A. C. (2006). *Developmental Dynamics: An Official Publication of the American Association of the Anatomists, 235*, 1152–1166.
Jones, N. C., Lynn, M. L., Gaudenz, K., Sakai, D., Aoto, K., Rey, J. P., et al. (2008). *Nature Medicine, 14*, 125–133.
Joubert, B. R., Håberg, S. E., Nilsen, R. M., Wang, X., Vollset, S. E., Murphy, S. K., et al. (2012). *Environmental Health Perspectives, 120*, 1425–1431.
Kaye, C. I., Martin, A. O., Rollnick, B. R., Rollnick, R., Nagatoshi, K., Israel, J., et al. (1992). *American Journal of Medical Genetics, 43*, 913–917.
Kelberman, D., Tyson, J., Chandler, D., McInerney, A., Slee, J., Albert, D., et al. (2001). *Human Genetics, 109*, 638–645.

Keller, R., Shih, J., Sater, A. K., & Moreno, C. (1992). *Developmental Dynamics*, *193*, 218–234.

Kennedy, A. E., Kandalam, S., Olivares-Navarrete, R., & Dickinson, A. J. G. (2017). *PLoS One*, *12*, e0185729.

Khoury, M. J., Adams, M. J., Jr., & Flanders, W. D. (1988). *American Journal of Human Genetics*, *42*, 89–95.

Kicheva, A., Bollenbach, T., Ribeiro, A., Valle Helena, P., Lovell-Badge, R., Episkopou, V., et al. (2014). *Science*, *345*, 1254927.

Kim, S., Lewis, A. E., Singh, V., Ma, X., Adelstein, R., & Bush, J. O. (2015). *PLoS Biology*, *13*, e1002122.

Kirby, M. L., Gale, T. F., & Stewart, D. E. (1983). *Science*, *220*, 1059–1061.

Kong, Y., Grimaldi, M., Curtin, E., Dougherty, M., Kaufman, C., White, R. M., et al. (2014). *Chemistry & Biology*, *21*, 488–501.

Kramer, A. G., Vuthiganon, J., & Lassiter, C. S. (2016). *Environmental Toxicology and Pharmacology*, *43*, 159–165.

Kulesa, P. M., & Fraser, S. E. (1998). *Developmental Biology*, *204*, 327–344.

Kurosaka, H., Mushiake, J., Saha, M., Wu, Y., Wang, Q., Kikuchi, M., et al. (2021). *Human Molecular Genetics*, *30*, 2383–2392.

Kurosaka, H., Wang, Q., Sandell, L., Yamashiro, T., & Trainor, P. A. (2017). *Human Molecular Genetics*, *26*, 1268–1279.

Kwon, H. J. (2019). *Journal of Developmental Biology*, 7.

Lammer, E. J., Chen, D. T., Hoar, R. M., Agnish, N. D., Benke, P. J., Braun, J. T., et al. (1985). *The New England Journal of Medicine*, *313*, 837–841.

Lange, S., Probst, C., Rehm, J., & Popova, S. (2018). *The Lancet Global Health*, *6*, e769–e776.

Le Douarin, N. M., Creuzet, S., Couly, G., & Dupin, E. (2004). *Development*, *131*, 4637–4650.

Le Douarin, N. M., & Teillet, M.-A. M. (1974). *Developmental Biology*, *41*, 162–184.

LeDouarin, N. M., & Kalcheim, C. (1999). *The neural crest*. Cambridge University Press.

Lerner-Ellis, J. P., Gradinger, A. B., Watkins, D., Tirone, J. C., Villeneuve, A., Dobson, C. M., et al. (2006). *Molecular Genetics and Metabolism*, *87*, 219–225.

Lerner-Ellis, J. P., Tirone, J. C., Pawelek, P. D., Dore, C., Atkinson, J. L., Watkins, D., et al. (2006). *Nature Genetics*, *38*, 93–100.

Li, R., Yang, Y.-G., Gao, Y., Wang, Z.-Q., & Tong, W.-M. (2012). *Cell Research*, *22*, 859–872.

Li, L., Zhu, G. Q., Meng, T., Shi, J. Y., Wu, J., Xu, X., et al. (2011). *American Journal of Medical Genetics. Part A*, *155A*, 2940–2948.

Lipinski, R. J., Hammond, P., O'Leary-Moore, S. K., Ament, J. J., Pecevich, S. J., Jiang, Y., et al. (2012). *PLoS One*, 7, e43067.

Liu, Z., Sun, H., Dai, J., Xue, X., Sun, J., & Wang, X. (2021). *Frontiers in Genetics*, *12*, 616329.

Lopez, E., Berenguer, M., Tingaud-Sequeira, A., Marlin, S., Toutain, A., Denoyelle, F., et al. (2016). *Journal of Medical Genetics*, *53*, 752–760.

Louryan, S., Heymans, O., & Goffard, J. C. (1995). *Surgical and Radiologic Anatomy*, *17*, 59–63.

Lovely, C. B., Fernandes, Y., & Eberhart, J. K. (2016). *Zebrafish*, *13*, 391–398.

Lovely, C., Rampersad, M., Fernandes, Y., & Eberhart, J. (2017). *Wiley Interdisciplinary Reviews. Developmental Biology*, *6*.

Mai, C. T., Isenburg, J. L., Canfield, M. A., Meyer, R. E., Correa, A., Alverson, C. J., et al. (2019). *Birth Defects Research*, *111*, 1420–1435.

Marjonen, H., Kahila, H., & Kaminen-Ahola, N. (2017). *Human Reproduction Open*, hox014.

Martinelli, M., Scapoli, L., Palmieri, A., Pezzetti, F., Baciliero, U., Padula, E., et al. (2006). *Human Mutation*, *27*, 294.

Martinelli, M., Scapoli, L., Pezzetti, F., Carinci, F., Carinci, P., Stabellini, G., et al. (2001). *American Journal of Medical Genetics, 98*, 357–360.

Maruko, E., Hayes, C., Evans, C. A., Padwa, B., & Mulliken, J. B. (2001). *The Cleft Palate-Craniofacial Journal, 38*, 15–19.

McCarthy, N., Wetherill, L., Lovely, C. B., Swartz, M. E., Foroud, T. M., & Eberhart, J. K. (2013). *Development, 140*, 3254–3265.

McKinney, M. C., McLennan, R., Giniunaite, R., Baker, R. E., Maini, P. K., Othmer, H. G., et al. (2020). *Developmental Biology, 461*, 184–196.

Millacura, N., Pardo, R., Cifuentes, L., & Suazo, J. (2017). *Public Health Nutrition, 20*, 2260–2268.

Mione, M. C., Cavanagh, J. F. R., Harris, B., & Parnavelas, J. G. (1997). *The Journal of Neuroscience, 17*, 2018.

Monahan, R., Seder, K., Patel, P., Alder, M., Grud, S., & O'Gara, M. (2001). *The Journal of the American Dental Association, 132*, 1402–1408.

Morris, J. K., Rankin, J., Garne, E., Loane, M., Greenlees, R., Addor, M.-C., et al. (2016). *British Medical Journal, 354*, i4721.

Morrison, J. A., McLennan, R., Wolfe, L. A., Gogol, M. M., Meier, S., McKinney, M. C., et al. (2017). *eLife, 6*, e28415.

Mossey, P. A., & Castilla, E. E. (2003). Global registry and database on craniofacial anomalies: Report of a WHO registry meeting on craniofacial anomalies. In *WHO Human Genetics Programme & WHO Meeting on International Collaborative Research on Craniofacial Anomalies*. Bauru, Brazil: World Health Organization.

Müller, F., & O'Rahilly, R. (2004). *Cells, Tissues, Organs, 177*, 2–20.

Nassif, A., Senussi, I., Meary, F., Loiodice, S., Hotton, D., Robert, B., et al. (2014). *Bone, 66*, 96–104.

Noden, D. M., & Trainor, P. A. (2005). *Journal of Anatomy, 207*, 575–601.

Nott, A., Watson, P. M., Robinson, J. D., Crepaldi, L., & Riccio, A. (2008). *Nature, 455*, 411–415.

Ohbayashi, N., & Eto, K. (1986). *Journal of Craniofacial Genetics and Developmental Biology, 2*, 41–44.

Otani, H., Tanaka, O., Naora, H., Yokoyama, M., Nomura, T., Kimura, M., et al. (1991). *Anatomischer Anzeiger, 172*, 1–9.

Passemard, S., Kaindl, A. M., & Verloes, A. (2013). Chapter 13—Microcephaly. In O. Dulac, M. Lassonde, & H. B. Sarnat (Eds.), *Handbook of Clinical Neurology* (pp. 129–141). Elsevier.

Pezzetti, F., Martinelli, M., Scapoli, L., Carinci, F., Palmieri, A., Marchesini, J., et al. (2004). *Human Mutation, 24*, 104–105.

Pierani, A., Brenner-Morton, S., Chiang, C., & Jessell, T. M. (1999). *Cell, 97*, 903–915.

Popova, S., Lange, S., Probst, C., Gmel, G., & Rehm, J. (2017). *The Lancet Global Health, 5*, e290–e299.

Poswillo, D. (1973). Oral Surgery, Oral Medicine, Oral. *Pathology, 35*, 302–328.

Rengasamy Venugopalan, S., Farrow, E., Sanchez-Lara, P. A., Yen, S., Lypka, M., Jiang, S., et al. (2019). *Orthodontics & Craniofacial Research, 22*, 163–167.

Reynolds, K., Zhang, S., Sun, B., Garland, M. A., Ji, Y., & Zhou, C. J. (2020). *Birth Defects Research, 112*, 1588–1634.

Rice, R., Cebra-Thomas, J., Haugas, M., Partanen, J., Rice, D. P. C., & Gilbert, S. F. (2017). *Scientific Reports, 7*, 12063.

Rice, R., Kallonen, A., Cebra-Thomas, J., & Gilbert, S. F. (2016). *Proceedings of the National Academy of Sciences of the United States of America, 113*, 5317–5322.

Rice, R., Riccio, P., Gilbert, S. F., & Cebra-Thomas, J. (2015). *Journal of Experimental Zoology. Part B, Molecular and Developmental Evolution, 324*, 208–220.

Rollnick, B. R., Kaye, C. I., Nagatoshi, K., Hauck, W., Martin, A. O., & Reynolds, J. F. (1987). *American Journal of Medical Genetics, 26*, 361–375.

Romitti, P. A., Lidral, A. C., Munger, R. G., Daack-Hirsch, S., Burns, T. L., & Murray, J. C. (1999). *Teratology*, *59*, 39–50.

Rubenstein, J. L. R., Shimamura, K., Martinez, S., & Puelles, L. (1998). *Annual Review of Neuroscience*, *21*, 445–477.

Rump, P., Jazayeri, O., van Dijk-Bos, K. K., Johansson, L. F., van Essen, A. J., Verheij, J. B. G. M., et al. (2016). *BMC Medical Genomics*, *9*, 7.

Sakai, D., Dixon, J., Achilleos, A., Dixon, M., & Trainor, P. A. (2016). Prevention of Treacher Collins syndrome craniofacial anomalies in mouse models via maternal antioxidant supplementation. *Nature Communications*, 7:10328.

Sakai, D., & Trainor, P. A. (2016). Face off against ROS: Tcof1/Treacle safeguards neuroepithelial cells and progenitor neural crest cells from oxidative stress during craniofacial development. *Development Growth & Differentiation*, *58*(7), 577–585.

Seidegård, J., & Ekström, G. (1997). *Environmental Health Perspectives*, *105*(Suppl. 4), 791–799.

Selleck, M. A., & Bronner-Fraser, M. (1995). *Development*, *121*, 525–538.

Shannon, S. R., Moise, A. R., & Trainor, P. A. (2017). *Wiley Interdisciplinary Reviews: Developmental Biology*, *6*.

Shannon, S. R., Yu, J., Defnet, A. E., Bongfeldt, D., Moise, A. R., Kane, M. A., et al. (2020). *Methods in Enzymology*, *637*, 367–418.

Sharp, G. C., Ho, K., Davies, A., Stergiakouli, E., Humphries, K., McArdle, W., et al. (2017). *Clinical Epigenetics*, *9*, 63.

Shi, M., Wehby, G. L., & Murray, J. C. (2008). *Birth Defects Research. Part C, Embryo Today: Reviews*, *84*, 16–29.

Shkoukani, M., Chen, M., & Vong, A. (2013). *Frontiers in Pediatrics*, *1*.

Simoes-Costa, M., & Bronner, M. E. (2016). *Science*, *352*, 1570–1573.

Smart, I. H. (1973). *Journal of Anatomy*, *116*, 67–91.

Sperber, G. H. (2002). Palatogenesis: Closure of the secondary palate. In D. F. Wyszynski (Ed.), *Cleft lip and palate: From origin to treatment* (pp. 14–24). New York: Oxford University Press.

Sperber, G. H., & Sperber, S. M. (2013). Embryogenetics of cleft lip and palate. In S. Berkowitz (Ed.), *Cleft lip and palate: Diagnosis and management* (pp. 3–33). Berlin Heidelberg, Berlin, Heidelberg: Springer.

Stiles, J., & Jernigan, T. L. (2010). *Neuropsychology Review*, *20*, 327–348.

Sun, D., Baur, S., & Hay, E. D. (2000). *Developmental Biology*, *228*, 337–349.

Suzuki, S., Marazita, M. L., Cooper, M. E., Miwa, N., Hing, A., Jugessur, A., et al. (2009). *American Journal of Human Genetics*, *84*, 406–411.

Swartz, M. E., Lovely, C. B., McCarthy, N., Kuka, T., & Eberhart, J. K. (2020). *Alcoholism, Clinical and Experimental Research*, *44*, 56–65.

Swartz, M. E., Wells, M. B., Griffin, M., McCarthy, N., Lovely, C. B., McGurk, P., et al. (2014). *Alcoholism, Clinical and Experimental Research*, *38*, 694–703.

Teng, T., Teng, C. S., Kaartinen, V., & Bush, J. O. (2022). *Development*, 149.

Timberlake, A. T., Griffin, C., Heike, C. L., Hing, A. V., Cunningham, M. L., Chitayat, D., et al. (2021). *Nature Communications*, *12*, 4680.

Trainor, P. A. (2005). *Brain, Behavior and Evolution*, *66*, 266–280.

Trainor, P. A. (2010). *American Journal of Medical Genetics. Part A*, *152A*, 2984–2994.

Trainor, P. A. (2016). *Current Topics in Developmental Biology*, *117*, 523–538.

Trainor, P. A., Dixon, J., & Dixon, M. J. (2009). *European Journal of Human Genetics*, *17*, 275–283.

Trainor, P., & Krumlauf, R. (2000). *Nature Cell Biology*, *2*, 96–102.

Trainor, P. A., & Krumlauf, R. (2001). *Current Opinion in Cell Biology*, *13*, 698–705.

Trainor, P. A., & Richtsmeier, J. T. (2015). *American Journal of Medical Genetics. Part A*, *167*, 1451–1454.

Trainor, P. A., Sobieszczuk, D., Wilkinson, D., & Krumlauf, R. (2002). *Development, 129*, 433–442.

Trainor, P. A., & Tam, P. P. (1995). *Development, 121*, 2569–2582.

Vastardis, H., Karimbux, N., Guthua, S. W., Seidman, J. G., & Seidman, C. E. (1996). *Nature Genetics, 13*, 417–421.

Vaziri Sani, F., Hallberg, K., Harfe, B. D., McMahon, A. P., Linde, A., & Gritli-Linde, A. (2005). *Developmental Biology, 285*, 490–495.

Veitch, E., Begbie, J., Schilling, T. F., Smith, M. M., & Graham, A. (1999). *Current Biology, 9*, 1481–1484.

Vento, A. R., Labrie, R. A., & Mulliken, J. B. (1991). *The Cleft Palate-Craniofacial Journal, 28*, 68–77.

Vieira, A. R., Cooper, M. E., Marazita, M. L., Castilla, E. E., & Orioli, I. M. (2008). *Brazilian Journal of Medical and Biological Research, 41*, 689–693.

Vlachos, A., Ball, S., Dahl, N., Alter, B. P., Sheth, S., Ramenghi, U., et al. (2008). *British Journal of Haematology, 142*, 859–876.

Wahl, S. E., Kennedy, A. E., Wyatt, B. H., Moore, A. D., Pridgen, D. E., Cherry, A. M., et al. (2015). *Developmental Biology, 405*, 108–122.

Wang, Q., Kurosaka, H., Kikuchi, M., Nakaya, A., Trainor, P. A., & Yamashiro, T. (2019). *Disease Models & Mechanisms, 12*.

Wang, Q., Xu, L., Miura, J., Saha, M. K., Uemura, Y., Sandell, L. L., et al. (2021). *Frontiers in Cell and Development Biology, 9*, 596838.

Warbrick, J. G. (1960). *Journal of Anatomy, 94*, 351–362.

Watt, K. E. N., Achilleos, A., Neben, C. L., Merrill, A. E., & Trainor, P. A. (2016). *PLoS Genetics, 12*, e1006187.

Watt, K. E. N., Macintosh, J., Bernard, G., & Trainor, P. A. (2022). *Seminars in Cell & Developmental Biology*.

Watt, K. E. N., Neben, C. L., Hall, S., Merrill, A. E., & Trainor, P. A. (2018). *Human Molecular Genetics, 27*, 2628–2643.

Watt, K. E. N., & Trainor, P. A. (2014). Chapter 17—Neurocristopathies: The etiology and pathogenesis of disorders arising from defects in neural crest cell development. In P. A. Trainor (Ed.), *Neural Crest Cells* (pp. 361–394). Boston: Academic Press.

Weaver, K. N., Watt, K. E. N., Hufnagel, R. B., Navajas Acedo, J., Linscott, L. L., Sund, K. L., et al. (2015). *American Journal of Human Genetics, 96*, 765–774.

Weinberg, S. M., Naidoo, S. D., Bardi, K. M., Brandon, C. A., Neiswanger, K., Resick, J. M., et al. (2009). *Orthodontics & Craniofacial Research, 12*, 271–281.

Williams, A. L., & Bohnsack, B. L. (2019). *Genesis, 57*, e23308.

Wu, T., Fallin, M. D., Shi, M., Ruczinski, I., Liang, K. Y., Hetmanski, J. B., et al. (2012). *Birth Defects Research Part A, Clinical and Molecular Teratology, 94*, 76–83.

Wu, Y., Kurosaka, H., Wang, Q., Inubushi, T., Nakatsugawa, K., Kikuchi, M., et al. (2022). *Journal of Dental Research, 101*, 686–694.

Wu, T., Liang, K. Y., Hetmanski, J. B., Ruczinski, I., Fallin, M. D., Ingersoll, R. G., et al. (2010). *Human Genetics, 128*, 401–410.

Xavier, D., Arif, Y., Murali, R., Kishore Kumar, S., Vipin Kumar, S., Tamang, R., et al. (2013). *Balkan Journal of Medical Genetics: BJMG, 16*, 49–54.

Yamaguchi, H., Kitami, K., Wu, X., He, L., Wang, J., Wang, B., et al. (2021). *Frontiers in Physiology, 12*.

Yang, L., Wang, H., Yang, L., Zhao, M., Guo, Y., Bovet, P., et al. (2022). *BMC Medicine, 20*, 4.

Yazdy, M. M., Honein, M. A., & Xing, J. (2007). *Birth Defects Research. Part A, Clinical and Molecular Teratology, 79*, 16–23.

Yong-Ping, L., Reichetzeder, C., Prehn, C., Yin, L. H., Chu, C., Elitok, S., et al. (2020). *The Journal of Steroid Biochemistry and Molecular Biology, 199*, 105591.

Yoon, H., Chung, I. S., Seol, E. Y., Park, B. Y., & Park, H. W. (2000). *Yonsei Medical Journal*, *41*, 477–484.
Yoshida, T., Vivatbutsiri, P., Morriss-Kay, G., Saga, Y., & Iseki, S. (2008). *Mechanisms of Development*, *125*, 797–808.
Young, N. M., Hu, D., Lainoff, A. J., Smith, F. J., Diaz, R., Tucker, A. S., et al. (2014). *Development*, *141*, 1059–1063.
Zbasnik, N., Dolan, K., Buczkowski, S. A., Green, R. M., Hallgrímsson, B., Marcucio, R. S., et al. (2022). *Developmental Dynamics*.
Zhang, S. J., Meng, P., Zhang, J., Jia, P., Lin, J., Wang, X., et al. (2018). *Genomics, Proteomics & Bioinformatics*, *16*, 354–364.
Zhang, Y., Pan, T., Zhong, X., & Cheng, C. (2014). *Tumour Biology*, *35*, 7063–7072.
Zhang, W., Venkataraghavan, S., Hetmanski, J. B., Leslie, E. J., Marazita, M. L., Feingold, E., et al. (2021). *Frontiers in Cell and Development Biology*, *9*, 621018.
Zhang, P., Wang, G., Lin, Z., Wu, Y., Zhang, J., Liu, M., et al. (2017). *Toxicology Letters*, *281*, 53–64.
Zhao, Z., & Reece, E. A. (2005). *Birth Defects Research Part B: Developmental and Reproductive Toxicology*, *74*, 383–391.
Zhao, R., & Trainor, P. A. (2022). *Seminars in Cell & Developmental Biology*. S1084-9521(22)00056-8.

CHAPTER SIX

Gene × environment associations in orofacial clefting

Mary L. Marazita*
Center for Craniofacial and Dental Genetics, University of Pittsburgh, Pittsburgh, PA, United States
Oral and Craniofacial Sciences, School of Dental Medicine, University of Pittsburgh, Pittsburgh, PA, United States
Human Genetics, School of Public Health, University of Pittsburgh, Pittsburgh, PA, United States
Clinical and Translational Science Institute, Department of Psychiatry, School of Medicine, University of Pittsburgh, Pittsburgh, PA, United States
*Corresponding author: e-mail address: marazita@pitt.edu

Contents

Abstract

This chapter reviews the evidence of gene × environment interactions (G × E) in the etiology of orofacial cleft birth defects (OFCs), specifically cleft lip (CL), cleft palate (CP), and cleft lip with or without cleft palate (CL/P). We summarize the current state of our understanding of the genetic architecture of nonsyndromic OFCs and the evidence that maternal exposures during pregnancy influence risk of OFCs. Further, we present possible candidate gene pathways for these exposures including metabolism of folates, metabolism of retinoids, retinoic acid receptor signaling, aryl hydrocarbon receptor signaling, glucocorticoid receptor signaling, and biotransformation and transport. We review genes in these pathways with prior evidence of association with OFCs, genes with evidence from prior candidate gene G × E studies, and genes identified from genome-wide searches specifically for identifying G × E. Finally, we suggest future directions for G × E research in OFCs.

Current Topics in Developmental Biology, Volume 152
ISSN 0070-2153
https://doi.org/10.1016/bs.ctdb.2022.10.006

1. Introduction

Orofacial clefts (OFCs), primarily cleft lip (CL), cleft palate (CP) and cleft lip with or without cleft palate (CL/P) are among the most common birth defects in all populations worldwide (Mossey, 2007; Mossey, Little, Munger, Dixon, & Shaw, 2009), with population and ethnic differences in birth prevalence; with worldwide incidence of approximately 1 per 700 newborns. Developmental origins of the affected structures, epidemiology, and familial patterns of CL/P and CP suggest that these two groups have separate etiologies (Fogh-Andersen, 1971; Fraser, 1955). OFCs represent a major public health problem due to the associated morbidity, mortality, and significant health care expenditures. On average, a child with an OFC initially faces feeding difficulties, undergoes 6 surgeries, spends 30 days in hospital, receives 5 years of orthodontic treatment, and participates in ongoing speech therapy, leading to an estimated total lifetime treatment cost of about $200,000 (Berk & Marazita, 2002; Semb, Brattstrom, Molsted, Prahl-Andersen, & Shaw, 2005; Wehby & Cassell, 2010). Furthermore, individuals born with an OFC have a higher infant mortality rate (particularly in developing countries where access to health care may be limited), higher mortality rates at all other stages of life, increased incidence of mental health problems, and higher risk for other disorders (notably breast, brain, and colon cancers) (Bille, Knudsen, & Christensen, 2005; Bille, Winther, et al., 2005; Christensen, Juel, Herskind, & Murray, 2004; Dietz et al., 2012; Menezes et al., 2009; Wehby & Cassell, 2010; Wehby et al., 2006; Zhu et al., 2002). The impact of OFCs is further complicated by ethnic differences in birth incidence—that is, about 1 in 500 live births for Asians and Native Americans, 1 in 1000 for Europeans, and 1 in 2500 for Africans and derived populations (Christensen, 1999; Cooper, Ratay, & Marazita, 2006; Mossey, 2007; Mossey et al., 2009)—as well as differences across populations in health care (Kantar et al., 2019).

Because OFC is a common birth defect with a chance of survival even in pre-surgical eras, there are references to OFC in myths, legends, art, and literature worldwide. OFC has been used for comic effect; e.g., Camille, a major character in the 1907 play *A Flea in Her Ear*, has cleft palate and repeatedly loses his silver obturator resulting in many misunderstandings (Feydeau, 1907). OFC has also been used as shorthand for either a saintly character (e.g., Hassan in 2003s *The Kite Runner* (Hosseini, 2003)), or for undesirable qualities as in the works of Shakespeare. In *Midsummer Night's*

Dream (Act V, Scene i (Shakespeare, 1595)), Oberon predicts all good things for the three central couples ("hare-lip" is an outmoded designation for CL):

> "And the blots of Nature's hand
> Shall not in their issue stand;
> Never mole, hare-lip, nor scar,"

The earliest explanations for the cause of OFCs focused on maternal exposures during pregnancy, resulting in a number of folklore explanations of how OFC arises, and how to prevent occurrence during pregnancy (Daack-Hirsch, 2010). Notable examples come from two populations with high birth prevalence of OFC: native South Americans and Asians.

Aztecs believed that eclipses occurred because a bite had been taken out of the moon, so that exposure to an eclipse during pregnancy could lead to OFC (a bite out of the infant's mouth). This Aztec explanation persisted throughout Central America, with the addition of exposure to a full moon. To prevent OFCs, Aztecs placed an obsidian knife on a pregnant woman's abdomen before going out at night; in Mexico a metal key or safety pin is used for protection.

In Asia there was a common belief that eating rabbit during pregnancy could lead to a "hare lip" (Cheng, 1990). Daack-Hirsch and colleagues surveyed modern-day Filipinos (Daack-Hirsch, 2010): the most frequent cause mentioned for OFC was "force to the fetal face," due to force applied to the pregnant abdomen due to a fall or other injury when the fetus had fingers or thumb in the oral cavity, thus causing the cleft. Other traditional explanations in Asia arose from Buddhist teachings that fate or bad karma lead to deformities, or as retributions for wrongdoings (Cheng, 1990). Similar attitudes prevailed in Europe, for example Shakespeare attributed "hare-lip" to the actions of a troublesome sprite (*King Lear (*Act III, Scene iv (Shakespeare, 1605)):

> "He gives the web and the pin,
> squints the eye, and makes the hare-lip;"

Notably, these and many other cultures also felt that OFC is familial, e.g., "in the blood" (Cheng, 1990; Daack-Hirsch, 2010), which we would still agree with today. Formal scientific interest in family patterns of OFCs began with published descriptive and observational family studies, with the first publication of a family with several affected members in 1757 (Trew, 1757). 100 years later Darwin (Darwin, 1875) cited a paper by Sproule (Sproule, 1863), mentioning "the transmission during a century of "hare-lip" with a cleft-palate" in his discussion of variation in plants and animals. Additional pre-1900 publications and OFC pedigrees were summarized by Rischbieth (Melnick, 1997; Rischbieth, 1910).

Once Mendel's research (Mendel, 1866) was re-discovered, translated, and championed by Bateson (Bateson, 1902, 1909; Melnick, 1997), there were debates between Bateson and Pearson (and the other members of the Galton Laboratory (Marazita, 2012; Melnick, 1997)). Of note, OFC was used as an example on both sides of the debate. The anti-Mendelian view of the Galton lab's conclusion that OFC was an expression of general "poor protoplasm" (Melnick, 1997; Rischbieth, 1910), would now be considered essentially a polygenic or multifactorial model for OFC. On the other side of the debate, Bateson included "hare-lip" as one of a group of "dominant hereditary diseases and malformations" (Bateson, 1909; Melnick, 1997).

The early years of the 20th century settled part of the debate in favor of Mendel's principal of unit inheritance, with the units now known as genes. However, both partners in this debate were partially correct in that today we consider OFC to be heterogeneous, with single genes of major effect (as per Darwin and Bateson's conclusions), potentially modified by polygenic background and environmental/behavioral factors (as per the Galton lab viewpoint).

In this chapter, we summarize the current understanding of nonsyndromic OFC genetic architecture, as well as other interacting aspects of OFC etiology, primarily maternal environmental exposures and behavioral factors during pregnancy that have evidence of association with OFCs. Finally, we summarize the evidence for GxE associations with such factors.

2. Genetic architecture of nonsyndromic OFCs

OFCs comprise a heterogeneous group of defects with both genetic and environmental factors playing important roles (Wyszynski, 2002), although understanding of these risk factors is incomplete. Syndromic forms of OFC are mostly due to simple Mendelian inheritance of coding variants in a gene, with little or no impact of environmental factors. Nonsyndromic OFCs—that is, isolated OFCs occurring in the absence of any other signs or symptoms characterizing a genetic syndrome—comprise the majority of OFCs, about 70–80% of CL/P and 50% of CP. Whereas OFCs found in genetic syndromes may be due to specific causal variants (usually coding variants), nonsyndromic OFCs are hypothesized to be etiologically complex, caused by the interplay of multiple genetic and environmental exposures, their interactions, and possibly stochastic events (Wyszynski, 2002). Heritability of nonsyndromic OFCs is high, estimated to be about 90% from twin studies (Grosen et al., 2011).

Despite substantial progress due to the tools made available by the Human Genome Project, our understanding of OFC etiology is incomplete. Genetic evidence, to date, supports the hypothesis that nonsyndromic OFCs have a complex genetic architecture. For example, whole exome sequencing screens of nonsyndromic OFCs find putatively causal coding variants in <10% of individuals (Basha et al., 2018). On the other hand, genome-wide association studies (GWAS) of millions of variants across the genome have cumulatively identified more than 50 associated genes/regions achieving a strict *p*-value threshold for genome-wide statistical significance (see reviews (Beaty, Marazita, & Leslie, 2016; Moreno Uribe & Marazita, 2021)).

These genes/regions have been identified across multiple independent GWASs for CL/P (Beaty et al., 2010; Birnbaum et al., 2009; Camargo et al., 2012; Gowans et al., 2016; Grant et al., 2009; Leslie, Carlson, et al., 2016; Mangold et al., 2010; Sun et al., 2015); and CP (Beaty et al., 2011; Butali et al., 2019; Leslie, Liu, et al., 2016)), and meta-analyses combining the separate studies (Carlson et al., 2019; Leslie, Carlson, Shaffer, Butali, et al., 2017, Leslie, Carlson, Shaffer, Buxo, et al., 2017; Ludwig et al., 2012). Cumulatively, these ~50 genes/regions explain about 20–25% of the heritability of nonsyndromic OFCs (Morris et al., 2020; Saleem, Zaib, Sun, & Fu, 2019). While these results showcase the great success that genome-wide approaches have had for investigating OFC genetics, more work is needed to fully catalog the genetic loci influencing risk. For example, the prevalence of OFC and some of the genes shown to be involved (Leslie, Carlson, et al., 2016; Ludwig et al., 2012; Marazita et al., 2009; Murray et al., 2012) differ across racial and ethnic groups (probably due to allele frequency differences); however, not all groups are well-represented in the existing literature. More work is needed to understand OFC genetics in underrepresented populations, particularly Africans.

Furthermore, the majority of OFC genetic association signals appear to be due to regulatory variants (Beaty et al., 2016; Leslie et al., 2015) (such as variants in/near *IRF6*, 8q24, *FOXE1*, and others (Beaty et al., 2016; Moreno Uribe & Marazita, 2021)), whereas only a few are due to causal coding variants (notably *GRHL3* for CP (Carlson et al., 2019; Eshete et al., 2018; Leslie, Liu, et al., 2016; Mangold et al., 2016)). The specific causal variants at these non-coding loci are quite difficult to determine from statistical associations in racially homogenous samples due to the high linkage disequilibrium (LD) at localized regions of the genome. More trans-ethnic association analyses across samples with differing LD patterns are needed to better map the causal variants at known loci.

3. Maternal exposures and behaviors during pregnancy influence OFC risk

Many studies have identified environmental exposures during pregnancy with the best supported including smoking and passive smoking, alcohol, and multivitamin/folate supplementation (see recent comprehensive reviews (Garland, Reynolds, & Zhou, 2020; Garland, Sun, et al., 2020; Ji et al., 2020; Reynolds et al., 2020)). A few more general characteristics, e.g., socio-economic status, have been suggested as OFC risk factors, but it is difficult to separate the combined effects of maternal nutrition and health from such broad measures of the environment (Mossey, 2007).

3.1 Smoking

Maternal smoking during pregnancy is the most replicated risk factor for OFCs, with strong enough evidence that the American College of OB/GYN included it in a recent report of the major risks of tobacco and nicotine exposure during pregnancy (ACOG Committee, 2020). An estimated 6.1% (95% confidence interval = 4.4–7.7%) of OFCs could be avoided by eliminating maternal smoking (Honein, Devine, Grosse, & Reefhuis, 2014; Little, Cardy, & Munger, 2004). Passive or second-hand exposure also appears to be a risk factor (Kummet et al., 2016).

Several studies have determined the risk of having a baby born with OFC due to maternal cigarette consumption. In general, the methodological approach used in these studies has been a case-control study design where mothers are interviewed regarding their smoking history and consumption level, classifying mothers according to the number of cigarettes consumed per day. Typically the smoking categories are none, 1–10, 11–20 and >21 cigarettes a day, or some slight modification of these dosage schemes. Although some prior studies showed negative results (Lieff et al., 1999; Werler, Lammer, Rosenberg, & Mitchell, 1990), most studies today agree that maternal smoking during pregnancy is a recognized risk factor for OFC. Further, maternal smoking is of great concern because of the increasing rate of smoking in early adults (Lantz, 2003) and young mothers in regions such as West-southern China where a recent study on ~4300 cases found that smoking was the second biggest risk during the first 3 months of pregnancy (Meng, Shi, Zheng, Wang, & Li, 2006).

Risk estimates from one of the largest studies published so far (Chung, Kowalski, Kim, & Buchman, 2000) in addition to three meta-analyses including 11 studies from 1966 to 1996 (Wyszynski, Duffy, & Beaty, 1997), 24 from 1974 to 2001 (Little et al., 2004) and 29 studies from 1966 to 2015 (Xuan et al., 2016) showed that smoking increases the odds of clefting by a factor in the range of 1.29–1.37 for CL/P and by 1.22–1.32 for CP. Also reported (Chung et al., 2000) were smoking intensity specific effects, with the odds for clefting estimated to increase by a factor of 1.50 (95% CI: 1.28–1.76), 1.55 (95% CI: 1.23–1.95) and 1.78 (95% CI: 1.22–2.59) with smoking 1–10, 11–20, and 21 or more cigarettes per day, respectively, compared to no smoking. The Chung et al. study (Chung et al., 2000), did not discriminate between CL/P and CP.

Finally, in addition to active smoking, passive smoking has also been implicated as an OFC risk factor via a meta-analysis and in population-based studies (Kummet et al., 2016; Sabbagh et al., 2015). Meta-analyses indicated that maternal passive smoking exposure results in a 1.5-fold increase in risk of nonsyndromic oral clefts (Sabbagh et al., 2015). Results from a population-based study showed that children of nonsmoking mothers exposed to passive smoking also had an increased risk (OR = 1.14, 95% CI: 1.02, 1.27). Moreover, OFC risk was further elevated among babies of smoking mothers who were exposed to passive smoke (OR = 1.51, 95% CI: 1.35, 1.70) (Kummet et al., 2016). Overall, these studies suggest that maternal smoking whether active or passive during the first trimester of pregnancy shows a moderate but consistent association with an increased risk of OFC.

3.2 Alcohol consumption

Evidence for maternal alcohol consumption as a risk factor for OFCs is less consistent than the evidence for smoking. Two recent studies including a meta-analysis of 9 studies from 1950 to 2019 (Yin, Li, Li, & Zou, 2019) and a large-scale Japanese national birth cohort study including 73,595 mothers (Kurita et al., 2020) failed to find significant differences between maternal drinking and not drinking with increased risk for CL/P or CP

(Kurita et al., 2020; Yin et al., 2019). However, other prior individual alcohol consumption studies (Munger et al., 1996; Romitti et al., 2007; Shaw & Lammer, 1999; Werler, Lammer, Rosenberg, & Mitchell, 1991) reported significant positive associations between the highest alcohol consumption category (≥4 drinks a month or ≥5 drinks per drinking occasion, i.e., "binge drinking") and CL/P with ORs in the ranges of 2.8–4.0. No increased risks were found for CP alone. Furthermore, one study utilizing pooled estimates from six population-based case-control studies indicated increased risk for CL only, with pooled ORs of 1.48 for mothers who drank an average of 5+ drinks/sitting. and 1.95 for mothers who drank at this level (i.e., 5+ drinks/siting) for 3 or more times per week (DeRoo et al., 2016).

3.3 Body mass index (over- and under-weight)

Interaction between smoking and body mass index (BMI) was assessed in a population-based study (Wehby et al., 2017) and showed a significant negative interaction where the risk of OFCs associated with maternal smoking is largest among underweight mothers and is strongest for CL alone and CP alone. Thus, for all clefts combined the odds ratio for smoking was 1.61 (95% confidence interval [CI]: 1.39–1.86) at BMI 17 (underweight), 1.47 (95% CI: 1.34–1.62) at BMI 22 (normal weight), 1.35 (95% CI: 1.22–1.48) at BMI 27 (overweight), 1.21 (95% CI: 1.04–1.41) at BMI 33 (obese), and 1.13 (95% CI: 0.92–1.38) at BMI 37 (very obese) (Wehby et al., 2017). This is particularly interesting because some studies show obesity as a separate risk factor for OFC.

The largest study of BMI in OFCs incorporated six large case-control studies from Northern Europe and the USA for analyses pooling the individual-level data (Kutbi et al., 2017). Cases included 4943 mothers of children with OFCs, and 10,592 controls (mothers of unaffected children). Pre-pregnancy maternal weight was classified by level of body mass index (BMI, kg/m^2). CP, both alone and with CL (CP ± CL), was associated with maternal pre-pregnancy obesity (BMI ≥ 35) compared with normal weight [adjusted odds ratio (aOR) = 1.36; 95% confidence interval (CI) = 1.16, 1.58]. CP+/-CL was also marginally associated with maternal underweight (aOR = 1.16; 95% CI = 0.98, 1.36). Cleft lip alone was not associated with BMI, suggesting that extremes of weight may have a specific effect on palatal development.

3.4 Prescription drug exposures

Several prescription medications have also been reported to increase OFC risk when taken during the first trimester, including folate antagonists and certain other drugs exhibiting anti-folate properties (Watkins, Meyer, Strauss, & Aylsworth, 2014).

Antiepileptic drugs: Offspring of women with epilepsy are at increased risk for "fetal anticonvulsant syndrome" which is a combination of intrauterine growth retardation, major congenital malformations (congenital heart defects, CL/P, neural tube defects), minor anomalies, cognitive dysfunction and infant mortality (Pennell, 2002). Therefore, several studies have assessed the association between antiepileptic drugs and OFC. Abrishamchian et al. (Abrishamchian, Khoury, & Calle, 1994) found an OR for CL/P of 3.78 for antiepileptic drugs (95% CI 1.65–7.88). Valproic acid and Phenytoin have been associated with CL/P (Pennell, 2002; Verrotti, Mencaroni, Castagnino, & Zaccara, 2015), but a larger effect of Valproic acid in increasing the risk of CP has been reported (Jackson, Bromley, Morrow, Irwin, & Clayton-Smith, 2016). The likely mechanism underlying this association might be the effect of the antiepileptic drug in reducing plasma folate levels and also through possible teratogenic effects caused by reactive intermediates such as free radicals and oxidative metabolites that get generated during the metabolism of these drugs (Pennell, 2002; Schwaninger et al., 1999).

Benzodiazepines are drugs prescribed to treat anxiety, induce sedation, and for the treatment of eclampsia and pre-eclampsia. Benodiazepines are thought to disrupt neurotransmitter mechanisms that regulate embryonic development. A meta-analysis of 23 studies on benzodiazepine first trimester exposure showed an increase risk of OFC (OR 1.79; 95%; CI: 1.13–2.82) (Dolovich et al., 1998), however, a more recent study in the Swedish Medical Birth Register failed to confirm this association (Marinucci et al., 2011).

Corticosteroids are drugs administered topically, systemically or by inhalation and are prescribed during pregnancy for their immunosuppressive and anti-inflammatory effects (Bandoli, Palmsten, Forbess Smith, & Chambers, 2017). Corticosteroids are also widely used to treat symptoms of diseases such as asthma, lupus and rheumatoid disorders. Several case-control studies have tested the effects of corticosteroids in the risk for OFCs. CL/P studies obtained significant ORs ranging from 1.9 to 8.9 (Carmichael & Shaw, 1999; Park-Wyllie et al., 2000; Pradat et al., 2003; Rodriguez-Pinilla & Martinez-Frias, 1998). Recent studies, however, have not been able to confirm this increased risk despite similar methodologies (Carmichael et al., 2007; Hviid & Molgaard-Nielsen, 2011; Skuladottir et al., 2014). For CP alone one recent study indicated a significant odds ratio of 5.3 (CI: 1.1–26.5) (Bandoli et al., 2017). The reduction in ORs from the earlier studies to the more recent ones has been attributed to factors such as a temporal trend toward shorter durations or lower doses of systemic corticosteroids in favor of alternative treatments. Overall, corticosteroid use may confer a modest increase in the risk for OFCs, however data are conflicting and it is unknown to what extent the underlying disease leading to the corticosteroid use also influences the risk of OFCs.

3.5 Multivitamin/folate consumption

Multivitamin (MV) supplementation during the periconceptual period (with an emphasis on folate) is also implicated (Wehby et al., 2013; Wehby & Murray, 2010) as reducing risk of OFCs. Further support comes from evidence that certain prescription medications including folate antagonists and certain other drugs exhibiting anti-folate properties (see section above) may increase risk of OFCs when taken during the first trimester (Watkins et al., 2014).

There are two types of studies evaluating the effects of MV supplementation in the prevention of OFCs: retrospective case-control studies and clinical trials. Retrospective case-control studies include those assessing self-reported vitamin use (Bower & Stanley, 1992; Hayes, Werler, Willett, & Mitchell, 1996; Itikala, Watkins, Mulinare, Moore, & Liu, 2001; Loffredo, Souza, Freitas, & Mossey, 2001; Shaw, Lammer, Wasserman, O'Malley, & Tolarova, 1995; Shaw et al., 2002; Werler, Hayes, Louik, Shapiro, & Mitchell, 1999), and those using serum biochemical measurements (Stoll, Dott, Alembik, & Koehl, 1999; van Rooij, Swinkels, Blom, Merkus, & Steegers-Theunissen, 2003; Wong et al., 1999). Some studies of self-reported MV use have shown consistent protective effects against OFCs, but others have been inconclusive (Gildestad et al., 2015). Similarly, when serum measurements indicate lower amounts of vitamins B6, B12 or hyperhomocysteinemia (i.e., high levels of homocysteine associated with deficiencies of vitamin B6, folic acid and vitamin B12) ORs range from of 2.9–5.3 indicating increased OFC risk. One population-based study in Norway (all births from 1999 to 2013 including nearly 900,000 pregnancies) showed a significant protective effect only for OFCs that occurred in combination with other malformations (adjusted RR 0·63; 95% CI: 0.45, 0.88) (Gildestad et al., 2015).

A recent meta-analysis for maternal folic acid supplementation, not MV, from 1958 to 2016 performed analysis of pooled OR data and returned a significant association with CL/P (OR = 0.73, 95% CI: 0.62–0.85) but no significant association for CP alone (OR = 0.75, 2 95% CI = 053–1.04) (Jahanbin, Shadkam, Miri, Shirazi, & Abtahi, 2018). Notably, this same study indicated that MV intake (not specifically folate) was associated with a significantly decreased risk for CL/P (OR = 0.65 95% CI = 0.55–0.80) and CP alone (OR = 0.69, 95% CI = 0.53–0.90) (Jahanbin et al., 2018). These results were similar to those from a previous meta-analysis (Johnson & Little, 2008) as well as a more recent systematic review of 64 MV studies (Zhou et al., 2020), indicating a synergistic effect of folic acid in combination with other vitamins rendering more effective prevention of OFCs. MV supplementation in recommended dosages (0.4–0.8 mg) reduces the risk of having a baby with OFC by 30–35% in women of reproductive age.

Early clinical trials of maternal MV supplementation (Czeizel, 1998; Czeizel, Dobo, & Vargha, 2004; Czeizel & Dudas, 1992; Li et al., 2012; Tolarova, 1982; Tolarova & Harris, 1995) yielded inconsistent findings. These conflicting results might be manifestations of the complexity in the etiology of OFCs and reflect the difficulty in designing studies to capture environment-specific effects. More recently, a prospective cohort study in China with 240,244 women enrolled indicated a reduction in CL/P with adjusted rate ratio (aRR) of 0.69 (95% CI: 0.55–0.87) with 0.4 mg of folic acid supplementation taken daily by women before their last menstrual period through pregnancy (Li et al., 2012). Finally, a randomized clinical trial to study the effects of high dosage folic acid supplementation for women who previously had a child with a cleft compared the usual dosage of 0.4 to a high dosage of 4 mg, and indicated a significant reduction in recurrence risk for CL/P from 2.9% in the 0.4 mg group to 0.8% in the 4 mg group, without compromising fetal growth (Wehby et al., 2013).

4. Gene × environment associations in OFC

Inconsistent results from the studies of exposures during pregnancy support the hypothesis that while environmental factors can influence OFC risk, they may also be interacting with an individual's genetic background resulting in potentially different outcomes for the same environmental exposures. Therefore, gene-by-environment (G × E) association studies are very important to tease out the pattern of risk that might be expected for specific individuals, and to better understand how genetic variants and environmental exposures act and interact to influence OFC risk. There are two major categories of studies to address G × E associations: candidate gene studies (chosen for a presumed relationship with specific environmental factors that modify OFC risk), and agnostic genome-wide studies of G × E associations (undertaken to discover new genetic variants associated with specific environmental factors). Genome-wide G × E studies require very large sample sizes for adequate power but are potentially less biased than candidate gene × environment studies. A combination of both approaches is proving fruitful.

4.1 Candidate gene × environment studies

Summarized here are statistically significant G × E interactions reported for several of the exposures summarized in the previous section (maternal smoking, alcohol consumption and vitamin supplementation). There have been numerous small G × E studies for each of these exposures as reviewed recently (Garland, Reynolds, & Zhou, 2020); note that the G × E results from different studies are not always consistent, in part because replicating statistical evidence across such studies is difficult and proving interaction exists remains a major challenge (Aschard, 2016).

Gene × smoking candidate gene studies have identified interactions associated with OFC risk in multiple populations for variants in the following genes: *TGFA*, *TGFB3*, *BMP4*, *MSX1*, *PHF8*, *PRDM16*, *AHRR*. Although there have been a variety of animal studies that have teased out ethanol teratogenicity there have been very few significant candidate gene × alcohol associations reported in humans. Given the importance of having a preventive measure for OFCs, and given the success of folate enrichment of grain products in preventing neural tube defects, there have been many gene × folate studies with folate pathway genes including, e.g., *SHTM1*, *MTR*, *MTHFR* (Reynolds et al., 2020; Wehby & Murray, 2010), most of which show evidence of association with OFCs, but inconsistent evidence of G × E interaction.

That maternal exposures during pregnancy such as smoking, alcohol, and multivitamin/folate influence OFC risk points to potential underlying biological pathways through which the exposures may operate, such as metabolism of folates, metabolism of retinoids, retinoic acid receptor signaling, aryl hydrocarbon receptor signaling, glucocorticoid receptor signaling, and biotransformation and transport (Garland, Reynolds, & Zhou, 2020; Garland, Sun, et al., 2020). These biological pathways may, in turn, be affected by genetic variation.

To illustrate the potential relevance of these specific pathways in OFC, Table 1 summarizes the above pathways, the key molecules plus key proteins and enzymes of each pathway, followed by significant associations in the literature between OFC and genes for the key components of the pathways. More work is needed to explore the underlying hypothesis that genetic variants in genes operating in these biological pathways moderate the effects of maternal exposure during pregnancy, but the fact that variants in relevant genes show significant association with OFC is promising.

4.2 Genome-wide G × E association scans

In contrast to candidate gene G × E studies where specific genes are picked to investigate with specific environmental exposures, genome-wide G × E association studies agnostically scan the genome for variants that interact with specific environments to influence OFC risk. To date there have been relatively few such genome-wide GxE scans for three main reasons: large sample sizes are necessary for statistical power, large numbers of genomic variants must be acquired for each of the participants, and, of course, the dataset must include evaluation of the prenatal exposures. To date there are two major datasets that have been utilized for published genome-wide G × E studies: the GENEVA data set of mostly European and Asian parent/child trios: 7507 total participants (Beaty et al., 2010), and the Pittsburgh Orofacial Cleft Dataset-1 (POFC1) of multi-ethnic OFC families of many structures plus controls with no known family history of craniofacial anomalies: 11,925 total participants. Each of these datasets was genotyped for Illumina SNP panels and imputed to a few million variants, and each has information recorded for smoking, alcohol, and MV/folate exposure before pregnancy and for each trimester of pregnancy. Both datasets are available through dbGaP (the GENEVA dbGaP accession number is phs000094.v1.p1 and for POFC1 is phs000774.v2.p1).

Table 2 summarizes the results from numerous genome-wide GxE analyses of GENEVA and POFC1. Several genome-wide significant GxE results were found, for example: *SLC2A9, WDR1, MUSK, ANK3, TBK1, ZNF236, PRL* with smoking; *ARHGEF10, SMC2, SLCO3A1, VGLL2* with alcohol, and *CASP9, ESRRG, RETREG1/FAM134B, BAALC* with MV/folate supplementation. Genome-wide analyses continue with these datasets, as well as upcoming new genome-wide datasets, some of which have whole genome sequencing as well as the SNP genotyping done for GWAS approaches. Much additional work will

Table 1 Candidates for G × E interaction analyses based on hypothesized etiological mechanisms for OFC-associated exposures (Garland, Reynolds, & Zhou, 2020; Garland, Sun, et al., 2020) plus genes in the pathways that have shown main effect associations with OFC.

Mechanism class	Key molecules	Key proteins and enzymes	Genes in the pathway that have been significantly associated with OFCs, with references[a]
Metabolism of folates	Folic acid, THF	Folate cycle: DHFR, SHMT (vit B6 cofactor), MTHFR	*SHTM1, MTR, MTHFR* (Reynolds et al., 2020; Wehby and Murray, 2010)
Metabolism of retinoids	Retinol, Retinoic Acid	Maternal and embryonic RBP4, STRA6, LRAT, RDHs	*RBP4, RARA* (Fan, Li, & Wu, 2007; Peanchitlertkajorn, Cooper, Liu, Field, & Marazita, 2003; Zhang et al., 2018)
Retinoic Acid receptor Signaling	Retinoic Acid	CRABP, RARs, RXRs, ATRA	*RARA, NOTCH1* (Reynolds et al., 2020; Wehby and Murray, 2010)
Aryl hydrocarbon receptor signaling	Aryl Hydrocarbon/Dioxin Receptor (AHR), PCBs, PCDFs, TCDD	AHR, bHLH/PAS transcription factors, ITE, CYP1A1, AHRR	*ARNT, AHRR* (Leslie, Carlson, Shaffer, Butali, et al., 2017; Leslie, Carlson, Shaffer, Buxo, et al., 2017; Linnenkamp, Raskin, Esposito, & Herai, 2020)
Glucocorticoid receptor signaling	Glucocorticoids (GCs), Cortisol	GC-mediated cellular processing, ECM, TGF-β2, β-catenin	*WNT, FGF, TGFB, BMP* (Ludwig et al., 2016; Suzuki et al., 2009; Wang, Shi, Zheng, Shi, & Jia, 2020)
Biotransformation and transport: Oxidation	Cytochrome P450s (CYPs), EHs, FMOs	CYP enzymes	*CYP1A1, CYP1A2, CYP1B1, CYP2E1, EPHX1* (Shi et al., 2007)
Biotransformation and transport: Conjugation	Covalently adding organic donors to, e.g., xenobiotics to detoxify	Detoxifying enzymes: NATs, GSTs, UGTs, SULTs	*NAT1, NAT2, GSTM1, GSTT1* (Lammer, Shaw, Iovannisci, & Finnell, 2005; Shi, Wehby, & Murray, 2008; Song et al., 2013)
Biotransformation and transport: Transport	Transporters in placenta, protect fetus from toxicants	ATP-binding cassette (ABC) family of proteins	*ABCA4/ARHGAP29, ABCB1* (Beaty et al., 2010; Wu-Chou et al., 2020)

[a]Genes in each class that have shown significant main effects association with OFCs along with representative recent publications.

Table 2 Genome-wide significant G × E interactions reported in genome-wide studies within the GENEVA dataset (Beaty et al., 2010) and/or the Pittsburgh Orofacial Cleft Dataset 1 (POFC1) (Leslie, Carlson, et al., 2016; Leslie, Liu, et al., 2016).

Locus	Associated gene	Maternal exposure	Phenotype	Ethnicity[a]	Dataset	References
1p36	*CASP9*	Multivitamin use	CL/P + CP	Multiple (pooled)	GENEVA + POFC1 trios	Zhang et al. (2021)
1q41	*ESRRG*	Multivitamin use	CL/P	European + Asian	GENEVA	Haaland et al. (2018)
4p16.1	*SLC2A9, WDR1*	Environmental/ passive smoke	CP	Asian	GENEVA	Wu et al. (2014)
5p15.1	*RETREG1/ FAM134B*	Multivitamin use	CL/P + CP	Multiple pooled	GENEVA +POFC1 trios	Zhang et al. (2021)
6p22.3	*PRL*	Smoking	CL/P	European	POFC1 cases/controls only	Carlson et al. (2022)
6q22.1	*VGLL2*	Alcohol	CL/P	European	POFC1 cases/controls only	Carlson et al. (2022)
8p23.3[b]	*ARHGEF10*[b]	Alcohol	CL/P	European	GENEVA	Haaland et al. (2019)
8q22.3	*BAALC*	Multivitamin use	CP	European + Asian	GENEVA	Beaty et al. (2011)
9q31.1	*SMC2*	Alcohol	CP	European + Asian	GENEVA	Beaty et al. (2011)
9q31.3	*MUSK*	Smoking	CL/P + CP	Multiple (pooled)	GENEVA + POFC1 trios	Zhang et al. (2021)
10q21.1[b]	*ANK3*[b]	Smoking	CL/P	European	GENEVA	Haaland et al. (2019)
12q14	*TBK1*	Smoking	CP	European + Asian	GENEVA	Beaty et al. (2011)
15q26.1	*SLCO3A1*	Alcohol	CL/P +CP	Multiple (pooled)	GENEVA + POFC1 trios	Zhang et al. (2021)
18q22	*ZNF236*	Smoking	CP	European + Asian	GENEVA	Beaty et al. (2011)

[a]Ethnicity in which significant G × E association was seen.
[b]Parent-of-origin-by-environment interaction method used.

need to be accomplished to understand how the environmental factors interact with variants in the genes summarized in Table 2 to appear as GxE interactions leading to increased risk of OFCs.

5. Summary and conclusions

Studies of the etiology of OFCs have come a long way from the observational studies and conclusions drawn by early civilizations such as the Aztecs, early family pattern investigations such as Darwin, Bateson and the Galton lab, up to the massive genome-wide studies of OFCs made possible by the tools developed by the Human Genome Project. The genome-wide approaches coupled with the environmental exposure data acquired over many years by investigators around the globe are making GxE interaction studies of OFCs ever more possible. Inconsistent results from the studies of exposures during pregnancy support the hypothesis that while environmental factors can influence OFC risk, they may also be interacting with an individual's genetic background resulting in potentially different outcomes for the same environmental exposures. Therefore, gene-by-environment association studies are very important to tease out the pattern(s) of risk that might be expected for specific individuals, and to better understand how genetic variants and environmental exposures act and interact to influence OFC risk.

References

Abrishamchian, A. R., Khoury, M. J., & Calle, E. E. (1994). The contribution of maternal epilepsy and its treatment to the etiology of oral clefts: A population based case-control study. *Genetic Epidemiology*, *11*, 343–351.

Aschard, H. (2016). A perspective on interaction effects in genetic association studies. *Genetic Epidemiology*, *40*, 678–688.

Bandoli, G., Palmsten, K., Forbess Smith, C. J., & Chambers, C. D. (2017). A review of systemic corticosteroid use in pregnancy and the risk of select pregnancy and birth outcomes. *Rheumatic Diseases Clinics of North America*, *43*, 489–502.

Basha, M., Demeer, B., Revencu, N., Helaers, R., Theys, S., Bou Saba, S., et al. (2018). Whole exome sequencing identifies mutations in 10% of patients with familial non-syndromic cleft lip and/or palate in genes mutated in well-known syndromes. *Journal of Medical Genetics*, *55*, 449–458.

Bateson, W. (1902). *Mendel's principles of heredity: A defense*. Cambridge, England: Cambridge University Press.

Bateson, W. (1909). *Mendel's principles of heredity*. Cambridge, England: Cambridge University Press.

Beaty, T. H., Marazita, M. L., & Leslie, E. J. (2016). Genetic factors influencing risk to orofacial clefts: today's challenges and tomorrow's opportunities. *F1000Research*, *5*, 2800.

Beaty, T. H., Murray, J. C., Marazita, M. L., Munger, R. G., Ruczinski, I., Hetmanski, J. B., et al. (2010). A genome-wide association study of cleft lip with and without cleft palate identifies risk variants near MAFB and ABCA4. *Nature Genetics*, *42*, 525–529.

Beaty, T. H., Ruczinski, I., Murray, J. C., Marazita, M. L., Munger, R. G., Hetmanski, J. B., et al. (2011). Evidence for gene-environment interaction in a genome wide study of nonsyndromic cleft palate. *Genetic Epidemiology*, *35*, 469–478.

Berk, N., & Marazita, M. (2002). The costs of cleft lip and palate: Personal and societal implications. In D. Wyszynski (Ed.), *Cleft lip and palate: From origin to treatment*. Oxford: Oxford University Press.

Bille, C., Knudsen, L. B., & Christensen, K. (2005). Changing lifestyles and oral clefts occurrence in Denmark. *The Cleft Palate-Craniofacial Journal*, *42*, 255–259.

Bille, C., Winther, J. F., Bautz, A., Murray, J. C., Olsen, J., & Christensen, K. (2005). Cancer risk in persons with oral cleft—A population-based study of 8,093 cases. *American Journal of Epidemiology*, *161*, 1047–1055.

Birnbaum, S., Ludwig, K. U., Reutter, H., Herms, S., Steffens, M., Rubini, M., et al. (2009). Key susceptibility locus for nonsyndromic cleft lip with or without cleft palate on chromosome 8q24. *Nature Genetics*, *41*, 473–477.

Bower, C., & Stanley, F. J. (1992). Dietary folate and nonneural midline birth defects: No evidence of an association from a case-control study in Western Australia. *American Journal of Medical Genetics*, *44*, 647–650.

Butali, A., Mossey, P. A., Adeyemo, W. L., Eshete, M. A., Gowans, L. J. J., Busch, T. D., et al. (2019). Genomic analyses in African populations identify novel risk loci for cleft palate. *Human Molecular Genetics*, *28*, 1038–1051.

Camargo, M., Rivera, D., Moreno, L., Lidral, A. C., Harper, U., Jones, M., et al. (2012). GWAS reveals new recessive loci associated with non-syndromic facial clefting. *European Journal of Medical Genetics*, *55*, 510–514.

Carlson, J. C., Anand, D., Butali, A., Buxo, C. J., Christensen, K., Deleyiannis, F., et al. (2019). A systematic genetic analysis and visualization of phenotypic heterogeneity among orofacial cleft GWAS signals. *Genetic Epidemiology*, *43*, 704–716.

Carlson, J. C., Shaffer, J. R., Deleyiannis, F., Hecht, J. T., Wehby, G. L., Christensen, K., et al. (2022). Genome-wide interaction study implicates VGLL2 and alcohol exposure and PRL and smoking in orofacial cleft risk. *Frontiers in Cell and Development Biology*, *10*, 621261.

Carmichael, S. L., & Shaw, G. M. (1999). Maternal corticosteroid use and risk of selected congenital anomalies. *American Journal of Medical Genetics*, *86*, 242–244.

Carmichael, S. L., Shaw, G. M., Ma, C., Werler, M. M., Rasmussen, S. A., Lammer, E. J., et al. (2007). Maternal corticosteroid use and orofacial clefts. *American Journal of Obstetrics and Gynecology*, *197*. 585 e581–587; discussion 683–584, e581–587.

Cheng, L. R. (1990). Asian-American cultural perspectives on birth defectrs: Focus on cleft palate. *The Cleft Palate-Craniofacial Journal*, *27*, 294–300.

Christensen, K. (1999). The 20th century Danish facial cleft population—Epidemiological and genetic-epidemiological studies. *The Cleft Palate-Craniofacial Journal*, *36*, 96–104.

Christensen, K., Juel, K., Herskind, A. M., & Murray, J. C. (2004). Long term follow up study of survival associated with cleft lip and palate at birth. *BMJ (Clinical Research Ed.)*, *328*, 1405.

Chung, K. C., Kowalski, C. P., Kim, H. M., & Buchman, S. R. (2000). Maternal cigarette smoking during pregnancy and the risk of having a child with cleft lip/palate. *Plastic and Reconstructive Surgery*, *105*, 485–491.

ACOG Committee. (2020). Tobacco and nicotine cessation during pregnancy: ACOG Committee Opinion, Number 807. *Obstetrics and Gynecology*, *135*, e221–e229.

Cooper, M. E., Ratay, J. S., & Marazita, M. L. (2006). Asian oral-facial cleft birth prevalence. *The Cleft Palate-Craniofacial Journal*, *43*, 580–589.

Czeizel, A. E. (1998). Periconceptional folic acid containing multivitamin supplementation. *European Journal of Obstetrics, Gynecology, and Reproductive Biology*, *78*, 151–161.

Czeizel, A. E., Dobo, M., & Vargha, P. (2004). Hungarian cohort-controlled trial of periconceptional multivitamin supplementation shows a reduction in certain congenital abnormalities. *Birth Defects Research. Part A, Clinical and Molecular Teratology*, *70*, 853–861.

Czeizel, A. E., & Dudas, I. (1992). Prevention of the first occurrence of neural-tube defects by periconceptional vitamin supplementation. *The New England Journal of Medicine*, *327*, 1832–1835.

Daack-Hirsch, S. (2010). Filipino explanatory models of cleft lip with or without cleft palate. *The Cleft Palate-Craniofacial Journal*, *47*, 122–133.

Darwin, C. (1875). *The variation of animals and plants under domestication. Vol. 1* (2nd Edition, Revised ed.). London, England: John Murray.

DeRoo, L. A., Wilcox, A. J., Lie, R. T., Romitti, P. A., Pedersen, D. A., Munger, R. G., et al. (2016). Maternal alcohol binge-drinking in the first trimester and the risk of orofacial clefts in offspring: A large population-based pooling study. *European Journal of Epidemiology*, *31*, 1021–1034.

Dietz, A., Pedersen, D. A., Jacobsen, R., Wehby, G. L., Murray, J. C., & Christensen, K. (2012). Risk of breast cancer in families with cleft lip and palate. *Annals of Epidemiology*, *22*, 37–42.

Dolovich, L. R., Addis, A., Vaillancourt, J. M., Power, J. D., Koren, G., & Einarson, T. R. (1998). Benzodiazepine use in pregnancy and major malformations or oral cleft: Meta-analysis of cohort and case-control studies. *BMJ (Clinical Research Ed)*, *317*, 839–843.

Eshete, M. A., Liu, H., Li, M., Adeyemo, W. L., Gowans, L. J. J., Mossey, P. A., et al. (2018). Loss-of-function GRHL3 variants detected in African patients with isolated cleft palate. *Journal of Dental Research*, *97*, 41–48.

Fan, G. Z., Li, Y. L., & Wu, P. A. (2007). Association between retinoic acid receptor alpha gene polymorphisms and nonsyndromic cleft lip with or without cleft palate susceptibility. *Zhonghua Yi Xue Za Zhi*, *87*, 396–398.

Feydeau, G. (1907). *La puce à l'oreille [a flea in her ear]*. London, England: NIck Herns Books Limited.

Fogh-Andersen, P. (1971). Epidemiology and etiology of clefts. *Birth Defects Original Article Series*, 7, 50–53.

Fraser, F. C. (1955). Thoughts on the etiology of clefts of the palate and lip. *Acta Genetica et Statistica Medica*, *5*, 358–369.

Garland, M. A., Reynolds, K., & Zhou, C. J. (2020). Environmental mechanisms of orofacial clefts. *Birth Defects Research*, *112*, 1660–1698.

Garland, M. A., Sun, B., Zhang, S., Reynolds, K., Ji, Y., & Zhou, C. J. (2020). Role of epigenetics and miRNAs in orofacial clefts. *Birth Defects Research*, *112*, 1635–1659.

Gildestad, T., Bjorge, T., Vollset, S. E., Klungsoyr, K., Nilsen, R. M., Haaland, O. A., et al. (2015). Folic acid supplements and risk for oral clefts in the newborn: A population-based study. *The British Journal of Nutrition*, *114*, 1456–1463.

Gowans, L. J., Adeyemo, W. L., Eshete, M., Mossey, P. A., Busch, T., Aregbesola, B., et al. (2016). Association studies and direct DNA sequencing implicate genetic susceptibility loci in the etiology of nonsyndromic orofacial clefts in sub-Saharan African populations. *Journal of Dental Research*, *95*, 1245–1256.

Grant, S. F., Wang, K., Zhang, H., Glaberson, W., Annaiah, K., Kim, C. E., et al. (2009). A genome-wide association study identifies a locus for nonsyndromic cleft lip with or without cleft palate on 8q24. *The Journal of Pediatrics*, *155*, 909–913.

Grosen, D., Bille, C., Petersen, I., Skytthe, A., Hjelmborg, J., Pedersen, J. K., et al. (2011). Risk of oral clefts in twins. *Epidemiology (Cambridge, Mass.)*, *22*, 313–319.

Haaland, O. A., Lie, R. T., Romanowska, J., Gjerdevik, M., Gjessing, H. K., & Jugessur, A. (2018). A genome-wide search for gene-environment effects in isolated cleft lip with or without cleft palate triads points to an interaction between maternal periconceptional vitamin use and variants in ESRRG. *Frontiers in Genetics*, *9*, 60.

Haaland, Ø. A., Romanowska, J., Gjerdevik, M., Lie, R. T., Gjessing, H. K., & Jugessur, A. (2019). A genome-wide scan of cleft lip triads identifies parent-of-origin interaction effects between ANK3 and maternal smoking, and between ARHGEF10 and alcohol consumption. *F1000Research*, *8*, 960.

Hayes, C., Werler, M. M., Willett, W. C., & Mitchell, A. A. (1996). Case-control study of periconceptional folic acid supplementation and oral clefts. *American Journal of Epidemiology*, *143*, 1229–1234.

Honein, M. A., Devine, O., Grosse, S. D., & Reefhuis, J. (2014). Prevention of orofacial clefts caused by smoking: Implications of the surgeon General's report. Birth defects research. *Birth Defects Research. Part A, Clinical and Molecular Teratology*, *100*, 822–825.

Hosseini, K. (2003). *The kite runner*. New York, NY: The Berkeley Publishing Group, a division of the Penguin Group.

Hviid, A., & Molgaard-Nielsen, D. (2011). Corticosteroid use during pregnancy and risk of orofacial clefts. *CMAJ*, *183*, 796–804.

Itikala, P. R., Watkins, M. L., Mulinare, J., Moore, C. A., & Liu, Y. (2001). Maternal multivitamin use and orofacial clefts in offspring. *Teratology*, *63*, 79–86.

Jackson, A., Bromley, R., Morrow, J., Irwin, B., & Clayton-Smith, J. (2016). In utero exposure to valproate increases the risk of isolated cleft palate. *Archives of Disease in Childhood. Fetal and Neonatal Edition*, *101*, F207–F211.

Jahanbin, A., Shadkam, E., Miri, H. H., Shirazi, A. S., & Abtahi, M. (2018). Maternal folic acid supplementation and the risk of oral clefts in offspring. *The Journal of Craniofacial Surgery*, *29*, e534–e541.

Ji, Y., Garland, M. A., Sun, B., Zhang, S., Reynolds, K., McMahon, M., et al. (2020). Cellular and developmental basis of orofacial clefts. *Birth Defects Research*, *112*, 1558–1587.

Johnson, C. Y., & Little, J. (2008). Folate intake, markers of folate status and oral clefts: Is the evidence converging? *International Journal of Epidemiology*, *37*, 1041–1058.

Kantar, R. S., Cammarata, M. J., Rifkin, W. J., Diaz-Siso, J. R., Hamdan, U. S., & Flores, R. L. (2019). Foundation-based cleft care in developing countries. *Plastic and Reconstructive Surgery*, *143*, 1165–1178.

Kummet, C. M., Moreno, L. M., Wilcox, A. J., Romitti, P. A., DeRoo, L. A., Munger, R. G., et al. (2016). Passive smoke exposure as a risk factor for oral clefts: A large international population-based study. *American Journal of Epidemiology*, *183*, 834–841.

Kurita, H., Motoki, N., Inaba, Y., Misawa, Y., Ohira, S., Kanai, M., et al. (2020). Maternal alcohol consumption and risk of offspring with congenital malformation: The Japan environment and Children's study. *Pediatric Research*, *90*, 479–486.

Kutbi, H., Wehby, G. L., Moreno Uribe, L. M., Romitti, P. A., Carmichael, S., Shaw, G. M., et al. (2017). Maternal underweight and obesity and risk of orofacial clefts in a large international consortium of population-based studies. *International Journal of Epidemiology*, *46*, 190–199.

Lammer, E. J., Shaw, G. M., Iovannisci, D. M., & Finnell, R. H. (2005). Maternal smoking, genetic variation of glutathione s-transferases, and risk for orofacial clefts. *Epidemiology (Cambridge, Mass.)*, *16*, 698–701.

Lantz, P. M. (2003). Smoking on the rise among young adults: Implications for research and policy. *Tobacco Control*, *12*, I60–I70.

Leslie, E. J., Carlson, J. C., Shaffer, J. R., Butali, A., Buxó, C. J., Castilla, E. E., et al. (2017). Genome-wide meta-analyses of nonsyndromic orofacial clefts identify novel associations between FOXE1 and all orofacial clefts, and TP63 and cleft lip with or without cleft palate. *Human Genetics*, *136*, 275–286.

Leslie, E. J., Carlson, J. C., Shaffer, J. R., Buxo, C. J., Castilla, E. E., Christensen, K., et al. (2017). Association studies of low-frequency coding variants in nonsyndromic cleft lip with or without cleft palate. *American Journal of Medical Genetics. Part A*, *173*, 1531–1538.

Leslie, E. J., Carlson, J. C., Shaffer, J. R., Feingold, E., Wehby, G., Laurie, C. A., et al. (2016). A multi-ethnic genome-wide association study identifies novel loci for non-syndromic cleft lip with or without cleft palate on 2p24.2, 17q23 and 19q13. *Human Molecular Genetics*, *25*, 2862–2872.

Leslie, E. J., Liu, H., Carlson, J. C., Shaffer, J. R., Feingold, E., Wehby, G., et al. (2016). A genome-wide association study of nonsyndromic cleft palate identifies an etiologic missense variant in GRHL3. *American Journal of Human Genetics*, *98*, 744–754.

Leslie, E. J., Taub, M. A., Liu, H., Steinberg, K. M., Koboldt, D. C., Zhang, Q., et al. (2015). Identification of functional variants for cleft lip with or without cleft palate in or near PAX7, FGFR2, and NOG by targeted sequencing of GWAS loci. *American Journal of Human Genetics*, *96*, 397–411.

Li, S., Chao, A., Li, Z., Moore, C. A., Liu, Y., Zhu, J., et al. (2012). Folic acid use and non-syndromic orofacial clefts in China: A prospective cohort study. *Epidemiology*, *23*, 423–432.

Lieff, S., Olshan, A. F., Werler, M., Strauss, R. P., Smith, J., & Mitchell, A. (1999). Maternal cigarette smoking during pregnancy and risk of oral clefts in newborns. *American Journal of Epidemiology*, *150*, 683–694.

Linnenkamp, B. D. W., Raskin, S., Esposito, S. E., & Herai, R. H. (2020). A comprehensive analysis of AHRR gene as a candidate for cleft lip with or without cleft palate. *Mutation Research, Reviews in Mutation Research*, *785*, 108319.

Little, J., Cardy, A., & Munger, R. G. (2004). Tobacco smoking and oral clefts: A meta-analysis. *Bulletin of the World Health Organization*, *82*, 213–218.

Loffredo, L. C., Souza, J. M., Freitas, J. A., & Mossey, P. A. (2001). Oral clefts and vitamin supplementation. *The Cleft Palate-Craniofacial Journal*, *38*, 76–83.

Ludwig, K. U., Ahmed, S. T., Böhmer, A. C., Sangani, N. B., Varghese, S., Klamt, J., et al. (2016). Meta-analysis reveals genome-wide significance at 15q13 for nonsyndromic clefting of both the lip and the palate, and functional analyses implicate GREM1 as a plausible causative gene. *PLoS Genetics*, *12*, e1005914.

Ludwig, K. U., Mangold, E., Herms, S., Nowak, S., Reutter, H., Paul, A., et al. (2012). Genome-wide meta-analyses of nonsyndromic cleft lip with or without cleft palate identify six new risk loci. *Nature Genetics*, *44*, 968–971.

Mangold, E., Bohmer, A. C., Ishorst, N., Hoebel, A. K., Gultepe, P., Schuenke, H., et al. (2016). Sequencing the GRHL3 coding region reveals rare truncating mutations and a common susceptibility variant for nonsyndromic cleft palate. *American Journal of Human Genetics*, *98*, 755–762.

Mangold, E., Ludwig, K. U., Birnbaum, S., Baluardo, C., Ferrian, M., Herms, S., et al. (2010). Genome-wide association study identifies two susceptibility loci for non-syndromic cleft lip with or without cleft palate. *Nature Genetics*, *42*, 24–26.

Marazita, M. L. (2012). The evolution of human genetic studies of cleft lip and cleft palate. *Annual Review of Genomics and Human Genetics*, *13*, 263–283.

Marazita, M. L., Lidral, A. C., Murray, J. C., Field, L. L., Maher, B. S., Goldstein McHenry, T., et al. (2009). Genome scan, fine-mapping, and candidate gene analysis of non-syndromic cleft lip with or without cleft palate reveals phenotype-specific differences in linkage and association results. *Human Heredity*, *68*, 151–170.

Marinucci, L., Balloni, S., Carinci, F., Locci, P., Pezzetti, F., & Bodo, M. (2011). Diazepam effects on non-syndromic cleft lip with or without palate: Epidemiological studies, clinical findings, genes and extracellular matrix. *Expert Opinion on Drug Safety*, *10*, 23–33.
Melnick, M. (1997). Cleft lip and palate etiology and its meaning in early 20th century England: Galton/Pearson vs. Bateson; polygenically poor protoplasm vs. Mendelism. *Journal of Craniofacial Genetics and Developmental Biology*, *17*, 65–79.
Mendel, G. (1866). Versuche uber pflanzenhybriden. *Verhandenlungen des Naturforschenden Vereines in Brunn*, *4*, 3–17.
Menezes, R., Marazita, M. L., Goldstein McHenry, T., Cooper, M. E., Bardi, K., Brandon, C., et al. (2009). AXIS inhibition protein 2, orofacial clefts and a family history of cancer. *Journal of the American Dental Association (1939)*, *140*, 80–84.
Meng, T., Shi, B., Zheng, Q., Wang, Y., & Li, S. (2006). Clinical and epidemiologic studies of nonsyndromic cleft lip and palate in China: Analysis of 4268 cases. *Annals of Plastic Surgery*, *57*, 264–269.
Moreno Uribe, L. M., & Marazita, M. L. (2021). Epidemiology, etiology and genetics of orofacial clefting. In P. Shetye, & T. L. Gibson (Eds.), *Cleft and craniofacial orthodontics* Wiley.
Morris, V. E., Hashmi, S. S., Zhu, L., Maili, L., Urbina, C., Blackwell, S., et al. (2020). Evidence for craniofacial enhancer variation underlying nonsyndromic cleft lip and palate. *Human Genetics*, *139*, 1261–1272.
Mossey, P. (2007). Epidemiology underpinning research in the aetiology of orofacial clefts. *Orthodontics & Craniofacial Research*, *10*, 114–120.
Mossey, P. A., Little, J., Munger, R. G., Dixon, M. J., & Shaw, W. C. (2009). Cleft lip and palate. *Lancet*, *374*, 1773–1785.
Munger, R. G., Romitti, P. A., Daack-Hirsch, S., Burns, T. L., Murray, J. C., & Hanson, J. (1996). Maternal alcohol use and risk of orofacial cleft birth defects. *Teratology*, *54*, 27–33.
Murray, T., Taub, M. A., Ruczinski, I., Scott, A. F., Hetmanski, J. B., Schwender, H., et al. (2012). Examining markers in 8q24 to explain differences in evidence for association with cleft lip with/without cleft palate between Asians and Europeans. *Genetic Epidemiology*, *36*, 392–399.
Park-Wyllie, L., Mazzotta, P., Pastuszak, A., Moretti, M. E., Beique, L., Hunnisett, L., et al. (2000). Birth defects after maternal exposure to corticosteroids: Prospective cohort study and meta-analysis of epidemiological studies. *Teratology*, *62*, 385–392.
Peanchitlertkajorn, S., Cooper, M. E., Liu, Y. E., Field, L. L., & Marazita, M. L. (2003). Chromosome 17: Gene mapping studies of cleft lip with or without cleft palate in Chinese families. *The Cleft Palate-Craniofacial Journal*, *40*, 71–79.
Pennell, P. B. (2002). Pregnancy in the woman with epilepsy: Maternal and fetal outcomes. *Seminars in Neurology*, *22*, 299–308.
Pradat, P., Robert-Gnansia, E., Di Tanna, G. L., Rosano, A., Lisi, A., Mastroiacovo, P., et al. (2003). First trimester exposure to corticosteroids and oral clefts. *Birth Defects Research. Part A, Clinical and Molecular Teratology*, *67*, 968–970.
Reynolds, K., Zhang, S., Sun, B., Garland, M. A., Ji, Y., & Zhou, C. J. (2020). Genetics and signaling mechanisms of orofacial clefts. *Birth Defects Research*, *112*, 1588–1634.
Rischbieth, H. (1910). Hare-lip and cleft palate. In K. Pearson (Ed.), *Treasury of human inheritance, part IV* (pp. 79–123). London, England: Dulau.
Rodriguez-Pinilla, E., & Martinez-Frias, M. L. (1998). Corticosteroids during pregnancy and oral clefts: A case-control study. *Teratology*, *58*, 2–5.
Romitti, P. A., Sun, L., Honein, M. A., Reefhuis, J., Correa, A., & Rasmussen, S. A. (2007). Maternal periconceptional alcohol consumption and risk of orofacial clefts. *American Journal of Epidemiology*, *166*, 775–785.

Sabbagh, H. J., Hassan, M. H., Innes, N. P., Elkodary, H. M., Little, J., & Mossey, P. A. (2015). Passive smoking in the etiology of non-syndromic orofacial clefts: A systematic review and meta-analysis. *PLoS One, 10*, e0116963.

Saleem, K., Zaib, T., Sun, W., & Fu, S. (2019). Assessment of candidate genes and genetic heterogeneity in human non syndromic orofacial clefts specifically non syndromic cleft lip with or without palate. *Heliyon, 5*, e03019.

Schwaninger, M., Ringleb, P., Winter, R., Kohl, B., Fiehn, W., Rieser, P. A., et al. (1999). Elevated plasma concentrations of homocysteine in antiepileptic drug treatment. *Epilepsia, 40*, 345–350.

Semb, G., Brattstrom, V., Molsted, K., Prahl-Andersen, B., & Shaw, W. C. (2005). The Eurocleft study: Intercenter study of treatment outcome in patients with complete cleft lip and palate. Part 1: Introduction and treatment experience. *The Cleft Palate-Craniofacial Journal, 42*, 64–68.

Shakespeare, W. (1595). *A Midsummer Night's Dream (V.i.2260–2265), OpenSourceShakespeare, an Experiment in Literary Technology*.

Shakespeare, W. (1605). *The Tragedy of King Lear (III.iv.1910-1913), OpenAccessShakespeare, an Experiment in Literary Technology*.

Shaw, G. M., & Lammer, E. J. (1999). Maternal periconceptional alcohol consumption and risk for orofacial clefts. *The Journal of Pediatrics, 134*, 298–303.

Shaw, G. M., Lammer, E. J., Wasserman, C. R., O'Malley, C. D., & Tolarova, M. M. (1995). Risks of orofacial clefts in children born to women using multivitamins containing folic acid periconceptionally. *Lancet, 346*, 393–396.

Shaw, G. M., Nelson, V., Carmichael, S. L., Lammer, E. J., Finnell, R. H., & Rosenquist, T. H. (2002). Maternal periconceptional vitamins: Interactions with selected factors and congenital anomalies? *Epidemiology, 13*, 625–630.

Shi, M., Christensen, K., Weinberg, C. R., Romitti, P., Bathum, L., Lozada, A., et al. (2007). Orofacial cleft risk is increased with maternal smoking and specific detoxification-gene variants. *American Journal of Human Genetics, 80*, 76–90.

Shi, M., Wehby, G. L., & Murray, J. C. (2008). Review on genetic variants and maternal smoking in the etiology of oral clefts and other birth defects. *Birth Defects Research. Part C, Embryo Today, 84*, 16–29.

Skuladottir, H., Wilcox, A. J., Ma, C., Lammer, E. J., Rasmussen, S. A., Werler, M. M., et al. (2014). Corticosteroid use and risk of orofacial clefts. *Birth Defects Research. Part A, Clinical and Molecular Teratology, 100*, 499–506.

Song, T., Wu, D., Wang, Y., Li, H., Yin, N., & Zhao, Z. (2013). Association of NAT1 and NAT2 genes with nonsyndromic cleft lip and palate. *Molecular Medicine Reports, 8*, 211–216.

Sproule, J. (1863). Hereditary nature of hare-lip. *British Medical Journal, 1*, 412.

Stoll, C., Dott, B., Alembik, Y., & Koehl, C. (1999). Maternal trace elements, vitamin B12, vitamin A, folic acid, and fetal malformations. *Reproductive Toxicology, 13*, 53–57.

Sun, Y., Huang, Y., Yin, A., Pan, Y., Wang, Y., Wang, C., et al. (2015). Genome-wide association study identifies a new susceptibility locus for cleft lip with or without a cleft palate. *Nature Communications, 6*, 6414.

Suzuki, S., Marazita, M. L., Cooper, M. E., Miwa, N., Hing, A., Jugessur, A., et al. (2009). Mutations in BMP4 are associated with subepithelial, microform, and overt cleft lip. *American Journal of Human Genetics, 84*, 406–411.

Tolarova, M. (1982). Periconceptional supplementation with vitamins and folic acid to prevent recurrence of cleft lip. *Lancet, 2*, 217.

Tolarova, M., & Harris, J. (1995). Reduced recurrence of orofacial clefts after periconceptional supplementation with high-dose folic acid and multivitamins. *Teratology, 51*, 71–78.

Trew, C. J. (1757). Sistens plura exempla palati deficientis. *Nova Acta Physico-Medica Academiae caesarae Leopoldion-Carolinae, 1*, 445–447.
van Rooij, I. A., Swinkels, D. W., Blom, H. J., Merkus, H. M., & Steegers-Theunissen, R. P. (2003). Vitamin and homocysteine status of mothers and infants and the risk of non-syndromic orofacial clefts. *American Journal of Obstetrics and Gynecology, 189*, 1155–1160.
Verrotti, A., Mencaroni, E., Castagnino, M., & Zaccara, G. (2015). Foetal safety of old and new antiepileptic drugs. *Expert Opinion on Drug Safety, 14*, 1563–1571.
Wang, Y., Shi, J., Zheng, Q., Shi, B., & Jia, Z. (2020). Gene-gene interactions between BMP4 and ARHGAP29 among non-syndromic cleft lip only (NSCLO) trios from western Han Chinese population. *International Journal of Clinical and Experimental Pathology, 13*, 295–301.
Watkins, S. E., Meyer, R. E., Strauss, R. P., & Aylsworth, A. S. (2014). Classification, epidemiology, and genetics of orofacial clefts. *Clinics in Plastic Surgery, 41*, 149–163.
Wehby, G. L., & Cassell, C. H. (2010). The impact of orofacial clefts on quality of life and healthcare use and costs. *Oral Diseases, 16*, 3–10.
Wehby, G. L., Castilla, E. E., Goco, N., Rittler, M., Cosentino, V., Javois, L., et al. (2006). Description of the methodology used in an ongoing pediatric care interventional study of children born with cleft lip and palate in South America [NCT00097149]. *BMC Pediatrics, 6*, 9.
Wehby, G. L., Felix, T. M., Goco, N., Richieri-Costa, A., Chakraborty, H., Souza, J., et al. (2013). High dosage folic acid supplementation, oral cleft recurrence and fetal growth. *International Journal of Environmental Research and Public Health, 10*, 590–605.
Wehby, G. L., & Murray, J. C. (2010). Folic acid and orofacial clefts: A review of the evidence. *Oral Diseases, 16*, 11–19.
Wehby, G. L., Uribe, L. M., Wilcox, A. J., Christensen, K., Romitti, P. A., Munger, R. G., et al. (2017). Interaction between smoking and body mass index and risk of oral clefts. *Annals of Epidemiology, 27*(103–107), e102.
Werler, M. M., Hayes, C., Louik, C., Shapiro, S., & Mitchell, A. A. (1999). Multivitamin supplementation and risk of birth defects. *American Journal of Epidemiology, 150*, 675–682.
Werler, M. M., Lammer, E. J., Rosenberg, L., & Mitchell, A. A. (1990). Maternal cigarette smoking during pregnancy in relation to oral clefts. *American Journal of Epidemiology, 132*, 926–932.
Werler, M. M., Lammer, E. J., Rosenberg, L., & Mitchell, A. A. (1991). Maternal alcohol use in relation to selected birth defects. *American Journal of Epidemiology, 134*, 691–698.
Wong, W. Y., Eskes, T. K., Kuijpers-Jagtman, A. M., Spauwen, P. H., Steegers, E. A., Thomas, C. M., et al. (1999). Nonsyndromic orofacial clefts: Association with maternal hyperhomocysteinemia. *Teratology, 60*, 253–257.
Wu, T., Schwender, H., Ruczinski, I., Murray, J. C., Marazita, M. L., Munger, R. G., et al. (2014). Evidence of gene-environment interaction for two genes on chromosome 4 and environmental tobacco smoke in controlling the risk of nonsyndromic cleft palate. *PLoS One, 9*, e88088.
Wu-Chou, Y. H., Chen, K. P., Lu, Y. C., Lin, Y. T., Chang, H. F., & Lo, L. J. (2020). The SNP rs560426 within ABCA4-ARHGAP29 locus and the risk of nonsyndromic oral clefts. *The Cleft Palate-Craniofacial Journal, 57*, 671–677.
Wyszynski, D. (2002). *Cleft lip and palate: From origin to treatment* (1st ed.). Oxford: Oxford University Press.
Wyszynski, D. F., Duffy, D. L., & Beaty, T. H. (1997). Maternal cigarette smoking and oral clefts: A meta-analysis. *The Cleft Palate-Craniofacial Journal, 34*, 206–210.
Xuan, Z., Zhongpeng, Y., Yanjun, G., Jiaqi, D., Yuchi, Z., Bing, S., et al. (2016). Maternal active smoking and risk of oral clefts: A meta-analysis. *Oral Surgery, Oral Medicine, Oral Pathology, Oral Radiology, 122*, 680–690.

Yin, X., Li, J., Li, Y., & Zou, S. (2019). Maternal alcohol consumption and oral clefts: A meta-analysis. *The British Journal of Oral & Maxillofacial Surgery*, *57*, 839–846.

Zhang, S. J., Meng, P., Zhang, J., Jia, P., Lin, J., Wang, X., et al. (2018). Machine learning models for genetic risk assessment of infants with non-syndromic orofacial cleft. *Genomics, Proteomics & Bioinformatics*, *16*, 354–364.

Zhang, W., Venkataraghavan, S., Hetmanski, J. B., Leslie, E. J., Marazita, M. L., Feingold, E., et al. (2021). Detecting gene-environment interaction for maternal exposures using case-parent trios ascertained through a case with non-syndromic orofacial cleft. *Frontiers in Cell and Development Biology*, *9*, 621018.

Zhou, Y., Sinnathamby, V., Yu, Y., Sikora, L., Johnson, C. Y., Mossey, P., et al. (2020). Folate intake, markers of folate status and oral clefts: An updated set of systematic reviews and meta-analyses. *Birth Defects Research*, *112*, 1699–1719.

Zhu, J. L., Basso, O., Hasle, H., Winther, J. F., Olsen, J. H., & Olsen, J. (2002). Do parents of children with congenital malformations have a higher cancer risk? A nationwide study in Denmark. *British Journal of Cancer*, *87*, 524–528.

CHAPTER SEVEN

Gene–environment interactions underlying the etiology of neural tube defects

Carlo Donato Caiaffa[a,†], Cristiane Sá Roriz Fonteles[b,†], Lei Yunping[a], and Richard H. Finnell[a,c,*]

[a]Center for Precision Environmental Health, Department of Molecular and Cellular Biology, Baylor College of Medicine, Houston, TX, United States
[b]Postgraduate Program in Dentistry, Faculty of Pharmacy, Dentistry and Nursing, Federal University of Ceara, Fortaleza, Brazil
[c]Departments of Molecular and Cellular Biology, Molecular and Human Genetics, and Medicine, Baylor College of Medicine, Houston, TX, United States
*Corresponding author: e-mail address: richard.finnell@bcm.edu

Contents

Abstract

Neural tube defects (NTDs) consist of severe structural malformations of the brain and spinal cord and are the second most common structural birth defect in humans, accounting for approximately 2700 affected pregnancies every year in the United States. These numbers are highly significant, considering that birth defects remain a leading cause

† Contributed equally to this manuscript.

Current Topics in Developmental Biology, Volume 152
ISSN 0070-2153
https://doi.org/10.1016/bs.ctdb.2022.10.007

of infant mortality in the United States, affecting approximately 120,000 babies born annually. Survivors of these congenital malformations face long-term disability and lifelong challenges imposed by severe physical burdens compromising the afflicted individual's overall quality of life. Clearly, birth defects, and especially NTDs remain a global public health challenge, and the source of significant financial repercussions for healthcare systems worldwide. In order to better understand the role gene–environment interactions play in the etiology of NTDs, this chapter provides an overview of NTD phenotypes and their embryonic origins, discusses the genetic landscape of NTDs as it is currently understood, with a focus on experimental models that best illustrate how environmental factors modulate individual susceptibility to these birth defects. As folic acid interventions have proven to be effective in reducing the prevalence of NTDs, the chapter ends with a discussion on the impact that maternal dietary status has on NTD prevalence from a population perspective.

1. Introduction

Neural tube defects (NTDs) result from failed closure of the neural tube, and represent the second most common type of human structural birth defects (Committee on Practice Bulletins-Obstetrics, 2017). Their estimated global prevalence is approximately 18.6 per 10,000 live births (Blencowe, Kancherla, Moorthie, Darlison, & Modell, 2018), while the prevalence of NTDs is 1–2 per 1000 births in the United States (Shaw, Jensvold, Wasserman, & Lammer, 1994; Wallingford, Niswander, Shaw, & Finnell, 2013). Worldwide differences in the prevalence of NTDs may be related to the complexity of their pathogenesis. There are 2300 to 2700 pregnancies complicated by an NTDs diagnosis in the United States annually, resulting in lifetime medical costs that exceed $560,000 per child or $1.68 billion per year nationwide.

Despite limited changes in treatment approaches to spina bifida (SB) over the past few decades, early diagnosis and improvement in clinical management has increased life expectancy of these patients. Thus, there are approximately 166,000 individuals with SB living in the United States, and over half of these patients are adults (Oakeshott et al., 2015; Rocque et al., 2015). NTDs are thought to have a multifactorial etiology derived from an interplay between specific genetic factors, such as genes that regulate folate one-carbon metabolism and planar cell polarity (Finnell et al., 2021; Wallingford et al., 2013), and specific environmental elements. It should be borne in mind that with respect to the etiology of NTDs, genetic factors tend to be highly complex and multiple variants differing between individual NTD cases, likely contribute to the occurrence of these malformations. As will be discussed later in this chapter, the periconceptional use of folic acid (FA) has been shown to be effective in reducing the prevalence of NTDs in most every population that has been studied to date (Berry et al., 1999; Canfield et al., 2005; Crider et al., 2022; De Wals et al., 2007). However, much remains unknown as to how embryos benefit from access to FA. It is also important to remember that there are significant numbers of NTDs that are not preventable by FA supplementation, with these FA-resistant NTDs occurring at a baseline rate of 5 per 10,000 live births (Heseker, Mason, Selhub, Rosenberg, & Jacques, 2009). We therefore maintain that NTDs remain a substantial public health problem even in countries with mandatory FA food fortification programs.

2. Neural tube defects phenotypes

There are a range of NTD subtypes that include both anterior and posterior defects. In anencephaly, the absence of the brain and the calvaria may be total or partial. Craniorachischisis results from failed elevation of the neural folds and this severe malformation is accompanied by exposure of neural tissue and bony defects along the length of the embryo. Encephaloceles are characterized by defects in the formation of the calvaria, presenting bone failure during the development of the skull, in addition to the herniation of both brain and meninges. It is uncertain whether to classify encephaloceles as NTDs, as it is most likely a post-neurulation defect. Iniencephaly is a rare form of NTD in which the cervico-occipital region presents dysraphism, which causes severe retroflexion of the junction between the neck and trunk. Anencephaly, craniorachischisis and iniencephaly are lethal malformations that present with a high frequency of associated multiple malformations, such as hydronephrosis, cardiovascular malformations, or cleft palate (David & Nixon, 1976). Failure of primary neurulation leads to open NTDs, including anencephaly, myelomeningocele (open spina bifida), and craniorachischisis. In open spina bifida, the lesion is not covered by skin, and defects can be observed in the thoracic vertebrae, as well as the herniation of the meninges and neural tissue from regions located at the posterior axis during neural tube development. In closed spina bifida, hernias also occur in the neural tissue and in the meninges associated with the spine, but the lesion is covered by skin, which covers the bony defect present in the lumbar region of the posterior vertebral arches (Copp et al., 2015).

3. Embryonic origin of neural tube defects

The process of embryonic development in animals with bilateral symmetry is traditionally divided into phases that occur after fertilization. During ontogeny, a cascade of canonical molecular events is set to modify cellular characteristics such as expression patterns, fate, and three-dimensional organization. The formation of the Central Nervous System (CNS) is a complex process known as neurulation that initiates during the early phases of embryonic development (Fig. 1). This process encompasses two consecutive and coordinated phases referred to as *primary* and *secondary* neurulation and begins approximately 21 days post-conception when the neurula undergoes morphogenetic changes that will shape what will become the CNS. Primary neurulation gives rise to the neural plate rostrally, while secondary neurulation determines the formation of the posterior neural tube. Primary and secondary neurulation are highly conserved processes in mammals, although there are some differences in the precise sequence of neural tube closure events (Copp, Greene, & Murdoch, 2003).

The neural plate originates from thickening of the dorsal surface ectoderm that folds and merges into the midline to create the neural tube, the structure that serves as the embryonic precursor for the brain and the spinal cord. In mammals, closure of the neural tube starts sequentially at different levels of the body axis (multisite closure). Primary neural tube closure is a discontinuous process that initiates at different points along the rostral-caudal axis, spreads bidirectionally in the rostral neural tube, extending caudally towards the spinal region. This multisite, progressive fusion which approximates zippering, proceeds along the spine and ultimately closing the posterior neuropore at the level of the second sacral segment, completing neurulation (Copp & Greene, 2010). In secondary neurulation, the transition of undifferentiated mesenchymal to epithelial cells allows formation of neural epithelium (Shimokita & Takahashi, 2011). The neural tube closes toward the caudal end of the embryo through a process that involves cavitation within the mesenchymal cell mass. The secondary neurulation or *canalization*, where a multipotential cellular population in the caudal eminence begins to differentiate into future neuronal cells, permits the formation of the spinal cord at the lower sacral and caudal levels. Therefore, the formation of a sacral caudal lumen occurs which will eventually join with the newly established primary lumen. This series of morphogenetic movements during neurulation are regulated by genetically encoded modules and mechanical forces generated by traction through cell-matrix adhesions and tension through cell-cell adhesions.

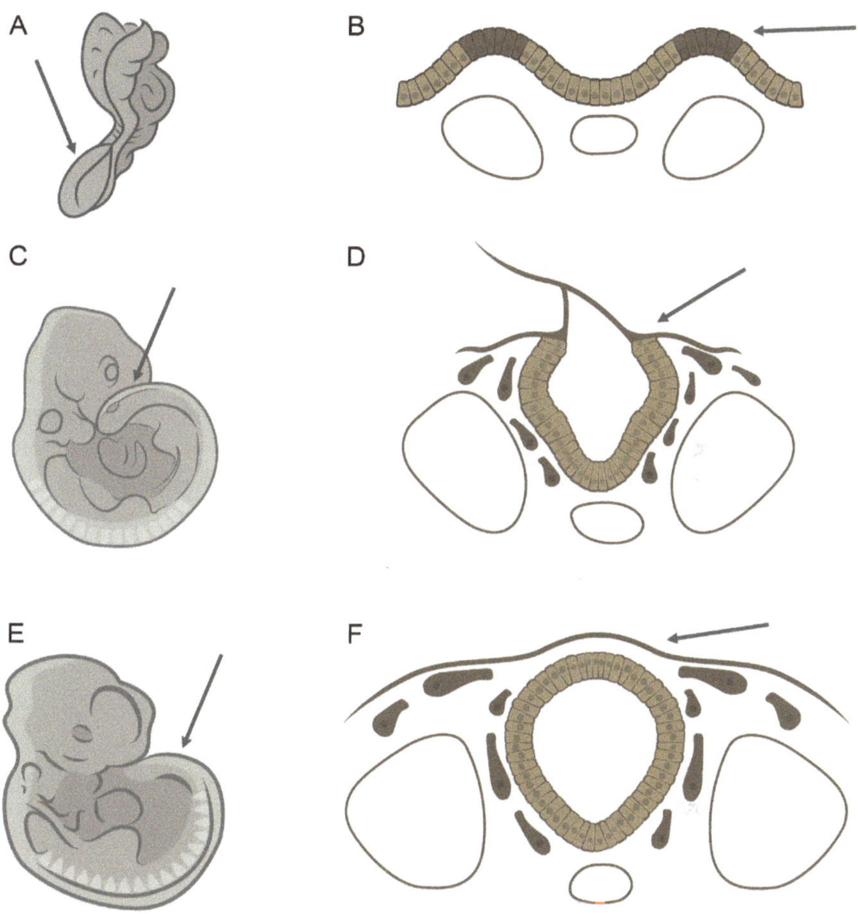

Fig. 1 Neural progenitor cells derived from the ectoderm will form the neural tube. (A) Dorsal view of an E8.5 mouse embryo showing the beginning of neural tube closure; (B) Transversal section at the dorsal region of an E8.5 mouse embryo showing the squamous surface ectoderm, the neural crest and the pseudostratified neuroepithelium; (C) Lateral view of an E9.5 mouse embryo showing the final stage of neural tube closure; (D) Transversal section at the dorsal region of an E9.5 mouse embryo highlighting two dorsolateral hinge points at the pseudostratified neuroepithelium converging towards a ventral hinge point. (E) Lateral view of an E10.5 mouse embryo showing a complete neural tube closure; (F) Transversal section at the dorsal region of an E10.5 mouse embryo presenting a closed neural tube surrounded by streams of migrating neural crest toward the notochord.

As the apical leading edges of the neural plate grow, the squamous surface ectoderm and the pseudostratified neuroepithelium remodel, forming *de novo* adhesions to create a continuous epithelium. As neurulation proceeds, protrusions consisting of long filopodia extensions and membrane ruffles, can be observed along the edges of both open neural folds, growing to guide neural tube closure. This series of tissue remodeling events are synchronized as the neural plate grows and bends at dorsolateral hinge points. The neuroepithelial cells within the median hinge point undergo relatively slow proliferation, with a 7 h cell cycle length while the dorsolateral neuroepithelium undergo rapid proliferation, with a 4 h cell cycle length. As the cell numbers increase to a greater extent in the dorsolateral neural plate, which is contacting the surface ectoderm, compared with the ventrally located neural plate cells, in contact with the paraxial mesoderm and notochord, there is an increase in cell density dorsolaterally compared with the more ventromedial neural plate.

3.1 The importance of cilia and adherens junctions for neural tube closure

As the neural folds elevate, the most apical cells from the neural plate coordinate closure together with the dorsally adjacent layer of non-neural ectodermal cells. The adhesion between the two nascent folds is initiated *via* activation of integrin β1 and is anchored to a zippering point rich in fibronectin, where novel cell-cell junctions must be established (Galea et al., 2017). A cellular structure resembling a rosette initiates the contacts between the two sides of the neural fold and focal anchorage at the fusion site promotes local constriction of the surface ectoderm, bringing the apex of the two neural folds together. A supracellular F-actin network forms a biomechanical structure at the zippering point, which includes an actin cable running along the neural fold tips. When the neural folds are close enough to reach the point of fusion, the actin cable forms a continuous ring around the neuropore, closing a zippering point and giving rise to a new caudal-to-rostral zippering point (Zhou et al., 2020). Many spina bifida cases correspond to this anatomic portion of the neural tube.

4. Genetic risk factors for neural tube defects

NTDs are thought to result from multiple factors with both genetic and environmental contributions (Blom, Shaw, den Heijer, & Finnell, 2006; Lundberg, Wing, Xiong, Zhao, & Finnell, 2003; Wolujewicz et al., 2021). Evidence supporting a genetic basis for NTDs is based on epidemiological studies, including the high concordance of NTDs observed between monozygotic twins (7.7%) when compared to like sex/dizygotic twins (4.0%) (Blencowe et al., 2018; Wallingford et al., 2013). The recurrence risk for NTDs is only 1 in 20, and even after two NTDs affected pregnancies the recurrence risk remains 10%, which strongly argues against a monogenic causation. There have been a number of case-control studies that focus on one or more candidate genes used to identify alleles suggestive of an association with increased NTD risk. Often times the candidate genes for interrogation were selected either on the basis of previously published mouse studies, or human studies involving very limited NTDs cohorts that focused on planar cell polarity (Kibar, Capra, & Gros, 2007; Lei et al., 2013; Lei et al., 2019) or WNT signaling pathway genes that showed some association with NTDs risk (Lei et al., 2015).

Mutations in well over 240 genes produce NTD phenotypes in mice (Harris & Juriloff, 2010), which demonstrates the involvement of numerous genetic pathways in the etiology of these malformations. It is also reasonable to consider these as candidate genes for larger human genome studies. Mouse studies revealed that the genetic background has a major impact on the penetrance of the NTDs phenotype in individuals with presumed NTDs causative mutations (Juriloff et al., 2001; Korstanje et al., 2008). While nearly all mouse null mutants (>90%) produce multiple affected structures with a high penetrance in homozygotes, it is important to note that some mutations cause NTDs in digenic, trigenic, and oligogenic combinations, which likely parallels the genetic etiology of NTDs in humans (Chen et al., 2018; Wang et al., 2018). We believe that human NTDs most often arise through an omnigenic interplay of deleterious genetic variants and environmental factors influencing the function of core pathways, such as one-carbon metabolism (Chen et al., 2018; Lemay et al., 2019; Wallingford et al., 2013). Despite considerable evidence of a contribution of genetic factors to NTDs in humans, from a clinical perspective, there are no known variants in actionable single genes that result in non-syndromic NTDs; therefore, they are of limited and express practical value in the management of high-risk pregnancies. The underlying genetic complexity of NTDs and the role of environmental interactions that we will now discuss make future efforts to reduce the prevalence of the FA resistant NTDs a very challenging proposition. Future prevention strategies targeting these resistant NTDs cases warrant advanced precision medicine approaches tailored to evaluate individual risks for mothers and infants, based upon the ever-increasing understanding of the genomic architecture underlying NTDs.

5. Environmental factors contributing to the etiology of neural tube defects

5.1 Arsenic

Arsenic is a global environmental risk factor for NTDs as well as other health concerns. Over 300 million people from 50 nations are at risk of arsenic exposure, primarily through drinking water, meat, seafood, fruit, and vegetables contamination (Sanyal, Bhattacharjee, Paul, & Bhattacharjee, 2020). The World Health Organization (WHO) safety standards recommend that arsenic levels in drinking water should be no >10 μg/L. However, in some countries, especially in south Asia, including Bangladesh where 42% of water wells exceed the WHO standard, and many other regions, the inorganic arsenic concentrations in drinking water are above 1000 μg/L, yielding levels that exceed by 100-fold the WHO standard (Chakraborti et al., 2010). Arsenic exists in both inorganic and organic forms. While inorganic arsenic such as sodium arsenite (AsIII) and calcium arsenate (AsV) exist in underground and drinking water, organic forms of arsenic, including methylarsonic, dimethylarsinic acid, arsanilic acid and arsenobetaine are most commonly found in seafood, fruits, and vegetables (Naujokas et al., 2013). In addition, arsenic has many significant agricultural applications as a pesticide or herbicide, being also used therapeutically for the treatment of multiple human diseases. Arsenic exposure is known to cause DNA damage, epigenetic changes, and gene expression disruptions. Multiple studies in both animal models and human cohorts were performed to explore the possible association between arsenic exposure and increased risk for NTDs.

The most widely used animal models for NTDs experiments involve mice, due to a comparable neural tube closure process to humans, and their well annotated genomes. Several studies explored arsenic exposure and NTDs using different highly inbred mouse strains. Wlodarczyk, Bennett, Calvin, and Finnell (1996) found that LM/Bc mice exposed to arsenate at 40 mg/kg by intraperitoneal injection on gestation days 7.5 and 8.5, resulted in a 90–100% response frequency of NTDs. Further investigation indicated that *Bcl2* and *P53* gene expression was upregulated by arsenate exposure, suggesting that arsenate exposure could inhibit cell proliferation in the developing embryos. Another study by the same group took advantage of the folate receptor beta (Folr2) mutant mouse model and tested the effect of arsenate on NTD development by evaluating gene expression following exposure to this heavy metal (Wlodarczyk et al., 2006). This study demonstrated that arsenate administration increased exencephaly rates in *Folr2* nullizygotes to 40% compared with 24% exencephaly phenotypes observed in arsenate exposed wildtype mice. When the Folr2 mice are maintained on a low folate diet (0.3 ppm FA) *in utero* exposure to arsenate by *Folr2* nullizygous increase frequency of NTDs affected mice to 64%, implying that multiple environmental risk factors could interact with genetic factors to increase risk of NTDs. Gefrides, Bennett, and Finnell (2002) investigated arsenic interactions with Splotch mice bearing a Pax3 mutation and found that arsenic increased the frequency of NTDs and embryonic death in the nullizygous Splotch embryos, while maternally administered folinic acid failed to ameliorate the teratogenic effect of this heavy metal. A partial rescue of arsenate induced NTDs was obtained experimentally using an insulin pellet and *N*-acetyl cysteine (NAC) in mice (Hill, Wlodarczyk, Mitchell, & Finnell, 2009). Beyond just mouse models, arsenate exposure also increased the risk of NTDs in rats (Beaudoin, 1974), hamsters (Carpenter, 1987) and chick embryos (Peterková & Puzanová, 1976).

Although several hundred million individuals are exposed to arsenic contaminated water and food, human epidemiological studies have not always demonstrated positive associations between arsenic exposure and increased NTDs incidence (Table 1). Various epidemiological studies collected different biological materials for arsenate exposure analyses, and these materials included toenails, hair, blood, plasma, and placental tissues. The studies often suffered from small sample sizes and only limited arsenic exposure, which led to negative association results between arsenate and NTDs (Ovayolu et al., 2020; Özel et al., 2019). Nonetheless, large cohort studies collected from highly arsenate contaminated regions, such as Bangladesh and the northern regions of China more often showed positive associations (Tindula et al., 2021; Wang et al., 2019). A case-control study in Texas focused on maternal arsenate exposure and birth defects revealed a trending risk of NTDs after arsenic exposure (Brender et al., 2006). Several studies conducted in Bangladesh have been very informative in establishing the association between arsenic and an increased risk of NTDs (Carter, Natiella, Baier, & Natiella, 1984; Kancherla et al., 2017; Mazumdar et al., 2015; Obrycki et al., 2019). One of the larger studies was performed in China and included 263 case-mothers who gave birth to an NTD affected child and 511 control-mothers. Arsenic analysis from hair samples demonstrated a crude odds ratio (OR) of 1.68 (95% CI: 1.24–2.27) and an adjusted OR of 1.32 (0.91–1.92) (Wang et al., 2019), indicating that arsenic interacted with some confounding factor to increase NTD risk in this cohort.

There are several suggested mechanisms by which *in utero* arsenic exposure can potentially increase the risk of NTDs. Arsenic exposure is known to inhibit both DNA methylation and DNA repair. Studies have shown that arsenic exposure decreases the activity of S-adenosyl methionine (SAM) as well as DNA methyltransferase (DNMT1 and DNMT3b) by 34.1% and 60.2%, respectively, thus inhibiting DNA methylation (Han, Song, Cui, Xia, & Ma, 2011). Gene inactivation studies in mice demonstrated that

Table 1 Association studies of arsenic and human Neural Tube Defects.

Number	Nation/ Country	Cases, n	Controls, n	Tissue/material	Associate OR (95%CI)	PMID	Journal	Year
1	Bangladesh	155	123	Toenail(mom)	1.74(1.26–2.42)	34358915	Environ Int	2021
2	Japanese	36	39	Amniotic fluid	No association	31743503	Congenit Anom	2020
3	Chinese	263	511	Hair (mom)	adj1.32(0.91–1.92); crude 1.68(1.24–2.27)	30807959	Environ Int	2019
4	Turkey	21	21	Blood	No association	29471703	J Matern Fetal Neonatal Med	2019
5	Turkey	100	70	Plasma	Yes	29199526	J Matern Fetal Neonatal Med	2019
6	Bangladesh	45	40	Toenail	2.19(sd3.96)	28387569	Epigenetics	2017
7	Bangladesh	54	55	(drinking) water	Yes	26250961	Birth Defects Res A Clin Mol Teratol	2015
9	China	80	50	Placental	0.88(0.43–1.78)	23164984	Reprod Toxicol	2013

inactivating the *Dnmt3b* gene produced fetuses with NTDs (Okano, Bell, Haber, & Li, 1999). Arsenic exposure results in the generation of reactive oxygen and nitrogen species (ROS and RNS), which can directly damage DNA and modify cysteine residues in proteins. Arsenic exposure was demonstrated to decrease *APE1*, *DNAligase 1 (LIG1)*, *OGG1*, *PARP1* and *DNA polymerase β* (*POLβ*) which play roles in base excision repair (BER) pathways (Osmond, Kunz, & Snow, 2010). Exposure to arsenic in drinking water was also found to reduce mRNA in *ERCC1*, *XPB*, and *XPE* gene expression, which are important for nucleotide excision repair (NER) (Andrew, Karagas, & Hamilton, 2003). In addition to BER and NER, exposure to arsenic was found to inhibit the DNA double-strand breaks (DSB) repair pathway. It was reported that arsenite inhibits H4K16Ac by binding to the zinc finger motif of two MYST family histone acetyltransferase TIP60 and hMOF, and arsenite was shown to inhibit H2BK120ub catalyzed by RNF20-RNF40 histone E3 ubiquitin ligase in a similar fashion, thereby diminishing the recruitment of BRCA1 and RAD51 to DSB sites for repair (Zhang et al., 2014). Therefore, arsenite could disrupt DSB repair by inhibiting histone epigenetic modifications, which leads to compact chromatin structures, unfavorable for DNA DSB repair (Tam, Price, & Wang, 2020). It was reported that *de novo* rare damaging variants were associated with NTDs cases. Thus, there are multiple gene-arsenic interactions that could potentially disrupt neural tube closure and increase the risk for NTDs.

There are other heavy metals which are associated with increasing the risk of NTDs in exposed embryos. These include lead (Pb), cadmium (Cd), and trace element copper (Cu). It was also shown that low plasma levels of trace elements zinc (Zn) and selenium (Se) were found to be risk factors for NTDs (Demir et al., 2019).

5.2 Anti-seizure medications

Anti-seizure medications (ASMs) are used to treat epilepsy and a number of other non-epileptic conditions (Spina & Perugi, 2004), with clinical efficacy in migraine prophylaxis, neuropathic pain, fibromyalgia, bipolar disorder, obesity and other disorders. The chronic use of these substances constitutes the mainstay treatment strategy to adequately control seizures and provide good therapeutic outcome to other conditions, raising serious concerns for pregnant women and those of childbearing age due to a significant increase in the risk of birth defects. Thus, treatment strategies should consider the benefits of preventing major maternal seizure events vs risks of major congenital malformations, prioritizing medications with lower risk of an adverse pregnancy outcome such as birth defects.

There is extensive literature linking first and second generation ASMs such as carbamazepine, phenobarbital, phenytoin, and valproic acid to an increased risk of birth defects (Tomson, Xue, & Battino, 2015). In addition, newer generation ASMs such as Lamotrigine (Lamictal, FDA-approved in 1994) and Topiramate (Topamax, FDA-approved in 1996), have also been associated with the induction of major congenital malformations in infants exposed *in utero*. Prospective cohorts derived from the European Registry of Antiepileptic Drugs and Pregnancy (EURAP), North American Anti-Epileptic Drug Pregnancy Registry (NAAPR), and United Kingdom of Great Britain (UK) and Ireland registries demonstrated a higher NTDs-prevalence among newborns in the presence of monotherapy with either valproic acid or carbamazepine (Tomson et al., 2015). In addition, combination therapies that include valproic acid were linked to the highest prevalence of major birth defects (Vajda et al., 2010), particularly NTDs, surpassing the unfavorable outcome of birth defects in children born from mothers exposed to other therapeutic regimens (Tomson & Battino, 2012). Therefore, clinical data with respect to NTDs demonstrates that fetuses exposed *in utero* to valproic acid are at a greater risk when compared to other commonly prescribed ASMs including lamotrigine, phenobarbital, phenytoin, levetiracetam, oxcarbazepine and topiramate. In fact, there is an approximately sevenfold greater risk of a child being affected by NTDs following prenatal VPA-exposure compared to other ASMs (Tanoshima et al., 2015).

5.2.1 Valproic acid

Valproic acid (VPA) was initially synthesized in 1882 (Tomson, Battino, & Perucca, 2016), but was only introduced for clinical use as an anticonvulsant in Europe and in the US market about one century later and remaining to this date as one of the most widely prescribed ASMs due to its efficacy and broad anticonvulsant spectrum (Romoli et al., 2019). After the first VPA associated NTD case was reported in the early 1980s (Dalens, Raynaud, & Gaulme, 1980); Robert and Guibaud (1982) reported in their epidemiological survey that out of 72 newborns affected by NTDs, 9 were born to epileptic mothers receiving VPA treatment. Thereafter, a considerable number of studies indicated that the use of VPA during pregnancy increased the NTD risk. Further work demonstrated the presence of a dose-dependent association between the use of VPA and birth defects (Omtzigt et al., 1992).

Clinical evidence demonstrates that VPA poses the greatest teratogenic risk among the clinically available ASMs (Tomson et al., 2018). It is also important to note that VPA exposure in pregnancy also increases the risk of other congenital malformations and neurodevelopment problem, including cleft palate, heart defects and autism spectrum disorders (Reynolds & Green, 2020), and studies indicate that VPA can interact with both genetic and environment factors to modify risk of NTDs. Despite previous attempts and decades of research, the underlying mechanisms by which VPA and other ASMs exert their teratogenic effects are complex and yet to be fully elucidated.

Studies using animal models have attempted to elucidate birth defect mechanisms caused by valproic acid and other ASMs, allowing for a better understanding of their teratogenic effects. Throughout the years, our laboratory has concentrated efforts in (1) designing experimental models to study the teratogenicity of VPA and VPA-analogs, (2) better understanding the mechanisms that underly VPA-teratogenicity, (3) identifying the genetic factors in mice that regulate susceptibility to VPA induced birth defects. In our earlier work (Finnell, Bennett, Karras, & Mohl, 1988), we demonstrated that although both VPA and the VPA-metabolite, 4-propyl-4-pentenoic acid induced NTDs in different inbred mouse strains, NTDs-susceptibility varied among strains under the same experimental conditions, demonstrating that different genetic backgrounds would account for the variation in the predisposition of inbred mouse strains to different NTDs-phenotypes. Subsequently, our group evaluated the expression of Pax-3, Emx-1, Emx-2, c-fos, c-jun, creb, bcl-2, p53, wee-1 during neural tube closure, to assess their impact on susceptibility to VPA induced teratogenicity. An overexpression of Emx-1, Emx-2, c-fos, c-jun, creb, bcl-2, p53 was observed, but gene expression ratio-based analysis favored bcl-2, an apoptosis regulator gene. Results suggested the altered expression of multiple genes may underly the plausible cause of VPA-induced malformations (Wlodarczyk, Craig, et al., 1996). Lundberg et al. (2004) performed linkage analysis and fine gene mapping to identify genetic susceptibility loci responsible for VPA sensitivity by using two inbred mouse strains with different sensitivity to VPA-induced NTDs, the SWV/Fnn, which was highly sensitive to VPA-induced NTDs and C57BL/6J, a VPA resistant strain. This work identified a 3.3 Mb region between *D7Mit285* (GRCm39/mm39, chr7:122085547–122085652) and *D7Mit101* (GRCm39/mm39, chr7: 125232211–125232299) that significantly associated with VPA-induced NTDs. Interesting genes mapped to this region include Tbx6 (T-box 6), Mapk3 (mitogen activated protein kinase 3),

Pkcβ (protein kinase C β subunit), all genes with potentially important roles in early embryonic development (Lundberg et al., 2004).

In 2001, two studies determined that VPA defines a novel class of histone deacetylase inhibitor (Göttlicher et al., 2001; Phiel et al., 2001). VPA can bind to the catalytic center of HDACs which inhibits HDAC activity both *in vitro* and *in vivo* (Göttlicher et al., 2001). Through HDAC inhibition, VPA activates Wnt-dependent gene expression, including β-Catenin, AP1 and Tcf/Lef dependent transcription (Phiel et al., 2001). It is well known that Wnt pathway signaling plays an important role in neural tube closure. Therefore, HDAC inhibition of Wnt pathway signaling could be a potential mechanism by which VPA increases the risk of NTDs. Given the pharmaceutical nature of VPA, it is highly plausible that enzymes and metabolites involved during its biotransformation may modify its effect on NTDs induction. Using unbiased, whole embryo metabolomic screening of mouse embryos from SWV/Fnn and C57BL/6J mice, Steele and colleagues identified metabolites of differential abundance between the two strains with and without VPA exposure. They found that lipid metabolism, carnitine metabolism, and a few specific amino acids (e.g., cysteine and methionine) differed significantly between these two strains. Hypotaurine was found to be significantly decreased in the susceptible (SWV/Fnn) strain regardless of VPA status. Maternal supplementation of hypotaurine to SWV/Fnn mice significantly reduced the occurrence of VPA-induced-NTDs by 33%, while L-carnitine and coenzyme-A decreased resorption rates but not VPA-induced NTDs (Steele et al., 2022).

Identifying the means to prevent VPA-induced birth defects is important due to its widespread therapeutic applications in clinical medicine. A clear understanding of the inherent mechanisms underlying VPA teratogenicity may open new venues in the design of compounds to safely treat seizure disorders while minimizing the inherent teratogenic risks. Several studies explored whether folate supplementation could rescue or decrease VPA induced NTDs using different models. The conclusions are inconsistent. Umur et al. (2012) used the chick as a model to study FA preventive effect on VPA teratogenic effect. They found that FA effectively prevented the teratogenic influence of VPA on chick embryos at neurulation stages by stopping a cascade of apoptosis before caspase 3 expression. Turgut, Kazan, Cakin, and Ozak (2019) also used chick embryos to investigate the effect of FA on preventing VPA teratogenicity. Although the number of embryos with NTDs decreased with concomitant FA administration, this difference did not reach statistical significance. Akimova et al. (2017) used

SWV mice to test FA supplementation effect and found that VPA-induced NTDs response frequencies was not diminished. Contrary to this finding, Bold and coworkers tested FA for its impact on embryonic development using ICR mice and found that FA (3 mg/kg/day) supplementation during gestational days 6–10 reduced the prevalence of VPA-induced NTDs (Bold, Sakata-Haga, & Fukui, 2018). Fathe, Palacios, and Finnell (2014) explored whether VPA could negatively affect folate receptors (Folr1, Folr2 and bovine folate binding protein) binding folate as a substrate competitor. They found that VPA at IC50 concentrations significantly reduces receptor affinity for folates. The investigators collected supernatant from HEK293T cells which were folate starved and then exposed to either FA or FA and VPA. As the VPA concentration was increased, the folate concentration in the supernatant decreased (Fathe et al., 2014). Although this study provides good evidence suggesting that FA supplementation is effective in preventing VPA induced NTDs in model organisms, there is no evidence that FA is effective in preventing VPA-induced NTDs in humans (Ban et al., 2015; Jentink, Bakker, Nijenhuis, Wilffert, & de Jong-van den Berg, 2010). More studies are required to find effective ways to prevent VPA-induced NTDs in humans using precision medicine paradigms.

5.3 Polycyclic aromatic hydrocarbons and neural tube defects

Polycyclic aromatic hydrocarbons (PAHs) are a group of semi-volatile organic compounds (SVOCs) that are present in crude oil and are formed during the incomplete combustion of gas, coal, garbage, creosote, and other organic substances, as well as from motor vehicle exhaust. Our laboratory, in collaboration with other investigators, have demonstrated an association between maternal placental PAH concentrations and an increased risk of NTDs to the offspring (Ren et al., 2011). We also demonstrated that PAHs (BaP and 1-Hydroxy-2-naphthoic acid) exposure of female mice increased NTDs incidence (Conde-Vancells et al., 2018; Lin et al., 2018). PAH-metabolites form reactive oxygen species (ROS) that can create DNA adduct formation (Wilk et al., 2013). Furthermore, accumulated cellular ROS can result in oxidative DNA damage, producing 8-hydroxy-2′-deoxyguanosine (8-OHdG), a biomarker for oxidative stress. The presence of 8-OHdG often results in GC-TA transversions. Accumulated ROS can also create DNA double-strand breaks (DSBs) (Woodbine, Brunton, Goodarzi, Shibata, & Jeggo, 2011). Recent studies showed that PAH exposure significantly altered genome-wide DNA methylation in NTDs fetus

(Wang et al., 2017). Specifically, a notable decrease in genomic DNA 5-hydroxymethylcytosine (5hmC) was found in NTDs which were exposed to PAHs, when compared to controls (Huang, Lin, Jin, Wang, & Ren, 2019).

Possible genes that interact with PAHs in risk regulation of NTDs include the PAH receptor, the aryl hydrocarbon receptor (AHR) and the cytochrome P450 enzymes, including CYP1A1, CYP1A2 and CYP1B, enzymes responsible for PAH metabolism. A case-control study performed by Wang and colleagues evaluated 534 case-mothers and 534 control-mothers and determined that the rs2855658 variant in the CYP1B1 gene can modify PAH-related risk of NTDs (Wang et al., 2014). AHR signaling is known to be regulated by aryl hydrocarbon nuclear translocator (ARNT) and transcription factor PAX3. Kozak et al. (Kozak, Abbott, & Hankinson, 1997) used homologous recombination technology to generate aryl hydrocarbon nuclear translocator (Arnt) null mice. In their study, homozygous (Arnt−/−) mice presented with NTDs, demonstrating the important role of Arnt in neural tube closure. Lin and colleagues found that mean methylation intensity in the body region of PAX3 in fetal neural tissues was positively associated with median concentrations of PAH in the maternal serum. They confirmed that BaP-induced NTDs were associated with hypermethylation of specific CpG sites within both the promoter and body region of Pax3 (Lin et al., 2019).

6. Organoid model systems to study neural tube defects

Ramon y Cajal established the cornerstones of modern neuroscience by providing the morphological evidence of free termination at the extremities of neuronal cells, late in the 19th century, when methodologies for neuro-histological studies were emerging for the first time (López-Muñoz, Boya, & Alamo, 2006). No other organs in the human body are as intricate and with such a high level of cellular diversity, connectivity, and morphology as the brain and the spinal cord. Still today, studying the nervous system remains a challenge, and much is yet to be elucidated on neural tube closure and development, function of specialized brain regions, and behavior of neuronal cells after environmental exposition to a plethora of compounds. Despite these limitations, much has been accomplished through technological advancements in this new era of digital biology. The technology provided by differentiating stem cells into 3D organoid structures is a breakthrough that allows the investigation of several aspects, not easily provided by animal models or conventional 2D cell culture systems, to understand how environmental exposure may affect the genetic networks driving neural tube closure and development, testing the activity of specific compounds on neuronal cells, or using high-throughput screenings to understand the effects of drugs before clinical trials (Wu, Peng, Finnell, & Zheng, 2021).

The nigrostriatal pathway is a brain dopaminergic circuit that connects the human mesencephalon to the prosencephalon. Renner et al. introduced an automated system for human midbrain organoid production, used to evaluate the effect of 84 compounds present in an NIH drug library, which includes pesticides, drugs, flame retardants, and also non-toxic controls. Their initial approach screened compounds presenting general neurotoxicity and specific dopaminergic toxicity, allowing the identification of the flame retardant 3,30,5,50-tetrabromobisphenol A (TBBPA) as a selective toxicant for dopaminergic neurons, and also identifying two other compounds with specific dopaminergic neuron toxicity: the organometallic compound methyl mercuric (II) chloride and the isoquinoline alkaloid berberine chloride. After comparing the automated system used with 2D cultures, the group detected differences in toxicity values in between conventional and automated systems, showing that the screening of toxicity using automated human midbrain organoids is feasible, and produces improved estimates of human toxicity when compared to the existing 2D cell culture systems (Renner et al., 2021).

Cadmium is a toxic heavy metal present in the environment and has no beneficial function in the human body. Oral exposure to Cadmium can cause kidney, liver, and brain damage in humans, while maternal exposure is associated with fetal growth restriction and congenital malformations. Yin, Zhu, Wang, and Qin (2018) developed brain organoids on a micropillar chip to study the effects of cadmium exposure. The device allowed high-throughput generation of brain organoids, and after cadmium treatment, these brain organoids exhibited increased cell death, impaired neural differentiation and abnormal neuron maturation indicating the potential causes for various learning and behavioral deficits observed after cadmium intoxication. Huang et al. (2021) also established a protocol to generate brain organoids to simulate the developmental neurotoxicity of maternal exposure to cadmium. On their experiments, when cerebral organoids were exposed to cadmium, apoptosis was induced in a dose-dependent manner, inhibiting neural progenitor cell proliferation. The group did not find a significant effect on neural differentiation but showed that cadmium exposure induced neuroinflammation and gene dysregulation, up-regulating mineral absorption pathways and impairing ciliogenesis.

7. Folate deficiencies and neural tube defects

Folate deficiency was known to be a risk factor for NTDs in the early 1950s when an increase in the incidence of NTDs was demonstrated as a result of exposure to the folate antagonist aminopterin in animal models and in human fetuses (Thiersch, 1952; Thiersch & Philips, 1950). Later, several landmark trials from 1960s to1980s supported these findings (Warkany, Beaudry, & Hornstein, 1959). In 1991, the Medical Research Council Vitamin Study published the first large randomized trial showing a strong effect of FA supplementation in preventing NTDs (MRC Vitamin Study Research Group, 1991). The trial was launched in 1983 and ended in 1991 after conclusive results were achieved. During this period, a total of 1817 women who had a previous pregnancy affected by a NTD were considered a high-risk group and were randomly allocated into one of the four study groups. A total of 1195 out of 1817 had a completed pregnancy with a known outcome status; 27 pregnancies resulted in NTDs, out of which 6 belonged to the 2 FA supplementation groups, and 21 integrated the 2 other remaining groups, representing a protective effect of 72%. Subsequently, several other large, randomized trials were conducted, including the US-China study (Berry et al., 1999), the Canadian study (De Wals et al., 2007), and the Hungary trial (Czeizel & Dudás, 1992), all supporting the conclusion that periconceptional folate supplementation could significantly decrease the prevalence of NTDs by 30–78%.

Despite multiple worldwide studies supporting evidence that FA supplementation during pregnancy can promote a reduction in the incidence of NTDs, the underlying mechanism by which FA prevents these congenital malformations is still not fully understood. Previous work has focused on folate transporter/receptors and/or folate metabolism research in order to understand how FA supplementation rescues NTDs. There are several known folate receptors or transporters, including the folate receptor alpha (FOLR1), reduced folate carrier (RFC; Slc19A1), the proton coupled folate transporter (PCFT; SLC46A1) and the mitochondrial folate transporter (MFT, SLC25A32). The FOLR1 has a very high affinity to FA and 5MTHF and is predominantly expressed in organs that require more folate, such as the brain, placenta, and kidney. The RFC1 is the major uptake route of reduced folates, universally expressed in most of cells along with PCFT, highly expressed in the duodenum and jejunum in humans. The Finnell group generated folate receptor alpha (Folr1) knock out mice and observed that Folr1 null mice presented with 100% penetrant NTDs (Piedrahita et al., 1999), while folinic acid supplementation could rescue normal neural tube closure in the Folr1 mutant mice by 23.5–77.8%, depending on the dose of folinic acid used (Spiegelstein et al., 2004). This research group also demonstrated that inactivation of the MFT gene produced embryos that were 100% exencephalic. FA treatment could not rescue Slc25a32 null mice from developing NTDs; however, formate, the mitochondrial folate metabolite was able to partially rescue Slc25a32 mutants (Kim et al., 2018). The PCFT null mice did not present with NTDs but showed anemia (Salojin et al., 2011), whereas RFC conventional knockout led to embryonic lethality at E6.5, but conditional knockout mice develop with NTDs (Toriyama, Toriyama, Wallingford, & Finnell, 2017).

In vertebrates, both folate and FA must be metabolized to the bioactive form. Initially, both folate and FA are reduced to tetrahydrofolate (THF) by dihydrofolate reductase (DHFR), and subsequently converted to a biologically active form, 5-methy-tetrahydrofolate (5mTHF), by serine-hydroxymethyltransferase (SHMT), and 5,10-methylenetetrahydrofolate reductase (MTHFR). Methylenetetrahydrofolate can also be metabolized to formyltetrahydrofolate by methylenetetrahydrofolate dehydrogenase 1 (MTHFD1). Folate metabolism pathway variants were extensively studied in relationship to risks for human NTDs, and several were found to be modestly associated with an increased risk of NTDs. These variants include MTHFR C677T (rs1801133), MTHFR A1298C (rs1801131), MTRR A66G (rs1801394), RFC1 A80G (rs1051226), MTHFD1 rs2236225 and MTHFD1L rs3832406 (Cai et al., 2019; Etheredge et al., 2012; Parle-McDermott et al., 2009; Zhang et al., 2013).

As folate plays important roles in DNA methylation as well as in the synthesis of nucleotide precursors, it was previously hypothesized that akin to folate status, DNA methylation status could possibly be associated with the etiology of NTDs. Wang and colleagues used an Infinium HumanMethylation 450 array to study methylation profiling from 10 NTD cases and 8 non-malformed controls. Their study revealed that hypermethylation of the tight junction pathway genes, CTNNA1 and MYH2 were associated with the risk of NTDs (Wang et al., 2017). In mice, genome-wide CpG methylation patterns were found to be significantly different in the cerebral tissue of offspring from wild-type dams exposed to an FA-fortified diet, compared with control diet (Barua et al., 2014). Further studies are warranted to better understand the link between FA status, changes in DNA methylation, and NTD prevalence.

Scientific evidence has guided public health measures to encourage that all women obtain adequate levels of FA in their diets. Increasing maternal FA intake is extremely important and can significantly reduce the prevalence of NTDs (Grosse, Berry, Mick Tilford, Kucik, & Waitzman, 2016). For the past 30 years, the Centers for Disease Control and Prevention (CDC), the Food and Drug Administration (FDA), the Health Resources and Services Administration, and the National Institutes of Health have advocated the use of FA supplementation at a daily dose of 400 μg to all women of reproductive age, in addition to consuming a diversified diet high in folate content (CDC, 1991, 1992; Crider et al., 2022). This measure is particularly important for women who have had a previous pregnancy affected by an NTD, due to high recurrency rates. Thus, in these cases daily doses of FA consumption increase to 10-fold the recommended dietary allowance, beginning 1 month before and continuing through the first 3 months of pregnancy.

8. Conclusions

NTDs arise from a complex set of interaction between genetic and environmental risk factors. Type, timing, strength, duration, and level of interaction between these factors dictate the impact on developmental signaling pathways, leading to abnormal embryonic development and the outcome of an NTD affected pregnancy. Some of the main environmental factors implicated in the etiology of NTDs include prenatal exposure to arsenic, antiseizure medications (e.g., valproic acid), polycyclic aromatic hydrocarbons. In addition, folate plays an important role in DNA methylation and nucleotide synthesis. Insufficient dietary folate intake significantly increases the risk of NTDs, preventable in most cases by adequate folate supplementation. While it is well established that periconceptional FA supplementation reduces the risk of NTDs, it is imperative that we figure out how this occurs. It is incredible that folate dose-response studies of NTDs in human populations do not exist and that the optimal maternal dose and target blood levels of folic acid for NTD prevention are unknown for a major public health intervention that is entering its fourth decade. The neural tube closure process of mice is analogous to humans, justifying why these models remain the most widely used models with which to study NTDs, in addition to their well annotated genomes. Despite the body of scientific knowledge that has been amassed throughout the years on the unfavorable gene and environment interactions that cause NTDs, we have yet to fully comprehend these malformations and make them truly preventable.

Acknowledgments

The authors acknowledge the financial support awarded on behalf of Professors Richard H. Finnell (CAPES-PRINT—88887.687698/2022-00) and Cristiane Sá Roriz Fonteles (CAPES-PRINT—88887.685501/2022-00) by the Ministry of Education of Brazil, CAPES Foundation, in order to conduct Senior Visiting Professorship at the Federal University of Ceara, and Baylor College of Medicine, respectively. The authors express their appreciation to Gabriel Tukeman, PhD candidate at the Baylor College of Medicine, for templates and editions of Fig. 1.

Disclosures

Dr. Finnell formerly held a leadership position with TeratOmic Consulting LLC, a now dissolved corporation. He also receives travel funds to attend quarterly board meetings of the Journal of Reproductive and Developmental Medicine.

References

Akimova, D., Wlodarczyk, B. J., Lin, Y., Ross, M. E., Finnell, R. H., Chen, Q., et al. (2017). Metabolite profiling of whole murine embryos reveals metabolic perturbations associated with maternal valproate-induced neural tube closure defects. *Birth Defects Research, 109*(2), 106–119.

Andrew, A. S., Karagas, M. R., & Hamilton, J. W. (2003). Decreased DNA repair gene expression among individuals exposed to arsenic in United States drinking water. *International Journal of Cancer, 104*(3), 263–268.

Ban, L., Fleming, K. M., Doyle, P., Smeeth, L., Hubbard, R. B., Fiaschi, L., et al. (2015). Congenital anomalies in children of mothers taking antiepileptic drugs with and without periconceptional high dose folic acid use: A population-based cohort study. *PLoS One, 10*(7), e0131130.

Barua, S., Kuizon, S., Chadman, K. K., Flory, M. J., Brown, W. T., & Junaid, M. A. (2014). Single-base resolution of mouse offspring brain methylome reveals epigenome modifications caused by gestational folic acid. *Epigenetics & Chromatin*, 7(1), 1–15.

Beaudoin, A. R. (1974). Teratogenicity of sodium arsenate in rats. *Teratology, 10*(2), 153–157.

Berry, R. J., Li, Z., Erickson, J. D., Li, S., Moore, C. A., Wang, H., et al. (1999). Prevention of neural-tube defects with folic acid in China. China-U.S. collaborative project for neural tube defect prevention. *The New England Journal of Medicine, 341*(20), 1485–1490.

Blencowe, H., Kancherla, V., Moorthie, S., Darlison, M. W., & Modell, B. (2018). Estimates of global and regional prevalence of neural tube defects for 2015: A systematic analysis. *Annals of the New York Academy of Sciences, 1414*(1), 31–46.

Blom, H. J., Shaw, G. M., den Heijer, M., & Finnell, R. H. (2006). Neural tube defects and folate: Case far from closed. *Nature Reviews. Neuroscience*, 7(9), 724–731.

Bold, J., Sakata-Haga, H., & Fukui, Y. (2018). Spinal nerve defects in mouse embryos prenatally exposed to valproic acid. *Anatomical Science International, 93*(1), 35–41.

Brender, J. D., Suarez, L., Felkner, M., Gilani, Z., Stinchcomb, D., Moody, K., et al. (2006). Maternal exposure to arsenic, cadmium, lead, and mercury and neural tube defects in offspring. *Environmental Research, 101*(1), 132–139.

Cai, C. Q., Fang, Y. L., Shu, J. B., Zhao, L. S., Zhang, R. P., Cao, L. R., et al. (2019). Association of neural tube defects with maternal alterations and genetic polymorphisms in one-carbon metabolic pathway. *Italian Journal of Pediatrics*, *45*(1), 37.

Canfield, M. A., Collins, J. S., Botto, L. D., Williams, L. J., Mai, C. T., Kirby, R. S., et al. (2005). Changes in the birth prevalence of selected birth defects after grain fortification with folic acid in the United States: Findings from a multi-state population-based study. *Birth Defects Research. Part A, Clinical and Molecular Teratology*, *73*(10), 679–689.

Carpenter, S. J. (1987). Developmental analysis of cephalic axial dysraphic disorders in arsenic-treated hamster embryos. *Anatomy and Embryology*, *176*(3), 345–365.

Carter, J. M., Natiella, J. R., Baier, R. E., & Natiella, R. R. (1984). Fibroblastic activities post implantation of cobalt chromium alloy and pure germanium in rabbits. *Artificial Organs*, *8*(1), 102–104.

Centers for Disease Control (CDC). (1991). Use of folic acid for prevention of spina bifida and other neural tube defects—1983-1991. *MMWR. Morbidity and Mortality Weekly Report*, *40*(30), 513–516.

Centers for Disease Control (CDC). (1992). Recommendations for the use of folic acid to reduce the number of cases of spina bifida and other neural tube defects. *MMWR. Recommendations and Reports: Morbidity and Mortality Weekly Report. Recommendations and Reports*, *41*(RR-14), 1–7.

Chakraborti, D., Rahman, M. M., Das, B., Murrill, M., Dey, S., Chandra Mukherjee, S., et al. (2010). Status of groundwater arsenic contamination in Bangladesh: A 14-year study report. *Water Research*, *44*(19), 5789–5802.

Chen, Z., Lei, Y., Cao, X., Zheng, Y., Wang, F., Bao, Y., et al. (2018). Genetic analysis of Wnt/PCP genes in neural tube defects. *BMC Medical Genomics*, *11*(1), 38.

Conde-Vancells, J., Vazquez-Chantada, M., McCollum, C. W., Bondesson, M., Kalasekar, S. M., Wlodarczyk, B. J., et al. (2018). Combining mouse embryonic stem cells and zebrafish embryos to evaluate developmental toxicity of chemical exposure. *Reproductive Toxicology (Elmsford, N.Y.)*, *81*, 220–228.

Copp, A. J., Adzick, N. S., Chitty, L. S., Fletcher, J. M., Holmbeck, G. N., & Shaw, G. M. (2015). Spina bifida. *Nature Reviews. Disease Primers*, *1*, 15007.

Copp, A. J., & Greene, N. D. (2010). Genetics and development of neural tube defects. *The Journal of Pathology*, *220*(2), 217–230.

Copp, A. J., Greene, N. D., & Murdoch, J. N. (2003). The genetic basis of mammalian neurulation. *Nature Reviews. Genetics*, *4*(10), 784–793.

Crider, K. S., Qi, Y. P., Yeung, L. F., Mai, C. T., Head Zauche, L., Wang, A., et al. (2022). Folic acid and the prevention of birth defects: 30 years of opportunity and controversies. *Annual Review of Nutrition*, *42*, 423–452.

Czeizel, A. E., & Dudás, I. (1992). Prevention of the first occurrence of neural-tube defects by periconceptional vitamin supplementation. *The New England Journal of Medicine*, *327*(26), 1832–1835.

Dalens, B., Raynaud, E. J., & Gaulme, J. (1980). Teratogenicity of valproic acid. *The Journal of Pediatrics*, *97*(2), 332–333.

David, T. J., & Nixon, A. (1976). Congenital malformations associated with anencephaly and iniencephaly. *Journal of Medical Genetics*, *13*(4), 263–265.

De Wals, P., Tairou, F., Van Allen, M. I., Uh, S. H., Lowry, R. B., Sibbald, B., et al. (2007). Reduction in neural-tube defects after folic acid fortification in Canada. *The New England Journal of Medicine*, *357*(2), 135–142.

Demir, N., Başaranoğlu, M., Huyut, Z., Değer, İ., Karaman, K., Şekeroğlu, M. R., et al. (2019). The relationship between mother and infant plasma trace element and heavy metal levels and the risk of neural tube defect in infants. *The Journal of Maternal-Fetal & Neonatal Medicine: The Official Journal of the European Association of Perinatal Medicine, the Federation of Asia and Oceania Perinatal Societies, the International Society of Perinatal Obstetricians*, *32*(9), 1433–1440.

Etheredge, A. J., Finnell, R. H., Carmichael, S. L., Lammer, E. J., Zhu, H., Mitchell, L. E., et al. (2012). Maternal and infant gene-folate interactions and the risk of neural tube defects. *American Journal of Medical Genetics. Part A, 158A*(10), 2439–2446.

Fathe, K., Palacios, A., & Finnell, R. H. (2014). Brief report novel mechanism for valproate-induced teratogenicity. *Birth Defects Research. Part A, Clinical and Molecular Teratology, 100*(8), 592–597.

Finnell, R. H., Bennett, G. D., Karras, S. B., & Mohl, V. K. (1988). Common hierarchies of susceptibility to the induction of neural tube defects in mouse embryos by valproic acid and its 4-propyl-4-pentenoic acid metabolite. *Teratology, 38*(4), 313–320.

Finnell, R. H., Caiaffa, C. D., Kim, S. E., Lei, Y., Steele, J., Cao, X., et al. (2021). Gene environment interactions in the etiology of neural tube defects. *Frontiers in Genetics, 12*, 659612.

Galea, G. L., Cho, Y. J., Galea, G., Molè, M. A., Rolo, A., Savery, D., et al. (2017). Biomechanical coupling facilitates spinal neural tube closure in mouse embryos. *Proceedings of the National Academy of Sciences of the United States of America, 114*(26), E5177–E5186. https://doi.org/10.1073/pnas.1700934114.

Gefrides, L. A., Bennett, G. D., & Finnell, R. H. (2002). Effects of folate supplementation on the risk of spontaneous and induced neural tube defects in splotch mice. *Teratology, 65*(2), 63–69.

Göttlicher, M., Minucci, S., Zhu, P., Krämer, O. H., Schimpf, A., Giavara, S., et al. (2001). Valproic acid defines a novel class of HDAC inhibitors inducing differentiation of transformed cells. *The EMBO Journal, 20*(24), 6969–6978.

Grosse, S. D., Berry, R. J., Mick Tilford, J., Kucik, J. E., & Waitzman, N. J. (2016). Retrospective assessment of cost savings from prevention: Folic acid fortification and Spina bifida in the U.S. *American Journal of Preventive Medicine, 50*(5 Suppl 1), S74–S80.

Han, Z. J., Song, G., Cui, Y., Xia, H. F., & Ma, X. (2011). Oxidative stress is implicated in arsenic-induced neural tube defects in chick embryos. *International journal of developmental neuroscience: The official journal of the International Society for Developmental Neuroscience, 29*(7), 673–680.

Harris, M. J., & Juriloff, D. M. (2010). An update to the list of mouse mutants with neural tube closure defects and advances toward a complete genetic perspective of neural tube closure. *Birth Defects Research. Part A, Clinical and Molecular Teratology, 88*(8), 653–669.

Heseker, H. B., Mason, J. B., Selhub, J., Rosenberg, I. H., & Jacques, P. F. (2009). Not all cases of neural-tube defect can be prevented by increasing the intake of folic acid. *The British Journal of Nutrition, 102*(2), 173–180.

Hill, D. S., Wlodarczyk, B. J., Mitchell, L. E., & Finnell, R. H. (2009). Arsenate-induced maternal glucose intolerance and neural tube defects in a mouse model. *Toxicology and Applied Pharmacology, 239*(1), 29–36.

Huang, Y., Dai, Y., Li, M., Guo, L., Cao, C., Huang, Y., et al. (2021). Exposure to cadmium induces neuroinflammation and impairs ciliogenesis in hESC-derived 3D cerebral organoids. *The Science of the Total Environment, 797*, 149043.

Huang, Y., Lin, S., Jin, L., Wang, L., & Ren, A. (2019). Decreased global DNA hydroxymethylation in neural tube defects: Association with polycyclic aromatic hydrocarbons. *Epigenetics, 14*(10), 1019–1029.

Jentink, J., Bakker, M. K., Nijenhuis, C. M., Wilffert, B., & de Jong-van den Berg, L. T. (2010). Does folic acid use decrease the risk for spina bifida after in utero exposure to valproic acid? *Pharmacoepidemiology and Drug Safety, 19*(8), 803–807.

Juriloff, D. M., Gunn, T. M., Harris, M. J., Mah, D. G., Wu, M. K., & Dewell, S. L. (2001). Multifactorial genetics of exencephaly in SELH/Bc mice. *Teratology, 64*(4), 189–200.

Kancherla, V., Ibne Hasan, M., Hamid, R., Paul, L., Selhub, J., Oakley, G., et al. (2017). Prenatal folic acid use associated with decreased risk of myelomeningocele: A case-control study offers further support for folic acid fortification in Bangladesh. *PLoS One, 12*(11), e0188726.

Kibar, Z., Capra, V., & Gros, P. (2007). Toward understanding the genetic basis of neural tube defects. *Clinical Genetics*, *71*(4), 295–310.

Kim, J., Lei, Y., Guo, J., Kim, S. E., Wlodarczyk, B. J., Cabrera, R. M., et al. (2018). Formate rescues neural tube defects caused by mutations in *Slc25a32*. *Proceedings of the National Academy of Sciences of the United States of America*, *115*(18), 4690–4695.

Korstanje, R., Desai, J., Lazar, G., King, B., Rollins, J., Spurr, M., et al. (2008). Quantitative trait loci affecting phenotypic variation in the vacuolated lens mouse mutant, a multigenic mouse model of neural tube defects. *Physiological Genomics*, *35*(3), 296–304.

Kozak, K. R., Abbott, B., & Hankinson, O. (1997). ARNT-deficient mice and placental differentiation. *Developmental Biology*, *191*(2), 297–305.

Lei, Y., Fathe, K., McCartney, D., Zhu, H., Yang, W., Ross, M. E., et al. (2015). Rare LRP6 variants identified in spina bifida patients. *Human Mutation*, *36*(3), 342–349.

Lei, Y., Kim, S. E., Chen, Z., Cao, X., Zhu, H., Yang, W., et al. (2019). Variants identified in PTK7 associated with neural tube defects. *Molecular Genetics & Genomic Medicine*, 7(4), e00584.

Lei, Y., Zhu, H., Duhon, C., Yang, W., Ross, M. E., Shaw, G. M., et al. (2013). Mutations in planar cell polarity gene SCRIB are associated with spina bifida. *PLoS One*, *8*(7), e69262.

Lemay, P., De Marco, P., Traverso, M., Merello, E., Dionne-Laporte, A., Spiegelman, D., et al. (2019). Whole exome sequencing identifies novel predisposing genes in neural tube defects. *Molecular Genetics & Genomic Medicine*, 7(1), e00467.

Lin, S., Ren, A., Wang, L., Huang, Y., Wang, Y., Wang, C., et al. (2018). Oxidative stress and apoptosis in benzo [a] pyrene-induced neural tube defects. *Free Radical Biology and Medicine*, *116*, 149–158.

Lin, S., Ren, A., Wang, L., Santos, C., Huang, Y., Jin, L., et al. (2019). Aberrant methylation of Pax3 gene and neural tube defects in association with exposure to polycyclic aromatic hydrocarbons. *Clinical Epigenetics*, *11*(1), 13.

López-Muñoz, F., Boya, J., & Alamo, C. (2006). Neuron theory, the cornerstone of neuroscience, on the centenary of the Nobel prize award to Santiago Ramón y Cajal. *Brain Research Bulletin*, *70*(4–6), 391–405.

Lundberg, Y. W., Cabrera, R. M., Greer, K. A., Zhao, J., Garg, R., & Finnell, R. H. (2004). Mapping a chromosomal locus for valproic acid-induced exencephaly in mice. *Mammalian Genome: Official Journal of the International Mammalian Genome Society*, *15*(5), 361–369.

Lundberg, Y. W., Wing, M. J., Xiong, W., Zhao, J., & Finnell, R. H. (2003). Genetic dissection of hyperthermia-induced neural tube defects in mice. *Birth Defects Research. Part A, Clinical and Molecular Teratology*, *67*(6), 409–413.

Mazumdar, M., Valeri, L., Rodrigues, E. G., Ibne Hasan, M. O., Hamid, R., Paul, L., et al. (2015). Polymorphisms in maternal folate pathway genes interact with arsenic in drinking water to influence risk of myelomeningocele. *Birth Defects Research. Part A, Clinical and Molecular Teratology*, *103*(9), 754–762.

Naujokas, M. F., Anderson, B., Ahsan, H., Aposhian, H. V., Graziano, J. H., Thompson, C., et al. (2013). The broad scope of health effects from chronic arsenic exposure: Update on a worldwide public health problem. *Environmental Health Perspectives*, *121*(3), 295–302.

No authors listed. (1991). Prevention of neural tube defects: results of the Medical Research Council Vitamin Study. MRC Vitamin Study Research Group. *Lancet (London, England)*, *338*(8760), 131–137.

Oakeshott, P., Reid, F., Poulton, A., Markus, H., Whitaker, R. H., & Hunt, G. M. (2015). Neurological level at birth predicts survival to the mid-40s and urological deaths in open spina bifida: A complete prospective cohort study. *Developmental Medicine and Child Neurology*, *57*(7), 634–638.

Obrycki, J. F., Lee, J. J., Kapur, K., Paul, L., Hasan, M., Mia, S., et al. (2019). A case-control analysis of maternal diet and risk of neural tube defects in Bangladesh. *Birth Defects Research*, *111*(14), 967–981.

Okano, M., Bell, D. W., Haber, D. A., & Li, E. (1999). DNA methyltransferases Dnmt3a and Dnmt3b are essential for de novo methylation and mammalian development. *Cell*, *99*(3), 247–257.

Omtzigt, J. G., Los, F. J., Grobbee, D. E., Pijpers, L., Jahoda, M. G., Brandenburg, H., et al. (1992). The risk of spina bifida aperta after first-trimester exposure to valproate in a prenatal cohort. *Neurology*, *42*(4 Suppl 5), 119–125.

Osmond, M. J., Kunz, B. A., & Snow, E. T. (2010). Age and exposure to arsenic alter base excision repair transcript levels in mice. *Mutagenesis*, *25*(5), 517–522.

Ovayolu, A., Ovayolu, G., Karaman, E., Yuce, T., Ozek, M. A., & Turksoy, V. A. (2020). Amniotic fluid levels of selected trace elements and heavy metals in pregnancies complicated with neural tube defects. *Congenital Anomalies*, *60*(5), 136–141.

Özel, Ş., Ozyer, S., Aykut, O., Çinar, M., Yılmaz, O. H., Caglar, A., et al. (2019). Maternal second trimester blood levels of selected heavy metals in pregnancies complicated with neural tube defects. *The Journal of Maternal-Fetal & Neonatal Medicine: The Official Journal of the European Association of Perinatal Medicine, the Federation of Asia and Oceania Perinatal Societies, the International Society of Perinatal Obstetricians*, *32*(15), 2547–2553.

Parle-McDermott, A., Pangilinan, F., O'Brien, K. K., Mills, J. L., Magee, A. M., Troendle, J., et al. (2009). A common variant in MTHFD1L is associated with neural tube defects and mRNA splicing efficiency. *Human Mutation*, *30*(12), 1650–1656.

Peterková, R., & Puzanová, L. (1976). Effect of trivalent and pentavalent arsenic on early developmental stages of the chick embryo. *Folia Morphologica*, *24*(1), 5–13.

Phiel, C. J., Zhang, F., Huang, E. Y., Guenther, M. G., Lazar, M. A., & Klein, P. S. (2001). Histone deacetylase is a direct target of valproic acid, a potent anticonvulsant, mood stabilizer, and teratogen. *The Journal of Biological Chemistry*, *276*(39), 36734–36741.

Piedrahita, J. A., Oetama, B., Bennett, G. D., van Waes, J., Kamen, B. A., Richardson, J., et al. (1999). Mice lacking the folic acid-binding protein Folbp1 are defective in early embryonic development. *Nature Genetics*, *23*(2), 228–232.

Practice Bulletin No 187. (2017). Neural tube defects. *Obstetrics and Gynecology*, *130*(6), e279–e290.

Ren, A., Qiu, X., Jin, L., Ma, J., Li, Z., Zhang, L., et al. (2011). Association of selected persistent organic pollutants in the placenta with the risk of neural tube defects. *Proceedings of the National Academy of Sciences of the United States of America*, *108*(31), 12770–12775.

Renner, H., Becker, K. J., Kagermeier, T. E., Grabos, M., Eliat, F., Günther, P., et al. (2021). Cell-type-specific high throughput toxicity testing in human midbrain organoids. *Frontiers in Molecular Neuroscience*, *14*, 715054.

Reynolds, E. H., & Green, R. (2020). Valproate and folate: Congenital and developmental risks. *Epilepsy & Behavior*, *108*, 107068.

Robert, E., & Guibaud, P. (1982). Maternal valproic acid and congenital neural tube defects. *Lancet (London, England)*, *2*(8304), 937.

Rocque, B. G., Bishop, E. R., Scogin, M. A., Hopson, B. D., Arynchyna, A. A., Boddiford, C. J., et al. (2015). Assessing health-related quality of life in children with spina bifida. *Journal of Neurosurgery. Pediatrics*, *15*(2), 144–149.

Romoli, M., Mazzocchetti, P., D'Alonzo, R., Siliquini, S., Rinaldi, V. E., Verrotti, A., et al. (2019). Valproic acid and epilepsy: From molecular mechanisms to clinical evidences. *Current Neuropharmacology*, *17*(10), 926–946.

Salojin, K. V., Cabrera, R. M., Sun, W., Chang, W. C., Lin, C., Duncan, L., et al. (2011). A mouse model of hereditary folate malabsorption: Deletion of the PCFT gene leads to systemic folate deficiency. *Blood*, *117*(18), 4895–4904.

Sanyal, T., Bhattacharjee, P., Paul, S., & Bhattacharjee, P. (2020). Recent advances in arsenic research: Significance of differential susceptibility and sustainable strategies for mitigation. *Frontiers in Public Health, 8*, 464.

Shaw, G. M., Jensvold, N. G., Wasserman, C. R., & Lammer, E. J. (1994). Epidemiologic characteristics of phenotypically distinct neural tube defects among 0.7 million California births, 1983-1987. *Teratology, 49*(2), 143–149.

Shimokita, E., & Takahashi, Y. (2011). Secondary neurulation: Fate-mapping and gene manipulation of the neural tube in tail bud. *Development, Growth & Differentiation, 53*(3), 401–410.

Spiegelstein, O., Mitchell, L. E., Merriweather, M. Y., Wicker, N. J., Zhang, Q., Lammer, E. J., et al. (2004). Embryonic development of folate binding protein-1 (Folbp1) knockout mice: Effects of the chemical form, dose, and timing of maternal folate supplementation. *Developmental Dynamics: An Official Publication of the American Association of the Anatomists, 231*(1), 221–231.

Spina, E., & Perugi, G. (2004). Antiepileptic drugs: Indications other than epilepsy. *Epileptic Disorders: International Epilepsy Journal With Videotape, 6*(2), 57–75.

Steele, J. W., Lin, Y. L., Chen, N., Wlodarczyk, B. J., Chen, Q., Attarwala, N., et al. (2022). Embryonic Hypotaurine levels contribute to strain-dependent susceptibility in mouse models of valproate-induced neural tube defects. *Frontiers in Cell and Development Biology, 10*, 832492.

Tam, L. M., Price, N. E., & Wang, Y. (2020). Molecular mechanisms of arsenic-induced disruption of DNA repair. *Chemical Research in Toxicology, 33*(3), 709–726.

Tanoshima, M., Kobayashi, T., Tanoshima, R., Beyene, J., Koren, G., & Ito, S. (2015). Risks of congenital malformations in offspring exposed to valproic acid in utero: A systematic review and cumulative meta-analysis. *Clinical Pharmacology and Therapeutics, 98*(4), 417–441.

Thiersch, J. B. (1952). Therapeutic abortions with a folic acid antagonist, 4-aminopteroylglutamic acid (4-amino P.G.A) administered by the oral route. *American Journal of Obstetrics and Gynecology, 63*(6), 1298–1304.

Thiersch, J. B., & Philips, F. S. (1950). Effect of 4-amino-pteroylglutamic acid (aminopterin) on early pregnancy. *Proceedings of the Society for Experimental Biology and Medicine. Society for Experimental Biology and Medicine (New York, N.Y.), 74*(1), 204–208.

Tindula, G., Mukherjee, S. K., Ekramullah, S. M., Arman, D. M., Biswas, S. K., Islam, J., et al. (2021). Parental metal exposures as potential risk factors for spina bifida in Bangladesh. *Environment International, 157*, 106800.

Tomson, T., & Battino, D. (2012). Teratogenic effects of antiepileptic drugs. *The Lancet. Neurology, 11*(9), 803–813.

Tomson, T., Battino, D., Bonizzoni, E., Craig, J., Lindhout, D., Perucca, E., et al. (2018). Comparative risk of major congenital malformations with eight different antiepileptic drugs: A prospective cohort study of the EURAP registry. *The Lancet. Neurology, 17*(6), 530–538.

Tomson, T., Battino, D., & Perucca, E. (2016). The remarkable story of valproic acid. *The Lancet. Neurology, 15*(2), 141.

Tomson, T., Xue, H., & Battino, D. (2015). Major congenital malformations in children of women with epilepsy. *Seizure, 28*, 46–50.

Toriyama, M., Toriyama, M., Wallingford, J. B., & Finnell, R. H. (2017). Folate-dependent methylation of septins governs ciliogenesis during neural tube closure. *FASEB Journal: Official Publication of the Federation of American Societies for Experimental Biology, 31*(8), 3622–3635.

Turgut, U., Kazan, S., Cakin, H., & Ozak, A. (2019). Valproic acid effect on neural tube defects is not prevented by concomitant folic acid supplementation: Early chick embryo model pilot study. *International journal of developmental neuroscience: The official journal of the International Society for Developmental Neuroscience, 78*, 45–48.

Umur, A. S., Selcuki, M., Bursali, A., Umur, N., Kara, B., Vatansever, H. S., et al. (2012). Simultaneous folate intake may prevent adverse effect of valproic acid on neurulating nervous system. *Child's Nervous System: Official Journal of the International Society for Pediatric Neurosurgery, 28*(5), 729–737.

Vajda, F. J., Hitchcock, A. A., Graham, J., O'Brien, T. J., Lander, C. M., & Eadie, M. J. (2010). The teratogenic risk of antiepileptic drug polytherapy. *Epilepsia, 51*(5), 805–810.

Wallingford, J. B., Niswander, L. A., Shaw, G. M., & Finnell, R. H. (2013). The continuing challenge of understanding, preventing, and treating neural tube defects. *Science (New York, N.Y.), 339*(6123), 1222002.

Wang, L., Li, Z., Jin, L., Li, K., Yuan, Y., Fu, Y., et al. (2014). Indoor air pollution and neural tube defects: Effect modification by maternal genes. *Epidemiology (Cambridge, Mass.), 25*(5), 658–665.

Wang, L., Lin, S., Zhang, J., Tian, T., Jin, L., & Ren, A. (2017). Fetal DNA hypermethylation in tight junction pathway is associated with neural tube defects: A genome-wide DNA methylation analysis. *Epigenetics, 12*(2), 157–165.

Wang, B., Zhang, Y., Dong, H., Gong, S., Wei, B., Luo, M., et al. (2018). Loss of Tctn3 causes neuronal apoptosis and neural tube defects in mice. *Cell Death & Disease, 9*(5), 520.

Wang, B., Zhu, Y., Yan, L., Zhang, J., Wang, X., Cheng, H., et al. (2019). Association of maternal chronic arsenic exposure with the risk of neural tube defects in Northern China. *Environment International, 126*, 222–227.

Warkany, J., Beaudry, P. H., & Hornstein, S. (1959). Attempted abortion with aminopterin (4-amino-pteroylglutamic acid); malformations of the child. *A.M.A. Journal of Diseases of Children, 97*(3), 274–281.

Wilk, A., Waligórski, P., Lassak, A., Vashistha, H., Lirette, D., Tate, D., et al. (2013). Polycyclic aromatic hydrocarbons-induced ROS accumulation enhances mutagenic potential of T-antigen from human polyomavirus JC. *Journal of Cellular Physiology, 228*(11), 2127–2138.

Wlodarczyk, B. J., Bennett, G. D., Calvin, J. A., & Finnell, R. H. (1996). Arsenic-induced neural tube defects in mice: Alterations in cell cycle gene expression. *Reproductive Toxicology (Elmsford, N.Y.), 10*(6), 447–454.

Wlodarczyk, B. J., Cabrera, R. M., Hill, D. S., Bozinov, D., Zhu, H., & Finnell, R. H. (2006). Arsenic-induced gene expression changes in the neural tube of folate transport defective mouse embryos. *Neurotoxicology, 27*(4), 547–557.

Wlodarczyk, B. C., Craig, J. C., Bennett, G. D., Calvin, J. A., & Finnell, R. H. (1996). Valproic acid-induced changes in gene expression during neurulation in a mouse model. *Teratology, 54*(6), 284–297.

Wolujewicz, P., Aguiar-Pulido, V., AbdelAleem, A., Nair, V., Thareja, G., Suhre, K., et al. (2021). Genome-wide investigation identifies a rare copy-number variant burden associated with human spina bifida. *Genetics in Medicine: Official Journal of the American College of Medical Genetics, 23*(7), 1211–1218.

Woodbine, L., Brunton, H., Goodarzi, A. A., Shibata, A., & Jeggo, P. A. (2011). Endogenously induced DNA double strand breaks arise in heterochromatic DNA regions and require ataxia telangiectasia mutated and Artemis for their repair. *Nucleic Acids Research, 39*(16), 6986–6997.

Wu, Y., Peng, S., Finnell, R. H., & Zheng, Y. (2021). Organoids as a new model system to study neural tube defects. *FASEB Journal: Official Publication of the Federation of American Societies for Experimental Biology, 35*(4), e21545.

Yin, F., Zhu, Y., Wang, Y., & Qin, J. (2018). Engineering brain organoids to probe impaired neurogenesis induced by cadmium. *ACS Biomaterials Science & Engineering, 4*(5), 1908–1915.

Zhang, T., Lou, J., Zhong, R., Wu, J., Zou, L., Sun, Y., et al. (2013). Genetic variants in the folate pathway and the risk of neural tube defects: A meta-analysis of the published literature. *PLoS One*, *8*(4), e59570.

Zhang, F., Paramasivam, M., Cai, Q., Dai, X., Wang, P., Lin, K., et al. (2014). Arsenite binds to the RING finger domains of RNF20-RNF40 histone E3 ubiquitin ligase and inhibits DNA double-strand break repair. *Journal of the American Chemical Society*, *136*(37), 12884–12887.

Zhou, C. J., Ji, Y., Reynolds, K., McMahon, M., Garland, M. A., Zhang, S., et al. (2020). Non-neural surface ectodermal rosette formation and F-actin dynamics drive mammalian neural tube closure. *Biochemical and Biophysical Research Communications*, *526*(3), 647–653. https://doi.org/10.1016/j.bbrc.2020.03.138.

CHAPTER EIGHT

Gene × environment interactions in autism spectrum disorders

Kimberly Keil-Stietz[a] and Pamela J. Lein[b,*]
[a]Department of Comparative Biosciences, University of Wisconsin-Madison, School of Veterinary Medicine, Madison, WI, United States
[b]Department of Molecular Biosciences, University of California, Davis, School of Veterinary Medicine, Davis, CA, United States
*Corresponding author: e-mail address: pjlein@ucdavis.edu

Contents

Abstract

There is credible evidence that environmental factors influence individual risk and/or severity of autism spectrum disorders (hereafter referred to as autism). While it is likely that environmental chemicals contribute to the etiology of autism via multiple

Current Topics in Developmental Biology, Volume 152
ISSN 0070-2153
https://doi.org/10.1016/bs.ctdb.2022.11.001

mechanisms, identifying specific environmental factors that confer risk for autism and understanding how they contribute to the etiology of autism has been challenging, in part because the influence of environmental chemicals likely varies depending on the genetic substrate of the exposed individual. Current research efforts are focused on elucidating the mechanisms by which environmental chemicals interact with autism genetic susceptibilities to adversely impact neurodevelopment. The goal is to not only generate insights regarding the pathophysiology of autism, but also inform the development of screening platforms to identify specific environmental factors and gene × environment (G × E) interactions that modify autism risk. Data from such studies are needed to support development of intervention strategies for mitigating the burden of this neurodevelopmental condition on individuals, their families and society. In this review, we discuss environmental chemicals identified as putative autism risk factors and proposed mechanisms by which G × E interactions influence autism risk and/or severity using polychlorinated biphenyls (PCBs) as an example.

1. Introduction

Autism spectrum disorder (ASD), hereafter referred to as autism, is a complex, multifactorial neurodevelopmental condition that is defined clinically by core deficits in social reciprocity and communication, restrictive interests and repetitive behaviors. The severity of core symptoms, the expression of co-morbidities, which include intellectual disability, anxiety, seizures, gastrointestinal disturbances, and immunological abnormalities, and the response to treatment vary considerably between individuals diagnosed with autism (Lord, Elsabbagh, Baird, & Veenstra-Vanderweele, 2018). The United States (U.S.) Center for Disease Control (CDC) estimated the prevalence of autism in U.S. children in 2021 to be 1 in 44, which is more than three times the CDC 2002 estimate of 1 in 150 U.S. children (https://www.autism-society.org). This trend of rising rates of autism, which dates back to the early 1990s, is a global occurrence not confined to the U.S. (Zeidan et al., 2022). Large-scale surveys estimate the current median worldwide prevalence of autism to be 1 in 100 children (Zeidan et al., 2022). When considered in the context of the tremendous costs that autism exacts on affected individuals, their families, and society (Hong et al., 2020; Leibson et al., 2020; Schofield et al., 2019), these statistics underscore the need to identify factors that confer autism risk and/or modify symptom severity.

Research on the etiology of autism has identified a strong hereditary component (Buxbaum & Hof, 2011; El-Fishawy & State, 2010; Geschwind, 2011). The candidate genes most strongly associated with autism encode proteins that regulate the patterning of synaptic connections in the developing brain (Delorme et al., 2013; Guang et al., 2018; Landrigan, Lambertini, & Birnbaum, 2012; Masini et al., 2020; Qiu, Aldinger, & Levitt, 2012; Stamou, Streifel, Goines, & Lein, 2013). However, genetic research has also shown that genes linked to autism rarely segregate in a simple Mendelian fashion (El-Fishawy & State, 2010) and that solely genetic causes account for approximately 10–30% of all autism cases (Ansel, Rosenzweig, Zisman, Melamed, & Gesundheit, 2016; Lai, Lombardo, & Baron, 2014; Masini et al., 2020). Moreover, there is incomplete concordance of autism in monozygotic twins (Herbert, 2010), and even in genetic syndromes strongly associated with autism, a significant percentage of carriers do not express autism phenotypes (Herbert, 2010; Levitt & Campbell, 2009; Masini et al., 2020). These observations are consistent with a model in which environmental factors act as modifiers of autism risk genes.

While the dramatic increase in autism rates over the past several decades is often presented as evidence that environmental factors are involved in autism, others have argued that the progressive rise in autism is the result of diagnostic substitution, or the labeling of individuals as autistic who previously would not have been so labeled, due to broadening of diagnostic criteria, coupled with increased awareness and improved detection of these disorders (Bishop, Whitehouse, Watt, & Line, 2008; Coo et al., 2008; Parner, Schendel, & Thorsen, 2008). However, studies that have tested this hypothesis found that diagnostic substitution accounted for less than half of new cases, indicating there has been a true increase in the number of individuals diagnosed with autism (Grether, Rosen, Smith, & Croen, 2009; Hertz-Picciotto & Delwiche, 2009; Hertz-Picciotto, Schmidt, & Krakowiak, 2018; King & Bearman, 2009). Consistent with these conclusions, two well-powered independent twin studies (Frazier et al., 2014; Hallmayer et al., 2011) and a third study of more than 14,000 Swedish children with autism (Sandin et al., 2014) concluded that environmental factors account for approximately 50–60% of autism cases. Collectively, these studies suggest that environmental factors can significantly influence autism susceptibility, thereby providing a plausible explanation for both the marked increase in autism rates and the significant clinical heterogeneity of autism.

In contrast to genetic risks, which are currently irreversible, environmental factors are modifiable risk factors. The identification of specific environmental factors that increase autism risk would enable the primary prevention of autism via public health policies aimed at reducing relevant environmental exposures in susceptible populations. Progress has been made in identifying environmental risk factors for autism, which include advanced paternal age at conception, complications during pregnancy, maternal diet, and prenatal exposure to psychotropic drugs (Bolte, Girdler, & Marschik, 2019; Carlsson, Molander, Taylor, Jonsson, & Bolte, 2020). Environmental chemical contaminants have also been proposed as autism risk factors (Landrigan, 2010; Landrigan et al., 2012). While experimental animal and predictive toxicology studies have identified environmental chemicals associated with autism-relevant phenotypes, it has been challenging to demonstrate this association in human studies (Carter & Blizard, 2016; Kalkbrenner, Schmidt, & Penlesky, 2014; Pelch, Bolden, & Kwiatkowski, 2019; Rock & Patisaul, 2018; Rossignol, Genuis, & Frye, 2014). A recent analysis of the relevant epidemiological literature concluded that overall, these studies were moderate in quality and of limited strength (Mari-Bauset et al., 2022). The authors identified numerous methodological challenges associated with epidemiological studies of environmental influences in autism that decreased quality and strength. These included difficulties in obtaining accurate measures of exposure, particularly for chemicals with short half-lives, such as some of the pesticides, phthalates and BPA; controlling for confounding factors, especially socioeconomic stressors that tend to co-vary with environmental exposures; and dealing with co-exposures, a not insignificant issue given the number of chemicals in the human chemosphere that vary in both composition and concentration across time and space. Epidemiologic studies of complex neurodevelopmental conditions face additional challenges in that these conditions are phenotypically and genetically heterogeneous. The former presents diagnostic challenges, while the latter likely creates a range of sensitivities to environmental risk factors (Pessah & Lein, 2008) that masks clear associations between exposure and diagnosis.

As a result of the difficulties associated with using epidemiological approaches to identify environmental chemicals that confer risk for autism, animal and cell-based models that demonstrate effects of environmental chemicals on outcomes with face validity to autism phenotypes are critically important for corroborating and informing human studies. Since autism is diagnosed based on behavioral criteria that are not readily translated to animal models, a key challenge has been identifying autism-relevant molecular and cellular phenotypes that can be quantified in experimental models, including cell-based and predictive toxicology models. An exciting advance in this regard is the convergence of genetic, histologic, *in vivo* imaging and functional data on altered patterns of synaptic connectivity as the biological basis underlying the behavioral phenotypes associated with autism (Alamdari, Sadeghi Damavandi, Zarei, & Khosrowabadi, 2022; Bourgeron, 2009; Geschwind & Levitt, 2007; Guang et al., 2018; Persichetti, Shao, Gotts, & Martin, 2022; Rubenstein & Merzenich, 2003; Zoghbi & Bear, 2012). As indicated earlier, candidate genes most strongly implicated in the causation of autism encode proteins that regulate the patterning of neuronal networks during development and influence the balance of excitatory to inhibitory signaling in the brain or E/I balance (Ansel et al., 2016; Bourgeron, 2009; Rubenstein & Merzenich, 2003; Stamou et al., 2013; Zoghbi & Bear, 2012). Histological analyses have identified abnormalities of dendritic and synaptic morphology (Penzes, Cahill, Jones, VanLeeuwen, & Woolfrey, 2011). For example, spine density was increased in Golgi-impregnated pyramidal neurons of Layer II in the frontal, temporal and parietal lobes and Layer V of the temporal lobe in autism (Hutsler & Zhang, 2010), suggesting skewed cortical synaptic activity towards excitation. Abnormal spine morphology was observed in the temporal and visual cortices of patients with Fragile-X, a genetic syndrome with a high incidence of autism (Irwin et al., 2001). Neuroimaging studies of children with autism have revealed both hypo- and hyper-connectivity in the brain of affected individuals (Baribeau & Anagnostou, 2013; Therien, Degre-Pelletier, Barbeau, Samson, & Soulieres, 2022). The available functional MRI data support a pattern of early brain overgrowth in the first few years of life followed by aberrant maturation in adolescence with impaired long-range connectivity but increased local and/or subcortical connectivity (reviewed in Baribeau & Anagnostou, 2013).

The differences in synaptic connectivity observed in individuals with autism may reflect influences of sex (Alaerts, Swinnen, & Wenderoth, 2016) and/or genotype (Dennis et al., 2011; Rudie et al., 2012; Scott-Van Zeeland et al., 2010). As an example of genotype, functional MRI studies of individuals expressing the autism risk variant of the Met Receptor Tyrosine Kinase (MET) gene revealed unique brain activity in response to social stimuli and altered connectivity in temporo-parietal brain regions, phenotypes that are exacerbated in individuals with autism (Eagleson, Xie, & Levitt, 2017; Rudie et al., 2012). These functional changes associated with MET genotype are not surprising considering that MET signaling is involved in regulating dendritic growth, spine formation and excitatory (glutamatergic) synaptogenesis during early development (Peng et al., 2016).

It has been proposed that the inherent imbalances in synaptic connectivity observed in autism present a genetic substrate that increases susceptibility to environmental insults that impact neurodevelopmental processes that determine synaptic connectivity, such as neuronal migration, apoptosis, elaboration of axons and dendrites, and activity-dependent refinements of synaptic connections (Belmonte & Bourgeron, 2006; Stamou et al., 2013). Perturbations of the spatiotemporal patterns and/or magnitude or extent of any of these neurodevelopmental events can modify the patterning of neural networks in the developing brain to give rise to behavioral phenotypes associated with autism (Belmonte & Bourgeron, 2006; Lein, Silbergeld, Locke, & Goldberg, 2005). Research findings from studies of transgenic animal models that express autism risk genes further support the hypothesis that perturbations of dendritic growth, synapse formation and/or synapse stabilization result in altered patterns of synaptic connectivity that coincide with autism-relevant behavioral phenotypes (Bourgeron, 2009; Zoghbi & Bear, 2012). How environmental and genetic risk factors interact to influence autism risk and/or severity remains an outstanding question.

In this review, we discuss current hypotheses regarding mechanisms by which environmental chemicals interact with genetic susceptibility factors to modify autism risk. These mechanisms are illustrated in the context of polychlorinated biphenyls (PCBs), an extensively studied class of environmental contaminants for which there is scientific consensus that they alter normal neurodevelopmental trajectories in the human brain (Pessah, Lein, Seegal, & Sagiv, 2019).

2. Environmental chemicals identified as putative autism risk factors

Early reports that *in utero* exposure to valproic acid or thalidomide during critical periods of development was associated with increased expression of autism-related traits in children (Rodier, Ingram, Tisdale, & Croog, 1997) raised significant interest regarding the role of chemical exposures in autism. Subsequent epidemiological studies reported an increased risk for autism associated with maternal use of various medications and drugs of abuse, including alcohol (Arndt, Stodgell, & Rodier, 2005; Christensen et al., 2013; Croen et al., 2011; Malm et al., 2012; Miller et al., 2005; Rodier et al., 1997; Rodier, Miller, & Brent, 2011), as well as prenatal or early postnatal exposure to diverse environmental chemicals. These epidemiological studies, as well as preclinical studies, have identified structurally and mechanistically diverse environmental chemicals as putative autism risk factors. These include legacy contaminants well-established to interfere with normal neurodevelopment, such as lead, mercury, and polychlorinated biphenyls (PCBs), as well as more contemporary contaminants, such as pesticides (organophosphorus and organochlorine pesticides, neonicotinoids and pyrethroids), flame retardants (polybrominated diphenyl ethers or PBDEs), plasticizers (phthalates and bisphenol A or BPA), and complex environmental mixtures, notably air pollution (Carter & Blizard, 2016; Kalkbrenner et al., 2014; Landrigan et al., 2012; Masini et al., 2020; Pelch et al., 2019).

The epidemiologic evidence linking environmental chemical exposures and autism was reviewed in Kalkbrenner et al. (2014). These authors identified 58 relevant articles in the peer-reviewed literature published prior to March 1, 2014, of which 32 met their inclusion criteria (individual-level data on autism, exposure measures pertaining to pregnancy or the 1st year of life, valid comparison groups, control for confounders and adequate sample sizes). These 32 articles included studies of autism and estimates of exposure to tobacco, air pollutants, volatile organic compounds and solvents, metals, pesticides, PCBs, PBDEs, PFCs, BPA and phthalates. Many of these chemicals have been shown to cause neurodevelopmental effects in experimental animal or cell-based models that have face validity to behavioral outcomes or cellular phenotypes associated with autism (reviewed in Kalkbrenner et al., 2014; Stamou et al., 2013). Some of these environmental exposures also showed associations with autism in the analysis of the epidemiologic literature by Kalkbrenner and colleagues. The most strongly and consistently associated were traffic-related air pollutants, some metals and the organophosphorus and organochlorine pesticides, with suggestive trends for some volatile organic compounds and phthalates (Kalkbrenner et al., 2014). Interestingly, neither tobacco smoke, which has been strongly associated with increased ADHD risk (Polanska, Jurewicz, & Hanke, 2012), nor alcohol showed a consistent association with increased risk for autism. The authors of this analysis reached the broad conclusion that with the possible exception of studies of tobacco and alcohol, the relevant publications available at the time were too limited in scope to either infer causality or to rule out the possibility that these or additional environmental chemicals confer risk for autism.

A later scoping review of the human and animal research was published in 2018, which aimed to identify and categorize primary research and reviews on the association between prenatal and early postnatal exposure to environmental chemicals and autism risk that were published through February 2018 (Pelch et al., 2019). Search criteria included studies that assessed exposure to environmental chemicals prior to 2 months of age in humans or 14 d in rodents and rodent studies that included at least one measurement of reciprocal social communicative behavior or repetitive and stereotyped behavior. The search identified 21,603 unique studies from which the authors selected 54 epidemiological studies, 46 experimental rodent studies, and 50 reviews as relevant. These studies covered 152 chemical exposures. The most frequently studied environmental chemicals in humans were particulate matter ($n = 14$), mercury ($n = 14$), nonspecific air pollution ($n = 10$) and lead ($n = 10$). Similarly, in rodents, mercury and lead were among the most studied environmental chemicals ($n = 6$ and 4, respectively) along with the organophosphorus pesticide, chlorpyrifos ($n = 9$). Similar to the conclusions reached by Kalkbrenner et al. (2014), Pelch et al. (2019) concluded that while research on this topic was rapidly expanding, the wide variability in study design and conduct, the exposures investigated and the outcomes assessed made it difficult to identify chemicals of concern, but based on the available literature, they recommended that chlorpyrifos, lead, and PCBs be systematically reviewed to assess their association with autism risk and severity.

A more recent narrative review of this literature published in 2020 (Cheroni, Caporale, & Testa, 2020) identified the following environmental chemicals as the most relevant in the context of autism risk: lead, methylmercury, air pollution, endocrine disruptors in general, phthalates, BPA, perfluoralkyl substances, and PCBs. In this review, we focus on the literature related to the association of PCBs with autism, but the reader is referred to recent reviews and meta-analysis of the literature regarding the association between autism phenotypes and either pesticides (Bertoletti et al., 2022; Maleki et al., 2022; Vester & Caudle, 2016) or air pollution (Cory-Slechta, Merrill, & Sobolewski, 2022; Fu, Guo, Cheung, & Yung, 2019; Yu et al., 2022).

3. Mechanisms by which gene × environment interactions influence autism risk

While it is now widely accepted that complex G × E interactions influence the etiology of autism, the mechanisms by which genetic and environmental factors interact to influence individual autism risk and/or severity remain speculative. Current hypotheses include (*i*) heritable deficits in xenobiotic metabolism, (*ii*) disruption of the gut microbiome, (*iii*) endocrine disruption, (*iv*) epigenetic alterations that alter gene expression, (*v*) inflammation and/or immune dysregulation, and (*vi*) convergence on signaling pathways that regulate brain development (Table 1). The scientific premise for each of these proposed mechanisms is briefly discussed in this section, with specific examples presented in subsequent sections focused on PCBs as putative autism risk factors. It is important to note that while there is evidence to support each of these G × E mechanisms in the determination of autism risk, at this time there is little evidence causally linking these mechanisms to autism. Establishing causality remains a significant goal of ongoing research.

3.1 Genetic polymorphisms in xenobiotic metabolism

Genetic mutations in enzymes that metabolize xenobiotics, including environmental chemicals, have been linked to increased autism risk (Mordaunt et al., 2019; Rossignol et al., 2014). In a genome-wide transcriptomic study of cord blood from autistic *vs.* neurotypical children, 172 genes were found to be differentially expressed between the two groups with several genes grouped into the "response to toxic substances" category (Mordaunt et al., 2019). The genes included those that encode members of the cytochrome P450 (CYP) family of enzymes, such as *CYP1A1*, that play critical roles in metabolizing diverse environmental chemicals (Nagai, Fukuno, Suzuki, & Konishi, 2016; Uwimana, Ruiz, Li, & Lehmler, 2019). Polymorphisms in CYP genes that influence CYP expression levels and/or activity have been shown to decrease detoxification of environmental chemicals or increase metabolic conversion of a non-toxic parent compound to a neurotoxic metabolite. Either functional consequence can effectively increase the neurotoxic potential of an environmental chemical.

Glutathione-dependent enzymes also play a key role in the detoxification of environmental chemicals, but this is not their only critical function: glutathione-mediated metabolism is critically important in maintaining intracellular redox homeostasis (Bjorklund et al., 2021; Liu, Tan, Song, & Song, 2020). Mothers who express genetic polymorphisms that decrease glutathione-mediated processes are at higher risk for having a child with autism, and these genetic polymorphisms occur with higher frequency in children with autism *vs.* neurotypical children (Rossignol et al., 2014). Consistent with these observations, glutathione S-transferase (GST)

Table 1 Proposed mechanisms for how genetic and environmental factors interact to influence autism risk and severity.

Mechanism	Rationale	References
Heritable deficits in xenobiotic metabolism	Genetic mutations in enzymes that metabolize environmental chemicals have been linked to increased autism risk. A decreased ability to detoxify environmental chemicals or an increased effectiveness in converting benign parent compounds to neurotoxic metabolites might increase the neurotoxicity of an environmental chemical contaminant.	Bjorklund et al. (2021) Herbert (2010) Rock and Patisaul (2018)
Disruption of the gut microbiome	Clinical and experimental data indicate that the gut microbiome regulates neurodevelopment; the gut microbiota in children with autism differs from that of neurotypical children. Toxicological studies show that the gut microbiome influences xenobiotic metabolism and disposition, and, conversely, environmental chemicals can alter the gut microbiome.	Balaguer-Trias, Deepika, Schuhmacher, and Kumar (2022) Bertotto, Catron, and Tal (2020) Needham et al. (2022) Rosenfeld (2017) Vuong et al. (2020) Yousefi et al. (2022)
Endocrine disruption	Endocrine signaling, including signaling by thyroid hormones, sex steroids and glucocorticoids, critically modulates neurodevelopment. The incidence of autism exhibits a pronounced sex bias, with a male:female ratio of 4:1. Diverse environmental chemicals have been shown to alter endocrine signaling via various mechanisms.	Demeneix (2019) Moosa, Shu, Sarachana, and Hu (2018) Mari-Bauset et al. (2022) Ramirez et al. (2022) Weiss (2012)
Epigenetic mechanisms	Epigenetic mechanisms are critically important in regulating brain development. Environmental chemicals can alter DNA methylation, histone acetylation and miRNA expression profiles, and these parameters are altered in at least some children with autism.	Keil and Lein (2016) Kong, Zhou, and Sun (2020) Park, Lee, Kim, and Yi (2022) Parra and Johnston (2022)
Inflammation and/or immune dysregulation	Crosstalk between nervous and immune systems is essential for normal neurodevelopment, and environmental chemicals can trigger systemic and neuroinflammation as well as alter immune function. Clinical and experimental data confirm neuroinflammation and immune dysregulation in autism.	Arambula and McCarthy (2020) McLellan, Kim, Bruce, Ramirez-Celis, and Van de Water (2022) Meltzer and Van de Water (2017) Sotgiu et al. (2020)
Convergence of G × E on signaling pathways that regulate brain development	Many genetic risk factors for autism converge on several major signaling pathways that critically regulate synaptic connectivity in the developing brain. Environmental chemicals have been identified that target these same signaling pathways.	Ansel et al. (2016) Ebert and Greenberg (2013) Pourtavakoli and Ghafouri-Fard (2022) Purushotham et al. (2022) Stamou et al. (2013)

detoxification activity was found to be lower in children with autism *vs.* neurotypical controls (Serajee, Nabi, Zhong, & Huq, 2004). It was also recently shown that imbalances in the glutathione redox system contribute to the pathogenesis of autism via not only increased oxidative stress, but also altered E/I balance in the brain (Bjorklund et al., 2021). Glutathione is involved in regulating the N-methyl-D-aspartate (NMDA) receptor, which mediates signaling by glutamate, the primary excitatory neurotransmitter in the brain. Autism-linked deficits in glutathione-related regulation of glutamate have been linked to E/I imbalance in the brain, and in extreme cases, to excitotoxic cell death of neurons (Bjorklund et al., 2021).

3.2 Disruption of the gut microbiome

The metabolism of environmental chemicals is also influenced by the gut microbiome (Li, Tian, et al., 2022; Lim et al., 2020; Zimmermann, Zimmermann-Kogadeeva, Wegmann, & Goodman, 2019), and changes in the composition of the gut microbiome have been shown to alter the disposition and concentration of environmental chemicals throughout the body, including the brain (Li, Tian, et al., 2022). Conversely, environmental chemicals can alter the gut microbiome (Gama, Neves, & Pereira, 2022; Lai, Chung, Li, Wan, & Wong, 2016; Li, Hefti, Marek, et al., 2022; Rude, Keogh, & Gareau, 2019). Given clinical and experimental data indicating that the gut microbiome regulates neurodevelopment and central nervous system function (Alam, Abdolmaleky, & Zhou, 2017; Muller et al., 2020; Ristori et al., 2019), it is widely posited that interactions between environmental chemicals and the gut microbiome during critical stages of neurodevelopment may alter neurodevelopmental trajectories to confer risk for autism. This is supported by reports that the gut microbiota in children with autism differs from that of neurotypical children (Kang et al., 2017; Liu et al., 2019; Mayer, Padua, & Tillisch, 2014; Sharon et al., 2019). However, data demonstrating that environmental chemical interactions with the gut microbiome are causally related to increased expression of autism phenotypes is largely lacking. Carefully designed experimental animal studies and prospective clinical studies are needed to better understand whether certain gut microbiome compositions are more likely to lead to adverse effects from environmental chemical exposures and whether therapeutic interventions targeting the gut microbiome could be employed to mitigate autism risk associated with G × E interactions.

3.3 Endocrine disruption

Diverse environmental chemicals have been classified as endocrine disruptors because they interfere with endogenous endocrine signaling either by mimicking endogenous hormones or by decreasing the production, transport or signaling of endogenous hormones (Mari-Bauset et al., 2022; Salazar, Villaseca, Cisternas, & Inestrosa, 2021; Sechman, Batoryna, Antos, & Hrabia, 2016; Takeuchi et al., 2011; Weiss, 2012). Since endocrine signaling influences neural circuitry in the developing brain via modulation of neurodevelopmental processes that regulate neuronal connectivity (Demeneix, 2019; Moosa et al., 2018; Ramirez et al., 2022), it is widely posited that endocrine disrupting chemicals interact with genetic factors that regulate endocrine signaling to alter neurodevelopmental trajectories.

One endocrine signaling system thought to be targeted by diverse environmental chemicals to cause developmental neurotoxicity is the thyroid hormone (TH) system. TH signaling is critical for normal neurodevelopment (Demeneix, 2019), and congenital hypothyroidism, if untreated, causes severe adverse neurodevelopmental outcomes (Oppenheimer & Schwartz, 1997; Rovet, 2014; Williams, 2008). While conflicting, there is some evidence that TH insufficiency increases autism risk (Salazar et al., 2021), and genetic mutations in TH receptors have been observed in children with autism (Kalikiri, Mamidala, Rao, & Rajesh, 2017). A large number of environmental chemicals have been shown to decrease circulating levels of TH and/or to interfere with TH signaling at the level of the receptor (Salazar et al., 2021). While these observations collectively support the hypothesis that environmental chemicals confer autism risk via modulation of TH signaling, evidence of a causal link is lacking, as are data demonstrating that expression of genetic mutations in TH receptors increase susceptibility to the neurotoxicity of environmental thyroid hormone disruptors.

Another observation that has promoted the hypothesis of endocrine signaling as a mechanisms by which G × E interactions confer autism risk is that autism disproportionately impacts males relative to females with a rate of diagnosis 3–4.5 times higher in males (Christensen et al., 2016; Zablotsky, Black, & Blumberg, 2017). This observation has led to the suggestion that sex steroids may be important in autism etiology. Sex steroid signaling determines the sexually dimorphic trajectories of brain development that produce sex differences in brain structure, synaptic connectivity and behavior (Collaer & Hines, 1995; McCarthy, 2016a, 2016b; McCarthy, Herold, & Stockman, 2018; Napolitano et al., 2022). Developmental surges of fetal testicular testosterone, which is converted to 17B-estradiol in the brain, are responsible for masculinizing the brain, and environmental chemicals that interfere with this developmental event can alter neural circuitry (McCarthy et al., 2018; McCarthy, Arnold, Ball, Blaustein, & De Vries, 2012; Napolitano et al., 2022). There is evidence of differential brain expression of genetic variants of the androgen receptor in autism compared to age- and sex-matched controls (Henningsson et al., 2009). Postmortem analysis has revealed decreased abundance of estrogen receptors in the brain of individuals with autism (Crider, Thakkar, Ahmed, & Pillai, 2014). Thus, environmental chemicals could contribute to autism risk by altering the spatiotemporal profiles and/or levels of sex hormones or their receptors to disrupt the normal sex specific patterning and development of the brain.

3.4 Epigenetic mechanisms

Epigenetic mechanisms are critically important in regulating brain development. DNA methylation patterns, chromatin remodeling, and miRNA expression profiles can be altered by environmental chemicals, and changes in these epigenetic modifiers have been documented in individuals with autism (Keil & Lein, 2016; Tremblay & Jiang, 2019). Perhaps the strongest evidence supporting epigenetic mechanisms as a means by which environmental chemicals might influence autism risk comes from studies of the DNA methylation binding protein, methyl-CpG binding protein 2 (MeCP2). Mutations in MeCP2 cause Rett syndrome, which has a high comorbidity of autism diagnosis, and mutations in MeCP2 have been identified in autism independent of Rett syndrome (Cukier et al., 2012; LaSalle & Yasui, 2009; Percy, 2011; Wen et al., 2017). MeCP2 is a transcriptional repressor regulated by miR132 that regulates dendritic and synaptic morphogenesis and function (Klein et al., 2007). Mice that lack MeCP2 have reduced expression of GABRB3, a gamma-aminobutyric acid (GABA) receptor important in determining the balance between excitatory and inhibitory neurotransmission (Samaco, Hogart, & LaSalle, 2005). Altered methylation of MeCP2 in the cord blood of mothers has been correlated with environmental chemical exposures (Eguchi et al., 2019; Pilsner et al., 2009).

The microRNA, miR132, plays an important role in synaptogenesis and connectivity (Xu, Li, Ji et al., 2019). In mice, genetic deletion of miR132 impaired performance in learning and memory tasks, and either deletion or overexpression of miR132 altered hippocampal transcription of genes related to synapse transmission, neuronal proliferation and learning and memory (Hansen et al., 2016). Brain levels of microRNAs have been reported to be elevated in autism (Abu-Elneel et al., 2008; Sarachana, Zhou, Chen, Manji, & Hu, 2010; Talebizadeh, Butler, & Theodoro, 2008), and environmental chemicals have been shown to alter brain expression of miR132 (Lesiak et al., 2014; Sima, Rossnerova, Simova, & Rossner Jr., 2021; Yu et al., 2021). Thus, microRNAs may represent a convergent target for genetic and environmental risk factors that alter synaptic connectivity.

There is evidence that epigenetic mechanisms influence xenobiotic metabolism. For example, comparative analyses of placental tissue from mothers who gave birth to a child later diagnosed with autism *vs.* placenta tissues from mothers whose children were neurotypical identified differentially methylated regions in a xenobiotic metabolism gene, *CYP2E1*, and the extent of differential methylation was further influenced by the mother's *CYP2E1* genotype (Zhu et al., 2019). Epigenetic mechanisms may also influence known genetic causes of autism. 15q11-q13 duplications are associated with a syndromic form of autism, and individuals with this duplication have lower levels of brain LINE-1 repetitive DNA methylation (Mitchell et al., 2012). The potential to alter gene expression across a wide array of genes important to neurodevelopment and neurotoxicity make epigenetic pathways a critical target of G $\times$ E interactions that modify autism risk and/or severity.

3.5 Inflammation and/or immune dysregulation

The immune system is inextricably linked with the central nervous system and plays a critical role in brain development and function (Ransohoff & Brown, 2012). For example, cytokines and microglia, the innate immune cells in brain, are necessary for establishing functional neural circuits in the developing brain, including those involved in social behavior (Deverman & Patterson, 2009; Garay & McAllister, 2010; Guedes, Ferreira, Costa, Cardoso, & Peca, 2022; Sullivan & Ciernia, 2022). While numerous cytokines have been implicated in modulating neurodevelopmental processes critical to determining synaptic connectivity, the proinflammatory cytokines interleukin (IL)-6 (Goines & Ashwood, 2013; Kummer, Zeidler, Kalpachidou, & Kress, 2021) and type II interferon, gamma (IFNγ) (Goines & Van de Water, 2010) are two key players. IL-6 is necessary for maintaining neural progenitor cells in mice (Bowen, Dempsey, & Vemuganti, 2011) and has been shown to promote neural differentiation in human induced pluripotent stem cells (Kummer et al., 2021; Xu et al., 2018). However, developmental timing is likely critical for these effects since IL-6 has also been reported to inhibit neural differentiation in hippocampal neurons from adult mice (Kong et al., 2019). IL-6 has also been reported to decrease dendrite number in cultured rat hippocampal neurons (Gadient, Lein, Higgins, & Patterson, 1998) and modulate dendrite growth in primary rat neurons in culture, promoting both dendrite retraction (Guo, Chandrasekaran, Lein, Kaplan, & Higgins, 1999) and increased dendrite number (Guo, Metzler-Northrup, Lein, Rueger, & Higgins, 1997) depending on cell type and culturing conditions. IL-6 also impacts synaptogenesis and connectivity, altering network activity in mouse hippocampal neurons, in part by reducing expression of glutamate receptors and calcium channel signaling (Vereyken, Bajova, Chow, de Graan, & Gruol, 2007). IFNγ also regulates dendritic growth and synaptogenesis, causing dendritic retraction and impairing synapse formation in rodent sympathetic and hippocampal neurons (Kim, Beck, Lein, & Higgins, 2002; Vikman, Owe-Larsson, Brask, Kristensson, & Hill, 2001). IFNγ can modulate synaptic activity in rat hippocampal neurons *in vitro*, causing marked increases in excitatory postsynaptic currents 48 h post exposure (Vikman et al., 2001). Similarly, microglia have been reported to play a critical role in regulating neural differentiation via engulfment of neural precursor cells and in pruning superfluous synapses in the developing brain (Guedes et al., 2022; Lebovitz, Ringel-Scaia, Allen, & Theus, 2018; Sullivan & Ciernia, 2022). Sex differences in microglial gene expression and function have been identified that play a key role in shaping sex-specific brain function and behavior. The levels of sex hormones, such as androgens, estrogens, and progesterone, vary in an age-dependent and sex-dependent manner, and microglia respond both directly and indirectly to changes in hormone levels by altering transcriptional gene expression, morphology, and function (Sullivan & Ciernia, 2022).

Human data support an association between the immune system and autism with a large body of evidence linking maternal or childhood infection with increased rates of autism (Krakowiak et al., 2017; Majerczyk, Ayad, Brewton, Saing, & Hart, 2022; Nielsen et al., 2022). A population-based study of children in Australia found that autoimmune disease in mothers or infection in children prior to age 2 were associated with increased odds of autism (Nielsen et al., 2022). Genetic studies have identified immune system genes differentially expressed in individuals with autism *vs.* neurotypical controls, and transcriptomic studies reported that pathways involved in immune function were significantly associated with autism diagnosis (Ansel et al., 2016). In a population-based case control study of children in California, levels of IL-1β and IL-4 from neonatal blood spots were associated with increased odds of autism (Krakowiak et al., 2017). Further IL-6 and IFNγ levels in mothers have been associated with elevated autism diagnosis in children (Majerczyk et al., 2022). Increased microglia cell activation has also been noted in postmortem analyses of brains from individuals with autism *vs.* neurotypical controls (Blagburn-Blanco, Chappell, De Biase, & DeNardo, 2022; Zhang, Wang, Li, Deng, & Shen, 2022). The role of the immune system in promoting masculinization of the brain likely provides a biological substrate for the sex bias in autism diagnosis among boys *vs.* girls (McCarthy, Nugent, & Lenz, 2017; McCarthy & Wright, 2017).

Given the importance of the immune system in establishing synaptic connectivity, and ample evidence that many environmental chemicals can disrupt immune function (Dietert, 2014; Eisner et al., 2022; Kreitinger, Beamer, & Shepherd, 2016; Winans, Humble, & Lawrence, 2011), it seems plausible that the immune system is a point of convergence for environmental and genetic factors to interact to confer risk for autism. However, as noted for other mechanisms of G × E interactions, to date, there is no evidence causally linking environmental chemical and genetic effects on immune function to increased risk of autism.

3.6 Convergence of genes and environment on neurodevelopmental signaling pathways

Many genetic risk factors for autism converge on signaling pathways that critically regulate synaptic connectivity in the developing brain (Delorme et al., 2013; Guang et al., 2018; Landrigan et al., 2012; Masini et al., 2020; Qiu et al., 2012). Therefore, these signaling pathways and the neurodevelopmental processes regulated by these signaling pathways represent molecular and cellular points of convergence for environmental and genetic risk factors for autism (Table 2) (Stamou et al., 2013).

A number of genes identified as autism risk genes regulate dendritogenesis or synapse formation, elimination or stabilization. These include genes for neuroligins (*NLG3*, *NLG4*) (Sudhof, 2008; Zoghbi & Bear, 2012), neurexins (*NRXN1*, *NRXN3*) (Sciacca et al., 2022), and contactin-associated protein-2 (*CNTNAP2*) (Dennis et al., 2011; Scott-Van Zeeland

Table 2 Autism-associated genes important for neuronal connectivity, dendritogenesis and synaptogenesis that may interact with environmental chemicals to increase autism risk.

Signaling pathway	Genes	References
Signaling pathways that regulate dendrite and/or synapse formation	• Met Receptor Tyrosine Kinase (MET) • Neuroligins (NLG3, NLG4) • Neurexins (NRXN1, NRXN3) • Contactin-associated protein-2 (CNTNAP2) • SH3 and multiple ankyrin repeat domains 3 (SHANK3) • Extracellular signal-regulated kinase (ERK) • Phosphatidylinositol 3-kinases (PI3K) • Mammalian target of rapamycin (mTOR)	Chaudry and Vasudevan (2022) Dennis et al. (2011) Eagleson et al. (2017) Guang et al. (2018) Huang et al. (2019) Lutz et al. (2021) Rudie et al. (2012) Sciacca, Marino, Vitaliti, Falsaperla, and Marino (2022) Scott-Van Zeeland et al. (2010) Sudhof (2008) Wan et al. (2022) Zoghbi and Bear (2012)
Convergence on calcium signaling pathways	• cAMP response element-binding protein (CREB) • Wingless-type MMTV integration site family, member 2 (Wnt2) • Ryanodine receptors (RyRs) • Voltage gated calcium channels (CACNA family) • Voltage gated sodium channels (SCN family) • Ligand gated chloride channels (GABRB family) • Calcium activated potassium channels (KCNM family)	Deidda, Bozarth, and Cancedda (2014) Marcantoni, Calorio, Hidisoglu, Chiantia, and Carbone (2020) Nanou and Catterall (2018) Panesar, Kennedy, Keil Stietz, and Lein (2020) Schmunk and Gargus (2013) Skafidas et al. (2014) Soueid et al. (2016) Wayman et al. (2006) Wayman et al. (2012) Wayman et al. (2012)

et al., 2010), all which have important roles in connecting and stabilizing presynaptic and postsynaptic components of the synapse. *CNTNAP2* mutations have been linked to increased autism risk, and expression of *CNTNAP2* gene variants are sufficient to alter connectivity in experimental models (Dennis et al., 2011; Scott-Van Zeeland et al., 2010). The SH3 and multiple ankyrin repeat domains 3 (*SHANK3*) gene has been identified as an autism risk gene (Huang et al., 2019; Lutz et al., 2021; Wan et al., 2022). SHANK3 functions to interact with neuroligin and neurexin complexes to regulate activity-dependent synaptogenesis (Craig & Kang, 2007; Stamou et al., 2013). ERK, PI3K and mTOR signaling pathways are also strongly implicated in autism (Chaudry & Vasudevan, 2022; Lutz et al., 2021) and are important in regulating dendritic growth and synaptogenesis (Levitt & Campbell, 2009).

Neuronal activity plays a significant role in sculpting neural circuits in the developing brain, and calcium signaling is the major mechanism linking neuronal activity to dendritogenesis and synaptogenesis in the developing brain (Pessah, Cherednichenko, & Lein, 2010; Wayman et al., 2006). Thus, it is not surprising that genes that either control calcium signaling or are regulated by calcium signaling, are predominant in the list of autism risk genes (Ansel et al., 2016; Nanou & Catterall, 2018; Nguyen, Medvedeva, Ayyagari, Schmunk, & Gargus, 2018; Pessah et al., 2010). Diverse environmental chemicals have also been shown to disrupt these same calcium signaling pathways, suggesting a "two hit" mechanism in which genetic and environmental factors converge to alter the fidelity of calcium signaling resulting in altered synaptic connectivity (Keil Stietz et al., 2021; Klocke & Lein, 2020; Panesar et al., 2020; Stamou et al., 2013).

Evidence that neuronal calcium signaling is disrupted in autism also derives from phenotypes observed in syndromic disorders that have a high incidence of autism. For example, Fragile × syndrome, which is caused by expansion of a CGG repeat (>200) in the 5′ non-coding portion of the fragile × mental retardation 1 gene (FMR1), is the most prevalent single gene disorder associated with increased risk for autism (Krueger & Bear, 2011; Leehey & Hagerman, 2012). FMR1 premutation (55–200 CGG repeats), which causes fragile X-associated tremor/ataxia syndrome marked by deficits in cognitive function, tremors and motor abilities in adults, is also associated with elevated risk of autism (Chonchaiya et al., 2012; Hagerman, Au, & Hagerman, 2011). Hippocampal neurons from FMR1 premutation mice displayed elevated cytoplasmic calcium concentrations and abnormal glutamatergic responses *in vitro* (Berman, Murray, Arque, Hunsaker, & Wenzel, 2012; Chen et al., 2010; Robin et al., 2017).

FMR1 premutation mouse neurons also exhibited increased frequency and duration of clustered burst firing compared to wild type controls (Cao, Hulsizer, et al., 2012; Cao et al., 2013). FMR1 premutation neurons derived from human iPSC cells similarly displayed altered calcium signaling under baseline conditions and in response to glutamatergic stimulation (Liu et al., 2012). Since calcium signaling is a major driver of dendritic morphogenesis, it is not surprising that mice with FMR1 premutations also exhibited deficits in neuronal cytoarchitecture and learning and memory behavior (Berman et al., 2012; Chen et al., 2010; Diep et al., 2012; Keil et al., 2019; Sethi et al., 2021).

Dendrites can structurally remodel in response to environmental factors and experience. This structural plasticity is mediated by calcium signaling triggered by NMDA receptor activation that drives the sequential activation of calmodulin-dependent protein kinase kinase (CaMKK), CaMKI, and MAPK ERK/MEK ERK kinases to enhance cAMP response element-binding protein (CREB) resulting in Wnt2 transcription and dendritic growth (Wayman et al., 2006). CREB also regulates expression of miR132 during neuronal maturation and synaptogenesis (Hansen et al., 2013). CREB dysfunction is implicated in autism (Barnby et al., 2005; Ngounou Wetie et al., 2015; Todd & Mack, 2001; Zheng et al., 2016), and downstream genes such as miR132 and Wnt pathway members have also been linked to autism or autism-related phenotypes (Bocchi et al., 2017; Kalkman, 2012; Li et al., 2016). Environmental chemicals are also known to alter neurodevelopment via modulation of CREB dependent mechanisms (Connolly & Kingsbury, 2010; Jang et al., 2012; Sethi, Keil, & Lein, 2018).

Ryanodine receptors (RyRs) are important for regulating calcium release from the endoplasmic reticulum (Pessah et al., 2010). Point mutations in *RYR* are prevalent in the human population with ~35% of individuals carrying one or more *RYR* variants (Kim et al., 2013) that increase sensitivity to some anesthetics to trigger malignant hyperthermia (Pessah et al., 2010). Mice engineered to express human *RYR* mutations displayed enhanced sensitivity to inhaled anesthetics, and their neurons exhibited elevated intracellular calcium concentrations (Barrientos et al., 2012; Yuen et al., 2012). Neurons from *RYR* transgenic mice also exhibited altered dendrite arborization and deficits in behavioral tasks (Keil et al., 2019; Sethi et al., 2021). RyR2 mutations have been linked to autism (Soueid et al., 2016) and environmental chemicals have been shown to interact with genetic mutations in RyRs to influence dendritic arborization and behavior in mouse models (Keil Stietz et al., 2021; Sethi et al., 2021).

Genes that encode calcium ion channels other than RyRs, as well as channels that conduct ions other than calcium, have also been implicated in autism (Ansel et al., 2016; Nanou & Catterall, 2018; Schmunk & Gargus, 2013). These include the voltage-gated calcium channels (CACNA family), voltage-gated sodium channels (SCN family), ligand-gated chloride channels (GABRB family) and, more recently, calcium activated potassium channels (KCNM family) (Schmunk & Gargus, 2013). Mutations in the L-type calcium channel Ca(v)1.2 cause Timothy syndrome, which is associated with high rates of autism (Splawski et al., 2004). Mutations in Cav1.2 and Cav1.3 are also implicated as autism risk factors independent of Timothy syndrome (Marcantoni et al., 2020; Nanou & Catterall, 2018). Timothy syndrome is thought to cause autistic phenotypes via altered neuronal connectivity downstream of dysregulated calcium signaling (Nanou & Catterall, 2018). In neurons, this mutation elevated intracellular calcium levels and altered expression of genes regulated by calcium signaling, including CAMKII and CREB (Pasca et al., 2011). In a mouse model of Timothy syndrome, brains had reduced numbers of cortical projecting neurons, deficits in neuronal differentiation and elevated production of catecholamines, all of which are autism-relevant phenotypes (Pasca et al., 2011). Similarly, *de novo* mutations in voltage-gated sodium channels (SCN2A) are implicated as autism risk factors (Sanders et al., 2012). SCN2A are important in regulating action potentials and play an important role in synaptic plasticity (Spratt et al., 2019).

E/I imbalance is implicated in the pathogenesis of autism (Rubenstein & Merzenich, 2003), and this can result from structural changes in neural circuits and/or altered expression of receptors for excitatory *vs.* inhibitory neurotransmitters. For example, GABA receptor mutations are associated with increased autism risk (Deidda et al., 2014), and levels of GABRB3 were found to be lower in brains from children with autism compared to neurotypical controls (Samaco et al., 2005). *GABRB3* is located in the chromosome 15q11-13 locus, which is implicated in syndromic forms of autism and is thought to increase susceptibility to environmental insults (Mitchell et al., 2012; Shelton, Hertz-Picciotto, & Pessah, 2012).

KCNM channels are also important for regulating neuronal excitability, and KCNMA1 mutations that reduce activity of the channel have been observed in individuals with autism (Du et al., 2005; Laumonnier et al., 2006; Schmunk & Gargus, 2013). KCNM channels have also been found to be regulated by Fragile × mental retardation protein (FMRP) via a KCNMB4 regulatory subunit and loss of FMRP reduced channel activity, which dysregulated neurotransmission (Deng et al., 2013). Mutations in the regulatory KCNMB4 subunit have been linked to autism, and KCNMB4 has been identified as one of the genes to be used in predictive panels of autism (Schmunk & Gargus, 2013; Skafidas et al., 2014).

4. PCBs as putative autism risk factors

PCBs belong to a larger grouping of environmental contaminants referred to as persistent organic pollutants (POPs) that have in common environmental persistence and resistance to metabolic degradation. PCBs are a family of 209 structurally similar congeners that vary according to the number and position of chlorine substitutions on the biphenyl backbone (Klocke, Sethi, & Lein, 2019). PCBs are classically divided into two groups based on their structure and binding affinity for the aryl hydrocarbon receptor (AhR), which is the canonical receptor for dioxin (2,3,7,8,-tetrachlorodibenzo-p-dioxin, TCDD): coplanar, dioxin-like PCBs bind the AhR with moderate to high affinity, whereas the non-coplanar, non-dioxin-like (NDL) PCBs have little to no affinity for the AhR (Klocke et al., 2019). A subset of NDL PCBs have RyR activity (Holland et al., 2017; Pessah et al., 2019; Sethi et al., 2019). Some PCB congeners are reported to interact with steroid hormone receptors (Bell, Dryden, Will, & Gore, 2018; Bell, Hart, & Gore, 2016; Dickerson, Cunningham, Patisaul, Woller, & Gore, 2011; Sechman et al., 2016; Takeuchi et al., 2011; Takeuchi, Anezaki, & Kojima, 2017) and TH receptors (Itoh et al., 2018; Pessah et al., 2019; Salazar et al., 2021; Zoeller, 2007) and, thus, are postulated to cause developmental neurotoxicity via endocrine disruption (Bell et al., 2018; Walker, Goetz, & Gore, 2014; Winneke et al., 2014). The NDL PCBs are thought to have the most significant impact on neurodevelopmental outcomes (Klocke & Lein, 2020) and, therefore, are highlighted in this review.

Despite a ban on production of PCBs in the U.S. in the 1970s and globally in the early 2000s (Klocke et al., 2019; Melymuk et al., 2022), legacy PCBs produced prior to the ban persist in the environment to this day. Sources of legacy PCBs include leachate from waste sites, degradation of old building materials and release from PCB-containing fluorescent light ballasts, hydraulic equipment and capacitors still in use. In addition, there are contemporary sources of PCBs, including the unintentional production of PCBs as byproducts of the modern manufacturing processes used to produce pigments found in paints, varnishes and caulking (Hannah, Megson, & Sandau, 2022; Herkert, Jahnke, & Hornbuckle, 2018; Hu & Hornbuckle, 2010; Liu et al., 2004), and the breakdown of higher chlorinated PCBs into lower chlorinated PCBs (Xiang, Xing, Liu, Qin, & Huang, 2020). Some of these contemporary PCBs were not present in the commercial mixtures produced prior to the ban, making contemporary PCB exposures unique compared to those that occurred prior to 1979. Understanding the neurotoxicological implications of PCBs not found in the legacy mixtures is an expanding area of research.

PCBs are ubiquitous contaminants readily detected in human tissues, livestock, fish, dairy products, water, soil, sediments and both indoor and outdoor air worldwide (Barmpas et al., 2019; Chen, Streifel, et al., 2017; Kania-Korwel & Lehmler, 2016; Li, Lim, et al., 2022; Sethi, Chen, Kass, & Puschner, 2017; Thompson & Boekelheide, 2013; Yu, Liu, Shu, Ma, & Pan, 2020). Because they are generally lipophilic, PCBs bioaccumulate up the food chain. Thus, humans are primarily exposed to PCBs via diet, although inhalation is becoming an increasingly important route of exposure, particularly to the lower chlorinated congeners (Ampleman et al., 2015; Carpenter, 2006; Chan-Hon-Tong, Charles, Forhan, Heude, & Sirot, 2013; Durand et al., 2008). In addition, PCBs cross the placenta and are present in breastmilk (Foerster et al., 2021; Kim et al., 2018). The National Health and Nutrition Examination Survey (NHANES) has documented widespread PCB exposure in women of child-bearing age in the U.S. (Thompson & Boekelheide, 2013). PCBs have been detected not only in serum of pregnant women but also in cord blood and brain tissue from young children (Baba et al., 2018; Eskenazi et al., 2017; Granillo et al., 2019; Hisada et al., 2014; Li, Lim, et al., 2022; Rosenquist et al., 2017). In one study, tissue levels of the NDL PCB congener, PCB 95, were found to be significantly higher in postmortem brain tissue from children with a genetic form of autism, 15q11-q13 duplication compared to neurotypical controls (Mitchell et al., 2012).

Reviews of the epidemiological evidence have almost universally concluded that the weight of evidence indicates that PCBs cause developmental neurotoxicity in humans; however, it is less clear whether PCBs are associated with increased risk for autism (Alampi et al., 2021; Bernardo et al., 2019; Lyall, Croen, Sjodin, et al., 2017; Mehri, Bashirian, Khazaei, & Jenabi, 2021; Pessah et al., 2019). For example, a recent narrative summary of the epidemiological literature published between 1990–2018 identified 29 peer-reviewed publications that fit inclusion criteria for sample size (>100), measurement of prenatal PCB levels in biological samples collected during pregnancy or delivery, and behavioral assessment of children at 3 years old or older (Pessah et al., 2019). The studies included in the analysis examined diverse cohorts from across North America, Asia, and Europe. The review identified 12 publications from 9 different cohorts that examined impacts of PCBs on cognitive function among children aged 3–11 years. Of these, most (8 of 12) found positive associations between PCB exposure and cognitive or language deficits in children (Pessah et al., 2019). A total of 17 studies from 11 different cohorts examined associations of prenatal PCBs with attention, behavioral regulation and social behavior among 3–12-year-old children (Pessah et al., 2019). The majority (10 of 17) reported PCB-related associations with impulse control, hyperactivity, and attention. Only two of the 17 studies examined PCBs in relation to social behavior and autistic traits, with one study reporting that total PCB levels were associated with fewer autistic traits (Nowack, Wittsiepe, Kasper-Sonnenberg, Wilhelm, & Scholmerich, 2015) and the other reporting congener-specific associations with autistic traits (Braun et al., 2014).

As acknowledged by the authors (Pessah et al., 2019), a limitation of the studies included in their review was that associations with specific PCB congeners or mechanism-based classes of congeners were not evaluated largely because most epidemiologic studies report total sum PCB levels or levels of "indicator" PCBs in biological samples. The significance of this limitation is underscored by a recent study suggesting that NDL PCBs with activity at the RyR, but not dioxin-like PCBs or total PCBs, are positively associated with autism (Granillo et al., 2019). The genetic substrate likely also influences the impact of PCBs on autism-relevant outcomes, as illustrated by a recent pilot study not included in this review that identified a trend towards a positive association between PCB 153 and autism in individuals with a deletion mutation in the gene encoding glutathione transferase but not in individuals who did not have this mutation (Bach et al., 2020). Also not included in the summary, because the age of the child at the time of autism diagnosis was not specified, was a 2017 study that found a positive association between PCB exposure and autism (Lyall, Croen, Sjodin, et al., 2017). Furthermore, a 2020 study reported an association between plasma PCB concentrations measured during pregnancy and increased incidence of autistic behaviors in children aged 3–4 years old when the data were analyzed using Bayesian predictive odds ratios (Bernardo et al., 2019). When considering these additional studies that have examined the impact of prenatal PCB on autism phenotypes together with the studies included in the review, most (five of six) report that PCBs increase the expression of autistic traits.

More recently, a meta-analysis was conducted of studies published through 2019 that reported odds ratios for PCB exposure and autism from case-control or cross-sectional studies (Mehri et al., 2021). The authors used random-effects models to examine the association among five different studies representing five independent cohorts from across North America and Finland using pooled odds ratios (ORs) and their 95% confidence intervals (CI). The pooled ORs indicated a significant association between PCB exposure during pregnancy and autism risk (OR, 1.80; 95% CI, 1.26–2.34) (Mehri et al., 2021). One of the studies included in the meta-analysis was a study from 546 mothers and infants in Canada that found children age 3–4 had increased social responsiveness scale (SRS) scores and increased odds of more autistic behavior in the highest plasma PCB concentration (sum of 24 PCB congeners) during pregnancy groups compared to the lowest PCB groups (Bernardo et al., 2019). A new study has since been conducted using the same cohort of 478 Canadian mothers and children but applying different regression models and found the sum of 4 PCBs (PCB 118, PCB 138, PCB 153 and PCB 180) was also associated with higher SRS scores (Alampi et al., 2021). These four PCBs were also identified in another study (Lyall, Croen, Sjodin, et al., 2017) included in the Mehri et al., 2021 meta-analysis that linked prenatal PCB exposure to autism risk. In a California cohort of 545 children with autism, mean levels of prenatal PCBs were higher in this group compared to general population controls, and autism risk was most elevated in the highest *vs.* lowest quartiles of prenatal PCB levels. Associated risk was especially prevalent for several individual congeners, including PCB 138/158, PCB 153, PCB 170 and PCB 180 that all had odds ratios of ~1.5 (Lyall, Croen, Sjodin, et al., 2017). Observations that specific congeners can differentially influence neurodevelopmental outcomes is an area of future study. This is particularly important given mechanistic data indicating that different subgroups of PCBs have different molecular targets (Klocke et al., 2019; Klocke & Lein, 2020). The implications are illustrated by the findings of a previously mentioned study of 104 mother-child pairs that found while total sum PCB levels were not associated with an increased risk of autism, prenatal levels of RyR-active congeners were linked to increased autism risk (Granillo et al., 2019). Thus, it is important to examine PCB links to autism phenotypes based on PCB structure and toxicological activity.

Clearly, these studies suggest that PCBs *per se* are not causing autism, but rather that PCBs increase risk for autism, likely by interacting with genetic susceptibilities. The questions then are which genes are PCBs interacting with to confer autism risk, and what are the mechanism(s) underlying G × PCB interactions? In a study of 206 autism susceptibility genes to determine which were targeted by chemicals included in the Comparative Toxicogenomics Database (CTD), over 1 million interactions were found with PCBs representing one of the major chemical classes of concern found to interact with autism risk genes (Carter & Blizard, 2016). In this study, the top three PCBs, PCB 153, PCB 138 and PCB 126, were found to affect a total 419, 569, and 1243 genes, respectively, and within those affected genes, 22, 20 and 24 were grouped as autism risk genes, respectively (Carter & Blizard, 2016). Below, we summarize the current literature related to how PCBs may interact with genetic risk factors to confer autism risk.

4.1 PCB interactions with xenobiotic metabolism

PCBs are metabolized by CYP enzymes, including *Cyp1a1* (Grimm et al., 2015; Liu et al., 2020). *Cyp1a1* is a transcriptional target of the AhR that is upregulated in the liver by PCB exposure in rodent models (Chubb et al., 2004), and the extent of *Cyp1a1* induction following PCB exposure is influenced by the gut microbiome (Lim et al., 2020). A transcriptomic study found that *CYP1A1* was downregulated in cord blood from individuals with autism *vs.* neurotypical controls, which would be predicted to decrease the rate and/or extent of PCB metabolism (Mordaunt et al., 2019; Thony, Auerbach, & Blau, 2000). This suggests that the population with autism, or at least a subgroup with decreased *CYP1A1* expression, may be more susceptible to PCB effects on neurodevelopment compared to neurotypical controls.

Glutathione-mediated transformation is another major route of PCB detoxification (Mandic-Maravic et al., 2019; Matelski & Van de Water, 2016). Glutathione S-transferase (GST) activity was found to be lower in children with autism *vs.* neurotypical controls (Serajee et al., 2004). In addition, genetic polymorphisms in the glutathione pathway were detected at a higher frequency in autism *vs.* typically developing children, and mothers who expressed these genetic polymorphisms were at higher risk for having a child diagnosed with autism (Rossignol et al., 2014). Importantly, GST genotypes have been shown to influence PCB concentrations in children with autism (Amen et al., 2022). In a study of 125 children from Pakistan with autism, blood concentrations of PCB 77, PCB 118, PCB 128 and PCB 153 were significantly higher in autism cases compared to neurotypical controls (Amen et al., 2022). Conversely, PCB 187 was higher in control blood than autism cases, while other PCB congeners analyzed were not different between populations (Amen et al., 2022). This study also examined the impact on PCB levels of expression of genetic polymorphisms in the glutathione transferase gene, *GSTT1*. PCB 66 and PCB 156 concentrations were higher, while PCB 81 and PCB 118 were lower in the *GSTT1* null genotype compared to the "wildtype" *GSTT1* genotype. Among the autism cases, PCB 44 and PCB 128 concentrations were higher, while PCB 167 was lower in null *vs.* wildtype *GSTT1* genotypes. Similar differences in PCB accumulation were also noted for another glutathione genotype, *GSTM1* (Amen et al., 2022). These findings highlight the role that xenobiotic metabolism may play in determining the PCB profile present in the brain of exposed individuals. Moreover, this study clearly demonstrates that genotypic variations in glutathione genes influence autism risk, underscoring the challenge of conducting epidemiological studies with sufficient power to study causal relationships between PCB exposures and autism risk.

PCB metabolism is also important to consider in light of emerging data that PCB metabolites can have detrimental effects on the central nervous system (Berghuis, Soechitram, Hitzert, Sauer, & Bos, 2013; Dreiem, Rykken, Lehmler, Robertson, & Fonnum, 2009; Hisada et al., 2014; Nomiyama et al., 2019; Sethi, Keil, et al., 2017; Takeuchi et al., 2011; Tehrani & Van Aken, 2014; Uwimana et al., 2019) and that hydroxylated PCB metabolites are detected in human brain samples (Li, Lim, et al., 2022). Hydroxylated metabolites of some PCB congeners inhibited TH-dependent dendritic growth in cultured rodent Purkinje neurons (Kimura-Kuroda, Nagata, & Kuroda, 2005, 2007), while other hydroxylated PCB metabolites induced TH-independent dendritic growth in Purkinje neurons (Kimura-Kuroda et al., 2007). PCB 11 and its hydroxylated and sulfated metabolites enhanced rat hippocampal and cortical dendritic and axonal complexity *in vitro*, although the sulfated metabolites were more potent than the parent compound (Sethi, Keil, et al., 2017). PCBs have also been shown to trigger apoptosis of hippocampal neurons *in vitro* and *in vivo* via oxidative stress (Howard, Fitzpatrick, Pessah, Kostyniak, & Lein, 2003; Yang & Lein, 2010), and hydroxylated PCB metabolites similarly induced apoptosis of cultured rat cerebellar granule cells via oxidative stress, in part through a mechanism involving ERK1/2 signaling (Dreiem et al., 2009). Thus, while the ultimate purpose of PCB metabolism is to facilitate elimination of PCBs from the body, the metabolites appear to have deleterious effects on critical neurodevelopmental processes implicated in the pathogenesis of autism. Moreover, these observations support the hypothesis that heritable deficits in metabolic pathways can exacerbate the developmental neurotoxicity of PCBs.

4.2 PCB interactions with the gut microbiome

Exposure to PCBs has recently been demonstrated to alter the gut microbiome in mice exposed throughout gestation and lactation to PCBs in the maternal diet. Specifically, in offspring exposed developmentally to the MARBLES (Markers of Autism Risk in Babies—Learning Early Signs) PCB mixture, which simulates the relative proportions of the twelve most abundant PCB congeners found in the serum of pregnant women living in Northern California who are at increased risk for having a child with autism (Hertz-Picciotto, Schmidt, et al., 2018), the beta diversity of the gut microbiome was significantly altered compared to vehicle controls (Rude et al., 2019). This change was further exacerbated by expression of heritable human mutations in genes that alter the fidelity of calcium signaling (CGG premutation plus gain-of-function RyR mutation) (Rude, Keogh, & Gareau, 2019). Adult mice exposed to PCBs were found to have altered gut microbiota characterized by changed alpha diversity and reduced beneficial bacteria, and these effects were exacerbated when combined with a high fat diet (Chi et al., 2018; Petriello, Hoffman, Vsevolozhskaya, Morris, & Hennig, 2018; Wahlang et al., 2021). Recently, it was demonstrated that not only does PCB exposure alter the gut microbiome, but also that the gut microbiome influences PCB metabolism and disposition. Thus, tissue levels of PCBs were higher in conventional mice with a normal microbiome than in germ-free mice that lack a microbiome following comparable PCB exposures (Li, Tian, et al., 2022). The gut microbiome also altered the gut–liver axis as evidenced by differential patterns of gene expression in the liver of conventional *vs.* germ-free mice following exposure to a human-relevant PCB mixture (Lim et al., 2020). Genes that were differentially regulated included *Cyp* genes and other genes involved in PCB metabolism. These findings revealed significant interplay between PCB exposures and the microbiome with the microbiome influencing neurotoxic responses and PCBs influencing microbial diversity in the gut. Whether and how these interactions influence autism risk and/or severity represent an intriguing area of future study.

4.3 PCBs as endocrine disruptors

PCBs have been shown to disrupt endocrine signaling via diverse mechanisms. In humans, PCB levels were reported to be inversely correlated with testosterone levels in adults (Goncharov et al., 2009), and maternal serum PCB levels were linked to altered testosterone and estradiol levels in cord blood of female and male newborns (Cao et al., 2008). PCB levels in maternal serum were also associated with changes in sex-typical behaviors in children age 6–8, generally increasing feminine scores in boys and decreasing feminine scores in girls (Winneke et al., 2014). In rats, developmental exposure to an estrogenic PCB mixture (the commercial mixture Aroclor 1221) changed the sexual differentiation of the adult brain, most notably causing masculinization or defeminization of the female brain (Dickerson et al., 2011; Walker et al., 2014). Thus, PCBs appear capable of influencing steroid hormone levels or signaling pathways, but whether these endocrine effects are causally linked to PCB effects on autism-relevant phenotypes is unknown.

Multiple animal studies have demonstrated that PCB exposures can decrease serum TH levels in pregnant dams and disrupt TH homeostasis in dams and offspring via effects on TH synthesis, transport, and metabolism (Zoeller, 2007). While similar relationships between PCBs and serum TH levels have been reported in humans, recent epidemiologic studies (Itoh et al., 2018) suggest that this may not be a consistent relationship in humans. Early experimental studies demonstrated a causal link between PCB-mediated decreases in serum TH levels and neurotoxic effects. Specifically, TH supplementation prevented motor and auditory deficits induced by developmental exposure to Aroclor 1254 in rats (Goldey & Crofton, 1998) and reversed PCB effects on oligodendrocyte maturation and myelination *in vitro* (Nave & Werner, 2014), both phenotypes associated with autism (Lord et al., 2018). However, experimental animal studies suggest that the cognitive deficits associated with developmental PCB exposure occur independent of decreased serum TH levels. For example, developmental exposures of rats to Aroclor 1254 at levels that significantly reduced maternal serum TH levels were not associated with learning and memory deficits or changes in neuronal progenitor cell proliferation and survival in exposed offspring (Zahalka, Ellis, Goldey, Stanton, & Lau, 2001). Conversely, developmental exposure to Aroclor 1254 at lower levels that did not significantly decrease serum TH levels caused significant performance deficits in the Morris water maze (Yang et al., 2009). Studies with individual PCB congeners also provide little support for the hypothesis that TH deficits mediate the cognitive effects of PCBs. For example, the NDL

PCBs 28, 118 and 153 produced similar deficits in spatial learning and memory, but their effects on serum T4 levels varied from a marked reduction to no effect. In contrast, DL PCBs 77 and 126 significantly reduced serum TH, but had few, if any, adverse effects on cognitive behavior (reviewed in Pessah et al., 2019). The inconsistencies observed across both human and rodent studies raise significant questions regarding a causal association between PCB effects on serum TH levels and cognitive behaviors. Fewer studies have focused on the effects of PCBs on sociability and repetitive behavior in rodent models (reviewed in Klocke & Lein, 2020). However, one study reported that developmental exposure to the MARBLES PCB mixture on behavioral phenotypes in mice were observed independent of altered serum TH levels (Sethi et al., 2021).

Alternatively, it has been proposed that the developmental neurotoxicity of PCBs is mediated by PCB interactions with TH receptors and/or the transcriptional machinery that regulates expression of TH-responsive genes. Early evidence in support of this hypothesis derived from a study showing that gestational exposure of rats to Aroclor 1254 in the maternal diet increased expression of TH-responsive genes in the fetal cortex despite significantly reducing maternal levels of serum TH, suggesting direct effects of PCBs on TH receptors in the fetal brain (reviewed in Zoeller, 2007). However, recent *in vitro* studies of PCBs abundant in the serum of pregnant women at increased risk for having a child diagnosed with autism found no significant agonistic or antagonistic interactions with canonical TH receptors expressed in a TH reporter cell line when exposed to these PCBs singly or in combination over a wide range of concentrations (Sethi et al., 2019). This same study also saw no effect of human-relevant hydroxylated or sulfated metabolites of PCB 11 and PCB 52 on TH receptor activity. These observations are consistent with an earlier study that failed to detect a direct interaction between Aroclor 1254 and TH receptors (Zoeller, 2007). The lack of convincing evidence demonstrating direct molecular interactions of PCBs with TH receptors has led to suggestions that PCBs may affect TH signaling via modulation of crosstalk between TH and other endocrine hormones and nuclear receptors (Kouidhi & Clerget-Froidevaux, 2018). However, to date, there are no experimental data directly linking PCB effects on TH signaling to effects of developmental PCB exposure on cognitive, social or repetitive behaviors. In contrast, evidence of causal relationships between PCB effects on serum TH levels and auditory and motor deficits and myelination suggest the possibility that interactions between PCBs and TH signaling may contribute to a subset of autism phenotypes.

4.4 Epigenetic effects of PCBs

In vitro and *in vivo* models have demonstrated that PCBs can alter epigenetic endpoints relevant to autism. One major target includes the family of genes responsible for maintaining or adding DNA methylation marks, the DNA methyltransferases (*Dnmts*) (Feng et al., 2010; Keil & Lein, 2016; Okano, Bell, Haber, & Li, 1999). PCB 153 decreased DNA methyltransferase activity in cultured embryos (Wu, Zhou, & Ohsako, 2006), a mixture of 15 PCBs reduced expression of *Dnmts* and methyl donor S-adenosylmethionine in the liver of developmentally exposed rats (Desaulniers et al., 2009), and developmental exposure to Aroclor 1221 increased *Dnmt* abundance in sexually dimorphic regions of the rat brain (Walker et al., 2014). These observations suggest that PCB effects on *Dnmt* expression and activity vary depending upon the timing of exposure, the PCB congeners and doses to which the system is exposed, and the target tissue. Evidence also suggests that PCBs alter global DNA methylation levels in the brain following developmental exposure. In a mouse model, developmental exposure to the MARBLES PCB mix caused significant brain hypermethylation compared to vehicle control, an effect unique to male offspring (Laufer et al., 2022). In addition to global changes, developmental exposure to the MARBLES PCB mix also resulted in differentially methylated genes within the brain and placenta of mouse offspring (Laufer et al., 2022). Several of the differentially methylated genes have been linked to PCB-induced developmental neurotoxicity, such as Wnt and calcium signaling molecules, as well as MeCP2 mutations associated with Rett syndrome and autism (Chen et al., 2015; Laufer et al., 2022).

These studies illustrate that PCBs can change DNA methylation in genes relevant to autism and are consistent with epidemiological evidence that PCB exposures are linked to alterations in the epigenome. For example, in a study of 399 Japanese women, serum levels of NDL PCBs were inversely associated with global DNA methylation levels in peripheral leukocytes (Itoh et al., 2014), which is similar to findings in another study of 89 Koreans (Kim et al., 2010) and 70 Greenlandic Inuit Indians (Rusiecki et al., 2008). Others have found levels of dioxin-like PCBs are associated with global DNA hypermethylation (Lind et al., 2013); thus, differences in PCB exposures with respect to congener profiles and levels and timing of exposure likely alter the influence of PCBs on the epigenetic landscape and make epidemiological studies extremely challenging. In addition to global changes, gene specific alterations have also been observed in humans exposed to PCBs. For example, altered methylation of MeCP2 in the cord blood of Japanese infants is correlated with environmental exposure to PCBs (Eguchi et al., 2019). Epigenetic mechanisms may also influence other known genetic causes of autism. For example, 15q11-q13 duplications are associated with a syndromic form of autism, and individuals with this duplication have lower levels of brain LINE-1 repetitive DNA methylation (Mitchell et al., 2012). Interestingly, these same individuals had higher levels of PCB 95 than typically developing controls, and this genetic mutation was the strongest predictor of PCB 95 levels (Mitchell et al., 2012). Considering the important role for epigenetic mechanisms in normal neurodevelopment and the epigenetic disruptions noted in autism, epigenetic modifications provide a convergent target by which PCBs may influence autism risk.

4.5 PCB effects on the immune system

Numerous studies in humans and both *in vivo* and *in vitro* experimental models have demonstrated that PCBs can modulate immune function (Kramer, Hikel, Adams, Hinds, & Moon, 2012; Liu, Lu, Zhong, Wang, & Xu, 2022). For example, it has been reported that developmental PCB exposures suppressed immune responses to diverse vaccines in children in Norway (Hochstenbach et al., 2012), the Netherlands (Stolevik et al., 2013; Weisglas-Kuperus et al., 2000), the Faroe Islands (Heilmann et al., 2010), Sweden (Glynn et al., 2008), and in Inuit infants from Nunavik, Quebec (Dallaire et al., 2004). In the Norwegian cohort, the immunosuppressive effects of PCBs coincided with transcriptional changes of several genes in cord blood related to immune function (Hochstenbach et al., 2012), and with increased infections (Stolevik et al., 2011). These associations are likely influenced by specific PCB congeners since increased rates of childhood infection in Swedish children were observed for PCB 28, PCB 52 and PCB 101, but not for other *mono-* and *di-ortho* substituted PCBs (Glynn et al., 2008). In mononuclear cells cultured from cord blood collected from children in Canada with relatively elevated PCB exposures via high dietary intake of PCB-contaminated fish, release of the proinflammatory cytokine tumor necrosis factor alpha (TNFα) was inversely associated with umbilical cord blood concentrations of PCBs (Bilrha et al., 2003). Gene expression analysis of human peripheral blood mononuclear cells incubated with various PCBs revealed differentially regulated gene expression in the presence of dioxin-like and NDL PCBs, with dioxin-like PCBs eliciting a greater number of changes

(Leijs et al., 2019). Genes that were downregulated were predominately those involved in immune response (Leijs et al., 2019). Rodent studies further strengthen the association of PCBs with altered immune/inflammation responses. In pregnant rats, exposure to Aroclor 1254 increased expression of IL-6, IFNγ and IL-2 in liver and serum (Xu, Guo, Li, Pan, & Ma, 2019). In another rat study, developmental exposure to PCB 126 via the dam increased ILl-1B and IFNγ in the serum of offspring (Ahmed, El-Gareib, & Shaker, 2018). Similarly, gestational exposure to Aroclor mixtures via the maternal diet resulted in differential cytokine expression within the brain of rat offspring and further exacerbated serum cytokine IL-1β and IL-6 increases upon challenge with the inflammatory agent, lipopolysaccharide (LPS) (Bell et al., 2018). The findings in rodent models are consistent with the epidemiological observations (Dietert, 2014).

G × E interaction studies in rodent models have been conducted to better understand the convergence between PCBs and autism risk genes on the immune response. The impact of developmental exposure to the MARBLES PCB mixture on brain and serum cytokine levels was quantified in mice engineered to express genetic mutations that alter the fidelity of calcium signaling (Keil et al., 2019), e.g., a RyR gain-of-function mutation and/or CGG premutation repeat expansions in the FMR1 gene (Matelski et al., 2020). Mutations in *RYR1* and *FMR1* are both linked to dysregulated calcium signaling and confer heightened sensitivity to environmental stressors in multiple cell types from mice (Barrientos et al., 2012; Cao et al., 2012, 2013) and affect immune function in humans and mice (Careaga et al., 2014; Marek et al., 2012; Vukcevic et al., 2013). Dose-dependent effects of PCB exposure on cytokine concentrations were observed in the serum, but not hippocampus, with 13 of 23 serum cytokines increased by PCB exposure. Differential effects of genotype were observed in the serum and hippocampus. Hippocampal cytokines were consistently elevated in RyR mutant mice and also in wildtype animals for some cytokines compared to CGG premutation and double mutant mice, while serum cytokines were mostly elevated in the mutant genotypes compared to the wildtype group. Males had elevated levels of 19 cytokines in the serum and 4 in the hippocampus compared to females, but there were also interactions between sex and genotype for 7 hippocampal cytokines. Only serum levels of the chemokine CCL5 showed an interaction between PCB dose, genotype, and sex. Serum levels of CCL5 generally increased with increasing PCB dose, and this effect was exacerbated in the mutant mouse models (Matelski et al., 2020).

These observations support the hypothesis that the genetic substrate influences PCB effects on cytokine expression. However, it remains unknown whether these interactions are causally related to autism risk. Another notable data gap is the effects of PCBs on microglia. Given the critical role that microglia play in sculpting neural circuitry in the developing brain (Sullivan & Ciernia, 2022), this would appear to be an important question that warrants investigation.

4.6 PCB interactions with genes that regulate calcium-dependent signaling

PCBs have been shown to modulate signaling pathways important for regulating dendritic arborization, synaptogenesis and synaptic connectivity (Klocke et al., 2019; Klocke & Lein, 2020). Dendritic morphology is a critical determinant of synaptic connectivity and changes in dendritic morphology (increased or decreased dendrite number, branching and/or spine density) contribute to the altered patterns of neuronal connectivity observed in many neurodevelopmental disorders, including autism (Alaerts et al., 2016; Cooper et al., 2017; Coskun, Loveland, Pearson, Papanicolaou, & Sheth, 2013; Keown et al., 2013; Khan et al., 2015). Various NDL PCBs have been shown to promote dendritic arborization and spine formation in hippocampal and cortical neurons both *in vitro* and *in vivo* (Keil, Sethi, & Lein, 2018; Keil Stietz et al., 2021; Lesiak et al., 2014; Sethi et al., 2018; Wayman, Yang, et al., 2012; Yang et al., 2009, 2014). These morphometric effects of NDL PCBs are largely mediated by calcium-dependent signaling pathways (reviewed in Panesar et al., 2020). Several NDL PCBs, exemplified by PCB 95 and PCB 136, trigger calcium signaling pathways via interactions with RyRs that enhance the open probability of these calcium ion channels embedded in the membrane of the endoplasmic reticulum, which effectively increases intracellular calcium concentrations (Pessah et al., 2010; Wayman, Bose, et al., 2012). These effects of PCB 95 and PCB 136 on RyR activity (Wayman, Yang, et al., 2012; Yang et al., 2014) trigger calcium-dependent signaling pathways, which in turn activate CREB-dependent transcriptional mechanisms (Lesiak et al., 2014; Wayman, Bose, et al., 2012) and mTOR-dependent translational mechanisms (Keil, Miller, Chen, et al., 2018) to stimulate dendritic growth and spine formation. The dendrite promoting effects of PCB 95 and 136 were blocked by pharmacological antagonism or siRNA knockdown of RyRs. PCB 66, a congener with physicochemical properties comparable to that of PCB 95 that lacks RyR activity, did not promote dendritic growth or spine formation (Lesiak et al., 2014; Wayman, Bose, et al., 2012; Yang et al., 2014). Interestingly, point mutations in the RyR that increase the sensitivity of these ion channels to PCB 95 in animal models (Kim et al., 2013; Ta & Pessah, 2007) have been identified in autism candidate gene studies (Lu & Cantor, 2012). Collectively, these observations suggest that RyRs represent a convergent target for G × PCB interactions that confer autism risk.

While PCB 95 and PCB 136 were found to enhance dendritic arborization via RyR-dependent mechanisms that converged on CREB-mediated transcription (Wayman, Bose, et al., 2012), a lower chlorinated contemporary PCB of concern, PCB 11, also has been found to promote dendritic growth in cultured hippocampal and cortical neurons via CREB-dependent mechanisms (Sethi et al., 2018) despite having negligible RyR activity (Sethi et al., 2019). The upstream calcium signaling molecules that link PCB 11 to CREB activation are as yet unknown. CREB dysfunction has been observed in autism (Barnby et al., 2005; Ngounou Wetie et al., 2015; Russo, Albert, & Bowman, 2017; Todd & Mack, 2001; Zheng et al., 2016), suggesting that heritable deficits in CREB signaling may be a more universal convergent target than RyRs in G × PCB interactions that increase autism risk.

PCB effects on dendritic growth have been demonstrated to coincide with autism-relevant behavioral phenotypes. Gestational and lactational exposure to Aroclor 1254 at 1 or 6 mg/kg/d (Yang et al., 2009) or to PCB 95 at 0.1, 1, but not 6, mg/kg/d (Wayman, Yang, et al., 2012) in the diet of rat dams increased dendritic complexity of pyramidal neurons in the hippocampus, cortex and cerebellum of offspring not subjected to training in the Morris Water maze. Experience-dependent dendritic growth associated with Morris water maze training was significantly diminished or even blocked in animals developmentally exposed to PCBs (Yang et al., 2009). PCB restriction of experience-dependent dendritic growth was associated with deficits in spatial learning and memory (Yang et al., 2009). Developmental exposure to PCB 95 also interfered with the synaptic plasticity and organization of the auditory cortex in rats, causing significant E/I imbalance (Kenet, Froemke, Schreiner, Pessah, & Merzenich, 2007). Establishing that RyR-dependent effects of PCBs on dendritic growth and synaptic plasticity are causally related to PCB effects on behavior is difficult since pharmacological or genetic deletion of RyR is embryolethal and in postnatal animals, cardiotoxic. However, a recent study using a larval zebrafish model to compare behavioral phenotypes following developmental exposures to PCBs with varying RyR activity ranging from little to very potent activity demonstrated that the structure-activity relationship (SAR) for RyR activity in zebrafish tissues predicted the SAR for PCB effects on behavior (Yaghoobi et al., 2022). This finding, coupled with observations that PCB causes comparable dose-response relationships for dendritic arborization and cognitive behavior in rodent models (Yang et al., 2009), provide strong support for a causal relationship between RyR-dependent effects of PCBs on dendritic growth and PCB-induced deficits in learning and memory.

Connectivity is influenced by not only the structural complexity of dendrites, but also by spine density (Chaudry & Vasudevan, 2022; Nanou & Catterall, 2018). In cultured primary rat hippocampal neurons and brain slices, PCB 95 significantly increased dendritic spine formation via RyR signaling and CREB-dependent upregulation of miR132 (Lesiak et al., 2014). The necessity of miR132 in this pathway is of particular interest since miR132 is also upregulated in the brain in response to valproic acid—a widely used model to induce autism phenotypes in mice (Hara et al., 2017). miR132 may be a convergent target for diverse environmental chemicals that confer autism risk. Interestingly, miR132 represses expression of MeCP2 (Klein et al., 2007), and dysfunction of MeCP2 is implicated in significant cognitive impairment, while, as mentioned earlier, both mutations and duplications of the *MECP2* gene are associated with Rett syndrome and autism (Cukier et al., 2012; LaSalle & Yasui, 2009; Percy, 2011). MiR132 has also been shown to interact with FMRP to regulate synapse formation (Edbauer et al., 2010).

Genetic mouse models engineered to express heritable human mutations that alter the fidelity of calcium signaling have been leveraged to test the hypothesis that genetic mutations converging on the same signaling pathways as PCBs increase susceptibility to the neurotoxic effects of PCBs. Mice that expressed a human gain-of-function mutation in the RyR, a CGG repeat expansion in the premutation range in the *Fmr1* gene, or both genetic mutations were exposed throughout gestation and lactation to varying doses of the MARBLES PCB mixture in the maternal diet. Morphometric and behavioral analyses of these mice revealed that dendritic morphogenesis and behavior were modulated by complex G × E interactions (Keil Stietz et al., 2021; Sethi et al., 2021). For example, in mouse cortical and hippocampal neurons, genotype influenced the dendritic response to PCBs in a dose, sex and brain region dependent manner. Expression of the RyR mutation or both the RyR and CGG expansion mutations increased dendritic complexity at lower doses of PCBs compared to the dose required to elicit the same dendritic response in wildtype neurons (Keil Stietz et al., 2021). Behavioral studies indicated that developmental exposure of wildtype males to the MARBLES PCB mixture phenocopied social behavior phenotypes observed in mice expressing heritable human mutations in calcium signaling, while expression of these mutations alleviated PCB effects on ultrasonic vocalizations and repetitive behavior, and modified the dose-response relationships and sex-dependent effects of PCBs on social behavior (Sethi et al., 2021). Additional studies using these same mouse models and PCB exposures indicated that G × E interactions influence the microbiome and intestinal physiology (Rude, Pusceddu, et al., 2019), as well as serum cytokine and chemokine levels (Matelski et al., 2020). Collectively, these observations support the hypothesis that heritable deficits in signaling pathways also impacted by environmental risk factors modify the neurotoxic response, which in turn influence autism risk.

4.7 PBDEs as autism risk factors

Polybrominated diphenyl ethers (PBDEs) were produced for use as flame retardants in diverse consumer products, including furniture and children's clothing. These persistent organic pollutants (POPs) are of growing concern as putative autism risk factors because of widespread human exposure and their structural and toxicological similarity to NDL PCBs (Dingemans, Kock, & van den Berg, 2016; Pessah et al., 2019). Developmental exposure to BDE-47, one of the more prevalent of the 209 possible PBDE congeners detected in human tissues, impaired the performance of MeCP2 mutant mice in the Morris Water Maze task (Woods et al., 2012). *In vitro* studies revealed that PBDEs altered calcium signaling in primary rodent neurons via RyR-dependent mechanisms (Kim et al., 2011), modulated glutamate receptor signaling, disrupted TH signaling and increased production of reactive oxygen species via effects on mitochondrial function (Costa, Pellacani, Dao, Kavanagh, & Roque, 2015; Costa, Tagliaferri, Roque, & Pellacani, 2016; Dingemans et al., 2016). PBDEs have also been shown to alter hepatic biotransformation and *CYP* expression, an effect that varied depending on the enterotype (conventional *vs.* germ-free) of the exposed animal (Li et al., 2017).

Curiously, in contrast to NDL PCBs, neither BDE 47 nor BDE 49, two congeners detected in the human gestational environment, promoted dendritic growth in primary rat hippocampal neurons (Chen, Lin, Dang, & Puschner, 2017). Rather both congeners delayed neuronal polarization resulting in significant inhibition of axonal outgrowth during the first few days *in vitro*. The axon inhibitory effects of these PBDE congeners occurred independent of cytotoxicity, and were blocked by pharmacological antagonism of RyR or siRNA knockdown of RyR2. These results demonstrate that the molecular and cellular mechanisms by which PBDEs interfere with neurodevelopment overlap are distinct from those of NDL PCBs. In addition, altered patterns of neuronal connectivity may contribute to the developmental neurotoxicity of PBDEs and increase autism risk in individuals with heritable autism risk factors that converge on calcium signaling and/or pathways that regulate neuronal connectivity (Table 2).

Epidemiological observations suggest that PBDE exposures are associated with reduced IQ and behavioral abnormalities (Pessah et al., 2019; Zhang et al., 2017). Associations between developmental PBDE exposures and abnormal executive function and attention control in young children, as well as impairments in externalizing behaviors (Vuong et al., 2018), have been reported. However, the literature regarding PBDE links to autism is conflicting. In one study of pregnant women, PBDE 28 levels were positively correlated with SRS (autistic behavior) scores in children aged 4–5, while levels of PBDE 85 were linked to fewer autistic behaviors (Braun et al., 2014). In another study of 100 children, PBDE levels were similar in children with autism *vs.* typically developing controls; however; there were no data regarding levels of PBDEs during prenatal development (Hertz-Picciotto et al., 2011). In a study of mother and child pairs in Northern California, maternal serum levels of PBDEs were lower in mothers with a child with autism; yet, sex may play a role in these associations as these effects were prominent in boys for whom there was an indication of increased autism risk, especially with PBDE 28 and PBDE 47, but not in girls (Lyall, Croen, Weiss, et al., 2017).

5. Evolving areas of research

5.1 Emerging chemicals of concern

Increased awareness of the potential influence of environmental exposures on autism risk and/or severity has driven interest in determining which of the 80,000+ chemicals in the human chemosphere pose the greatest autism risk. This in turn has stimulated the development of experimental models for testing the impacts of environmental chemicals on phenotypes of relevance to autism. Data from these studies combined with exposure monitoring data have identified several chemical classes of emerging concern, notably per- and polyfluoroalkyl substances (PFAS), including perfluorooctane sulfonate (PFOS).

In preclinical rodent models, lactational exposure to PFOS decreased learning and memory function and elevated glutamate and GABA levels in the hippocampus of adult mice (Mshaty et al., 2020). Developmental PFOS exposure also caused deficits in motor function, changed short-term and long-term plasticity in Purkinje cells, and altered levels of proteins involved in neurotransmission, including glutamate receptors (Ninomiya et al., 2022). Postnatal exposure to perfluorohexane sulfonate (PFHxS), perfluorooctanoic acid (PFOA) and PFOS altered adult behavior and cognitive function in mice, in part by modulating cholinergic signaling in the brain (Johansson, Fredriksson, & Eriksson, 2008; Viberg, Lee, & Eriksson, 2013). While PFOS are less well studied than PCBs and other legacy pollutants, there is increasing evidence that PFOS are worth investigating as putative autism risk factors.

Epidemiological data suggest that certain types of PFAS, such as PFOA, increase autism risk in a non-monotonic (U-shaped) dose-related manner, while mixtures of PFAS show an inverse dose relationship with autism (Skogheim et al., 2021). PFAS were found in serum of women in California with a child age 2–5, and longer breastfeeding was linked to lower serum levels in the mothers, indicating that lactation may be a source of PFAS in children (Kim, Bennett, Calafat, Hertz-Picciotto, & Shin, 2020). In a study of mother-child pairs in the CHARGE study (CHildhood Autism Risk from Genetics and Environment), PFHxS and PFOS were higher in the serum of women who had a child with autism *vs.* typically developing controls (Shin, Bennett, Calafat, Tancredi, & Hertz-Picciotto, 2020). In 173 mother-child pairs from the MARBLES study, prenatal PFOA and perfluorononanoate (PFNA) levels were positively associated with autism risk, while PFHxS showed a negative association with autism risk (Oh, Bennett, Calafat, et al., 2021). In other studies, serum PFAS levels in children aged 5 and 7 were associated with behavioral problems at age 7 (Oulhote, Steuerwald, Debes, Weihe, & Grandjean, 2016). In 140 mother-child pairs from the MARBLES study, prenatal PFOA was inversely associated with scores for the Early Learning Composite, which is a measure of motor, visual and language function in children aged 2–3 (Oh, Schmidt, Tancredi, et al., 2021). However, two other studies that sampled maternal sera during early to mid-pregnancy found no or inverse associations between PFAS and autism diagnosis (Liew et al., 2015; Lyall et al., 2018).

The different outcomes observed between these studies likely result from differences in the study populations, including, location, year and stage of pregnancy at time of collection; differences in methodologies for measuring exposures; as well as well differences in clinical criteria used to diagnose autism. Despite the inconsistent findings, the available data suggest that PFAS warrant further investigation to better understand how much of a risk they pose to genetically susceptible individuals and to elucidate their mechanism(s) of developmental neurotoxicity to identify which genetic substrates are most susceptible.

5.2 Influence of environmental risk factors on autism co-morbidities

Children diagnosed with autism often suffer from several comorbid conditions, including GI complications as well as bladder dysfunction (von Gontard, Baeyens, Van Hoecke, Warzak, & Bachmann, 2011). The role of G × E interactions in inducing these comorbid conditions is currently understudied and discussed in a recent review (Panesar et al., 2020). As mentioned earlier, experimental evidence suggests that PCB exposures alter the gut microbiome (Li et al., 2017; Lim et al., 2020), which would be expected to alter GI function. PCBs may also contribute to GI dysfunction via mechanisms independent of effects on the gut microbiome. For example, developmental exposure to the MARBLES PCB mixture altered intestinal permeability and expression of inflammatory mediators in wild type mice, and these effects were exacerbated in mice with heritable mutations in calcium signaling (Li et al., 2017; Rude, Keogh, & Gareau, 2019). Interestingly, developmental exposure to the MARBLES PCB mixture influenced the ability of adult mice to urinate (Kennedy et al., 2022). Mechanistic studies of these observations are ongoing but suggest two testable possibilities: (*i*) PCBs directly affect peripheral target organs such as the intestine and bladder; and (*ii*) PCBs alter intestinal and bladder function indirectly via effects on the autonomic nervous system that innervates these organs. The latter hypothesis is derived from data linking disturbances in the autonomic nervous system to GI, bladder, and immune dysfunction (Matteoli & Boeckxstaens, 2013; Muller et al., 2020; Roy & Green, 2019). The peripheral nervous system may be a convergence point through which environmental factors such as PCBs influence the manifestation of autism comorbidities. How the genetic substrate ties into these phenotypes is currently not understood, but the mechanisms detailed above for the central nervous system likely play a role in the peripheral nervous system.

6. Concluding remarks

A significant contribution from environmental factors in determining autism risk and severity is consistent with both the rapid increase in autism rates and the clinical heterogeneity that is the hallmark of this neurodevelopmental condition. However, this phenotypic heterogeneity combined with the complex genetic substrate of autism significantly increases the challenge of identifying specific environmental factors that confer increased risk for autism as well as the genetic susceptibilities that are particularly vulnerable to these environmental factors. To close this significant data gap, we endorse the suggestion that research efforts focus on leveraging experimental models to identify mechanisms by which environmental factors interact with genetic factors to increase the risk of adverse neurodevelopmental outcomes of relevance to autism. Such mechanistic insights can then be used to develop rapid throughput screens for potential environmental risk factors, which in turn will inform and focus epidemiological studies.

The success of this strategy is predicated on choosing appropriate model systems. While a number of transgenic animals (mice, and more recently, rats) expressing autism-linked genetic mutations have been or are being developed, the costs associated with using these models for developmental neurotoxicity exposure studies limits their utility for early mechanistic and screening approaches (Lein et al., 2005). Such models do, however, serve an important role in confirming putative G × E interactions identified in simpler systems. But for initial mechanistic and screening studies, we propose the use of *in vitro* models (such as primary neuronal cell cultures or human iPSC-derived neural systems) or simple systems-based models (such as *C. elegans* or *D. rerio*) that recapitulate the neurodevelopmental processes disrupted in autism. Ongoing research demonstrates that such models can be adapted for rapid throughput screening of multiple culture wells simultaneously using, for example, high content imaging systems to analyze dendritic morphology and synaptic density (Harrill, Robinette, & Mundy, 2011), microelectrode arrays to measure network activity as a measure of functional connectivity (Johnstone et al., 2010; Robinette, Harrill, Mundy, & Shafer, 2011), the FLIPR® to assess spontaneous calcium oscillations (Cao, Hammock, et al., 2012) and "omic" technologies to identify changes in molecular and biochemical targets (Modafferi et al., 2021). *In silico* approaches that leverage databases, such as the Comparative

Toxicogenomics Database, to interrogate chemical influences on the hundreds of autism susceptibility genes identified to date (Carter & Blizard, 2016) would also be an effective high-throughput approach for identifying potential autism-relevant G × E interactions.

Another consideration in studies of G × E interactions in autism is cross-talk between signaling systems and G × E mechanisms that influence neuronal connectivity, as well as between chemicals in mixtures (Fig. 1). For example, endocrine signaling regulates gene transcription of proinflammatory cytokines, and soluble mediators of the innate immune system, such as cytokines, can feed back on the brain to regulate endocrine signaling (Irwin & Cole, 2011). The immune system is also strongly influenced by the gut microbiome, and emerging evidence suggests that the microbiome influences not only peripheral immune calls, but also microglia in the brain (Lebovitz et al., 2018). There is also evidence that microbes within the human gut may play a significant role in the regulation of the brain via their influence on the production of inflammatory cytokines, antimicrobial peptides and neurotransmitters that affect the epigenome (Alam et al., 2017; Yousefi et al., 2022).

Clearly, work is urgently needed to better predict which combination of autism genes and environmental exposures pose the greatest autism risk. The fact that chemical exposures are more readily controlled than genetic

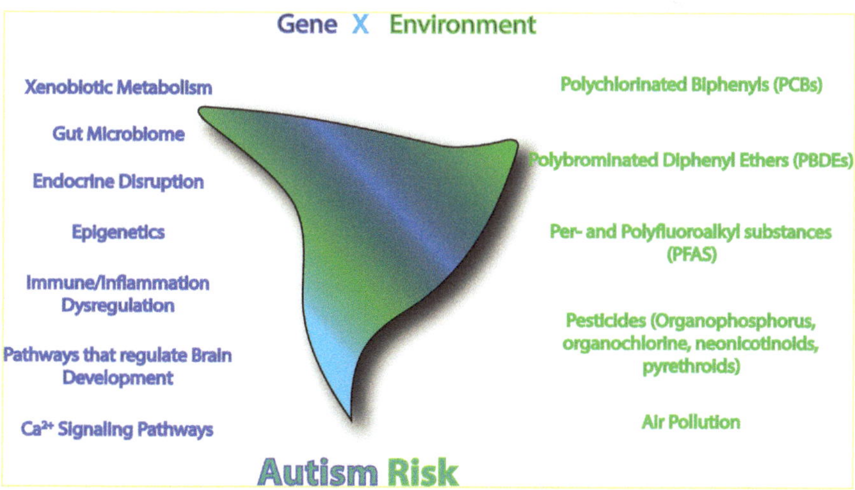

Fig. 1 Autism risk is likely determined by cross talk between multiple signaling pathways, mechanisms of G × E interactions and interactions between structurally and mechanistically diverse chemicals in mixtures.

factors to prevent or mitigate the expression of autism-related traits coupled with the significant toll that autism places on autistic children, their families and society, provides a compelling reason to engage in this endeavor.

Acknowledgments

This work was supported by funding from the National Institute of Environmental Health Sciences (grants R01 ES014901 to PJL, R00 ES029537 to KPKS, and P30 ES023513 to UC Davis) and the Eunice Kennedy Shriver National Institute of Child Health and Human Development (grant P50 HD103526 to the UC Davis MIND Institute).

References

Abu-Elneel, K., Liu, T., Gazzaniga, F. S., Nishimura, Y., Wall, D. P., Geschwind, D. H., et al. (2008). Heterogeneous dysregulation of microRNAs across the autism spectrum. *Neurogenetics*, *9*, 153–161.

Ahmed, R. G., El-Gareib, A. W., & Shaker, H. M. (2018). Gestational 3,3',4,4',5-pentachlorobiphenyl (PCB 126) exposure disrupts fetoplacental unit: Fetal thyroid-cytokines dysfunction. *Life Sciences*, *192*, 213–220.

Alaerts, K., Swinnen, S. P., & Wenderoth, N. (2016). Sex differences in autism: A resting-state fMRI investigation of functional brain connectivity in males and females. *Social Cognitive and Affective Neuroscience*, *11*, 1002–1016.

Alam, R., Abdolmaleky, H. M., & Zhou, J. R. (2017). Microbiome, inflammation, epigenetic alterations, and mental diseases. *American Journal of Medical Genetics. Part B, Neuropsychiatric Genetics*, *174*, 651–660.

Alamdari, S. B., Sadeghi Damavandi, M., Zarei, M., & Khosrowabadi, R. (2022). Cognitive theories of autism based on the interactions between brain functional networks. *Frontiers in Human Neuroscience*, *16*, 828985.

Alampi, J. D., Lanphear, B. P., Braun, J. M., Chen, A., Takaro, T. K., Muckle, G., et al. (2021). Association between gestational exposure to toxicants and autistic behaviors using Bayesian quantile regression. *American Journal of Epidemiology*, *190*, 1803–1813.

Amen, N. E., Eqani, S., Bilal, K., Ali, N., Rajeh, N., Adelman, D., et al. (2022). Molecularly tracing of children exposure pathways to environmental organic pollutants and the Autism Spectrum Disorder Risk. *Environmental Pollution*, *315*, 120381.

Ampleman, M. D., Martinez, A., DeWall, J., Rawn, D. F., Hornbuckle, K. C., & Thorne, P. S. (2015). Inhalation and dietary exposure to PCBs in urban and rural cohorts via congener-specific measurements. *Environmental Science & Technology*, *49*, 1156–1164.

Ansel, A., Rosenzweig, J. P., Zisman, P. D., Melamed, M., & Gesundheit, B. (2016). Variation in gene expression in autism spectrum disorders: an extensive review of transcriptomic studies. *Frontiers in Neuroscience*, *10*, 601.

Arambula, S. E., & McCarthy, M. M. (2020). Neuroendocrine-immune crosstalk shapes sex-specific brain development. *Endocrinology*, *161*(6), bqaa055. https://doi.org/10.1210/endocr/bqaa055.

Arndt, T. L., Stodgell, C. J., & Rodier, P. M. (2005). The teratology of autism. *International Journal of Developmental Neuroscience*, *23*, 189–199.

Baba, T., Ito, S., Yuasa, M., Yoshioka, E., Miyashita, C., Araki, A., et al. (2018). Association of prenatal exposure to PCDD/Fs and PCBs with maternal and infant thyroid hormones: The Hokkaido Study on Environment and Children's Health. *Sci Total Environ*, *615*, 1239–1246.

Bach, M. A., Samms-Vaughan, M., Hessabi, M., Bressler, J., Lee, M., Zhang, J., et al. (2020). Association of polychlorinated biphenyls and organochlorine pesticides with autism spectrum disorder in Jamaican children. *Research in Autism Spectrum Disorder*, *76*, 101587.
Balaguer-Trias, J., Deepika, D., Schuhmacher, M., & Kumar, V. (2022). Impact of contaminants on microbiota: linking the gut-brain axis with neurotoxicity. *International Journal of Environmental Research and Public Health*, *19*.
Baribeau, D. A., & Anagnostou, E. (2013). A comparison of neuroimaging findings in childhood onset schizophrenia and autism spectrum disorder: A review of the literature. *Frontiers in Psychiatry*, *4*, 175.
Barmpas, M., Vakonaki, E., Tzatzarakis, M., Sifakis, S., Alegakis, A., Grigoriadis, T., et al. (2019). Organochlorine pollutants' levels in hair, amniotic fluid and serum samples of pregnant women in Greece. A cohort study. *Environmental Toxicology and Pharmacology*, *73*, 103279.
Barnby, G., Abbott, A., Sykes, N., Morris, A., Weeks, D. E., Mott, R., et al. (2005). Candidate-gene screening and association analysis at the autism-susceptibility locus on chromosome 16p: Evidence of association at GRIN2A and ABAT. *American Journal of Human Genetics*, *76*, 950–966.
Barrientos, G. C., Feng, W., Truong, K., Matthaei, K. I., Yang, T., Allen, P. D., et al. (2012). Gene dose influences cellular and calcium channel dysregulation in heterozygous and homozygous T4826I-RYR1 malignant hyperthermia-susceptible muscle. *The Journal of Biological Chemistry*, *287*, 2863–2876.
Bell, M. R., Dryden, A., Will, R., & Gore, A. C. (2018). Sex differences in effects of gestational polychlorinated biphenyl exposure on hypothalamic neuroimmune and neuromodulator systems in neonatal rats. *Toxicology and Applied Pharmacology*, *353*, 55–66.
Bell, M. R., Hart, B. G., & Gore, A. C. (2016). Two-hit exposure to polychlorinated biphenyls at gestational and juvenile life stages: 2. Sex-specific neuromolecular effects in the brain. *Molecular and Cellular Endocrinology*, *420*, 125–137.
Belmonte, M. K., & Bourgeron, T. (2006). Fragile X syndrome and autism at the intersection of genetic and neural networks. *Nature Neuroscience*, *9*, 1221–1225.
Berghuis, S. A., Soechitram, S. D., Hitzert, M. M., Sauer, P. J., & Bos, A. F. (2013). Prenatal exposure to polychlorinated biphenyls and their hydroxylated metabolites is associated with motor development of three-month-old infants. *Neurotoxicology*, *38*, 124–130.
Berman, R. F., Murray, K. D., Arque, G., Hunsaker, M. R., & Wenzel, H. J. (2012). Abnormal dendrite and spine morphology in primary visual cortex in the CGG knock-in mouse model of the fragile X premutation. *Epilepsia*, *53*(Suppl. 1), 150–160.
Bernardo, B. A., Lanphear, B. P., Venners, S. A., Arbuckle, T. E., Braun, J. M., Muckle, G., et al. (2019). Assessing the relation between plasma PCB concentrations and elevated autistic behaviours using bayesian predictive odds ratios. *International Journal of Environmental Research and Public Health*, *16*(3), 457. https://doi.org/10.3390/ijerph16030457.
Bertoletti, A. C. C., Peres, K. K., Faccioli, L. S., Vacci, M. C., Mata, I. R. D., Kuyven, C. J., et al. (2022). Early exposure to agricultural pesticides and the occurrence of autism spectrum disorder: A systematic review. *Rev Paul Pediatr*, *41*, e2021360.
Bertotto, L. B., Catron, T. R., & Tal, T. (2020). Exploring interactions between xenobiotics, microbiota, and neurotoxicity in zebrafish. *Neurotoxicology*, *76*, 235–244.
Bilrha, H., Roy, R., Moreau, B., Belles-Isles, M., Dewailly, E., & Ayotte, P. (2003). In vitro activation of cord blood mononuclear cells and cytokine production in a remote coastal population exposed to organochlorines and methyl mercury. *Environmental Health Perspectives*, *111*, 1952–1957.
Bishop, D. V., Whitehouse, A. J., Watt, H. J., & Line, E. A. (2008). Autism and diagnostic substitution: evidence from a study of adults with a history of developmental language disorder. *Developmental Medicine and Child Neurology*, *50*, 341–345.

Bjorklund, G., Dosa, M. D., Maes, M., Dadar, M., Frye, R. E., Peana, M., et al. (2021). The impact of glutathione metabolism in autism spectrum disorder. *Pharmacological Research*, *166*, 105437.

Blagburn-Blanco, S. V., Chappell, M. S., De Biase, L. M., & DeNardo, L. A. (2022). Synapse-specific roles for microglia in development: New horizons in the prefrontal cortex. *Frontiers in Molecular Neuroscience*, *15*, 965756.

Bocchi, R., Egervari, K., Carol-Perdiguer, L., Viale, B., Quairiaux, C., De Roo, M., et al. (2017). Perturbed Wnt signaling leads to neuronal migration delay, altered interhemispheric connections and impaired social behavior. *Nature Communications*, *8*, 1158.

Bolte, S., Girdler, S., & Marschik, P. B. (2019). The contribution of environmental exposure to the etiology of autism spectrum disorder. *Cellular and Molecular Life Sciences*, *76*, 1275–1297.

Bourgeron, T. (2009). A synaptic trek to autism. *Current Opinion in Neurobiology*, *19*, 231–234.

Bowen, K. K., Dempsey, R. J., & Vemuganti, R. (2011). Adult interleukin-6 knockout mice show compromised neurogenesis. *Neuroreport*, *22*, 126–130.

Braun, J. M., Kalkbrenner, A. E., Just, A. C., Yolton, K., Calafat, A. M., Sjodin, A., et al. (2014). Gestational exposure to endocrine-disrupting chemicals and reciprocal social, repetitive, and stereotypic behaviors in 4- and 5-year-old children: the HOME study. *Environmental Health Perspectives*, *122*, 513–520.

Buxbaum, J. D., & Hof, P. R. (2011). The emerging neuroscience of autism spectrum disorders. *Brain Research*, *1380*, 1–2.

Cao, Z., Hammock, B. D., McCoy, M., Rogawski, M., Lein, P., & Pessah, I. (2012). Tetramethylenedisulfotetramine
alters Ca2+ dynamics in cultured hippocampal neurons: Mitigation by NMDA blockade and GABAA receptor positive modulation. *Toxicological Sciences*, *130*(2), 362–372. https://doi.org/10.1093/toxsci/kfs244.

Cao, Z., Hulsizer, S., Cui, Y., Pretto, D. L., Kim, K. H., Hagerman, P. J., et al. (2013). Enhanced asynchronous Ca(2+) oscillations associated with impaired glutamate transport in cortical astrocytes expressing Fmr1 gene premutation expansion. *The Journal of Biological Chemistry*, *288*, 13831–13841.

Cao, Z., Hulsizer, S., Tassone, F., Tang, H. T., Hagerman, R. J., Rogawski, M. A., et al. (2012). Clustered burst firing in FMR1 premutation hippocampal neurons: amelioration with allopregnanolone. *Human Molecular Genetics*, *21*, 2923–2935.

Cao, Y., Winneke, G., Wilhelm, M., Wittsiepe, J., Lemm, F., Furst, P., et al. (2008). Environmental exposure to dioxins and polychlorinated biphenyls reduce levels of gonadal hormones in newborns: results from the Duisburg cohort study. *International Journal of Hygiene and Environmental Health*, *211*, 30–39.

Careaga, M., Rose, D., Tassone, F., Berman, R. F., Hagerman, R., & Ashwood, P. (2014). Immune dysregulation as a cause of autoinflammation in fragile X premutation carriers: Link between FMR1 CGG repeat number and decreased cytokine responses. *PLoS One*, *9*, 1–8. https://doi.org/10.1371/journal.pone.0094475.

Carlsson, T., Molander, F., Taylor, M. J., Jonsson, U., & Bolte, S. (2020). Early environmental risk factors for neurodevelopmental disorders—A systematic review of twin and sibling studies. *Development and Psychopathology*, 1–48.

Carpenter, D. O. (2006). Polychlorinated biphenyls (PCBs): Routes of exposure and effects on human health. *Reviews on Environmental Health*, *21*, 1–23.

Carter, C. J., & Blizard, R. A. (2016). Autism genes are selectively targeted by environmental pollutants including pesticides, heavy metals, bisphenol A, phthalates and many others in food, cosmetics or household products. *Neurochemistry International*.

Chan-Hon-Tong, A., Charles, M. A., Forhan, A., Heude, B., & Sirot, V. (2013). Exposure to food contaminants during pregnancy. *Sci Total Environ*, *458-460*, 27–35.

Chaudry, S., & Vasudevan, N. (2022). mTOR-dependent spine dynamics in autism. *Frontiers in Molecular Neuroscience*, *15*, 877609.

Chen, L., Chen, K., Lavery, L. A., Baker, S. A., Shaw, C. A., Li, W., et al. (2015). MeCP2 binds to non-CG methylated DNA as neurons mature, influencing transcription and the timing of onset for Rett syndrome. *Proceedings of the National Academy of Sciences of the United States of America*, *112*, 5509–5514.

Chen, X., Lin, Y., Dang, K., & Puschner, B. (2017). Quantification of polychlorinated biphenyls and polybrominated diphenyl ethers in commercial cows' milk from California by gas chromatography-triple quadruple mass spectrometry. *PLoS One*, *12*, e0170129.

Chen, H., Streifel, K. M., Singh, V., Yang, D., Mangini, L., Wulff, H., et al. (2017). From the cover: BDE-47 and BDE-49 inhibit axonal growth in primary rat hippocampal neuron-glia co-cultures via ryanodine receptor-dependent mechanisms. *Toxicological Sciences*, *156*, 375–386.

Chen, Y., Tassone, F., Berman, R. F., Hagerman, P. J., Hagerman, R. J., Willemsen, R., et al. (2010). Murine hippocampal neurons expressing Fmr1 gene premutations show early developmental deficits and late degeneration. *Human Molecular Genetics*, *19*, 196–208.

Cheroni, C., Caporale, N., & Testa, G. (2020). Autism spectrum disorder at the crossroad between genes and environment: contributions, convergences, and interactions in ASD developmental pathophysiology. *Molecular Autism*, *11*, 69.

Chi, Y., Lin, Y., Zhu, H., Huang, Q., Ye, G., & Dong, S. (2018). PCBs-high-fat diet interactions as mediators of gut microbiota dysbiosis and abdominal fat accumulation in female mice. *Environmental Pollution*, *239*, 332–341.

Chonchaiya, W., Au, J., Schneider, A., Hessl, D., Harris, S. W., Laird, M., et al. (2012). Increased prevalence of seizures in boys who were probands with the FMR1 premutation and co-morbid autism spectrum disorder. *Human Genetics*, *131*, 581–589.

Christensen, D. L., Baio, J., Van Naarden Braun, K., Bilder, D., Charles, J., Constantino, J. N., et al. (2016). Prevalence and characteristics of autism spectrum disorder among children aged 8 years—Autism and developmental disabilities monitoring network, 11 sites, United States, 2012. *MMWR Surveillance Summaries*, *65*, 1–23.

Christensen, J., Gronborg, T. K., Sorensen, M. J., Schendel, D., Parner, E. T., Pedersen, L. H., et al. (2013). Prenatal valproate exposure and risk of autism spectrum disorders and childhood autism. *JAMA*, *309*, 1696–1703.

Chubb, L. S., Andersen, M. E., Broccardo, C. J., Legare, M. E., Billings, R. E., Dean, C. E., et al. (2004). Regional induction of CYP1A1 in rat liver following treatment with mixtures of PCB 126 and PCB 153. *Toxicologic Pathology*, *32*, 467–473.

Collaer, M. L., & Hines, M. (1995). Human behavioral sex differences: a role for gonadal hormones during early development? *Psychological Bulletin*, *118*, 55–107.

Connolly, S., & Kingsbury, T. J. (2010). Caffeine modulates CREB-dependent gene expression in developing cortical neurons. *Biochemical and Biophysical Research Communications*, *397*, 152–156.

Coo, H., Ouellette-Kuntz, H., Lloyd, J. E., Kasmara, L., Holden, J. J., & Lewis, M. E. (2008). Trends in autism prevalence: diagnostic substitution revisited. *Journal of Autism and Developmental Disorders*, *38*, 1036–1046.

Cooper, R. A., Richter, F. R., Bays, P. M., Plaisted-Grant, K. C., Baron-Cohen, S., & Simons, J. S. (2017). Reduced hippocampal functional connectivity during episodic memory retrieval in autism. *Cerebral Cortex*, *27*, 888–902.

Cory-Slechta, D. A., Merrill, A., & Sobolewski, M. (2022). Air pollution-related neurotoxicity across the life span. *Annual Review of Pharmacology and Toxicology*.

Coskun, M. A., Loveland, K. A., Pearson, D. A., Papanicolaou, A. C., & Sheth, B. R. (2013). Functional assays of local connectivity in the somatosensory cortex of individuals with autism. *Autism Research*, *6*, 190–200.

Costa, L. G., Pellacani, C., Dao, K., Kavanagh, T. J., & Roque, P. J. (2015). The brominated flame retardant BDE-47 causes oxidative stress and apoptotic cell death in vitro and in vivo in mice. *Neurotoxicology*, *48*, 68–76.

Costa, L. G., Tagliaferri, S., Roque, P. J., & Pellacani, C. (2016). Role of glutamate receptors in tetrabrominated diphenyl ether (BDE-47) neurotoxicity in mouse cerebellar granule neurons. *Toxicology Letters*, *241*, 159–166.

Craig, A. M., & Kang, Y. (2007). Neurexin-neuroligin signaling in synapse development. *Current Opinion in Neurobiology*, *17*, 43–52.

Crider, A., Thakkar, R., Ahmed, A. O., & Pillai, A. (2014). Dysregulation of estrogen receptor beta (ERbeta), aromatase (CYP19A1), and ER co-activators in the middle frontal gyrus of autism spectrum disorder subjects. *Molecular Autism*, *5*, 46.

Croen, L. A., Connors, S. L., Matevia, M., Qian, Y., Newschaffer, C., & Zimmerman, A. W. (2011). Prenatal exposure to beta2-adrenergic receptor agonists and risk of autism spectrum disorders. *Journal of Neurodevelopmental Disorders*, *3*, 307–315.

Cukier, H. N., Lee, J. M., Ma, D., Young, J. I., Mayo, V., Butler, B. L., et al. (2012). The expanding role of MBD genes in autism: identification of a MECP2 duplication and novel alterations in MBD5, MBD6, and SETDB1. *Autism Research*, *5*, 385–397.

Dallaire, F., Dewailly, E., Muckle, G., Vezina, C., Jacobson, S. W., Jacobson, J. L., et al. (2004). Acute infections and environmental exposure to organochlorines in Inuit infants from Nunavik. *Environmental Health Perspectives*, *112*, 1359–1365.

Deidda, G., Bozarth, I. F., & Cancedda, L. (2014). Modulation of GABAergic transmission in development and neurodevelopmental disorders: investigating physiology and pathology to gain therapeutic perspectives. *Frontiers in Cellular Neuroscience*, *8*, 119.

Delorme, R., Ey, E., Toro, R., Leboyer, M., Gillberg, C., & Bourgeron, T. (2013). Progress toward treatments for synaptic defects in autism. *Nature Medicine*, *19*, 685–694.

Demeneix, B. A. (2019). Evidence for prenatal exposure to thyroid disruptors and adverse effects on brain development. *Eur Thyroid J*, *8*, 283–292.

Deng, P. Y., Rotman, Z., Blundon, J. A., Cho, Y., Cui, J., Cavalli, V., et al. (2013). FMRP regulates neurotransmitter release and synaptic information transmission by modulating action potential duration via BK channels. *Neuron*, *77*, 696–711.

Dennis, E. L., Jahanshad, N., Rudie, J. D., Brown, J. A., Johnson, K., McMahon, K. L., et al. (2011). Altered structural brain connectivity in healthy carriers of the autism risk gene, CNTNAP2. *Brain Connectivity*, *1*, 447–459.

Desaulniers, D., Xiao, G. H., Lian, H., Feng, Y. L., Zhu, J., Nakai, J., et al. (2009). Effects of mixtures of polychlorinated biphenyls, methylmercury, and organochlorine pesticides on hepatic DNA methylation in prepubertal female Sprague-Dawley rats. *International Journal of Toxicology*, *28*, 294–307.

Deverman, B. E., & Patterson, P. H. (2009). Cytokines and CNS development. *Neuron*, *64*, 61–78.

Dickerson, S. M., Cunningham, S. L., Patisaul, H. B., Woller, M. J., & Gore, A. C. (2011). Endocrine disruption of brain sexual differentiation by developmental PCB exposure. *Endocrinology*, *152*, 581–594.

Diep, A. A., Hunsaker, M. R., Kwock, R., Kim, K., Willemsen, R., & Berman, R. F. (2012). Female CGG knock-in mice modeling the fragile X premutation are impaired on a skilled forelimb reaching task. *Neurobiology of Learning and Memory*, *97*, 229–234.

Dietert, R. R. (2014). Developmental Immunotoxicity, Perinatal Programming, and Noncommunicable Diseases: Focus on Human Studies. *Adv Med*, *2014*, 867805.

Dingemans, M. M., Kock, M., & van den Berg, M. (2016). Mechanisms of action point towards combined PBDE/NDL-PCB risk assessment. *Toxicological Sciences*, *153*, 215–224.

Dreiem, A., Rykken, S., Lehmler, H. J., Robertson, L. W., & Fonnum, F. (2009). Hydroxylated polychlorinated biphenyls increase reactive oxygen species formation and induce cell death in cultured cerebellar granule cells. *Toxicology and Applied Pharmacology*, *240*, 306–313.

Du, W., Bautista, J. F., Yang, H., Diez-Sampedro, A., You, S. A., Wang, L., et al. (2005). Calcium-sensitive potassium channelopathy in human epilepsy and paroxysmal movement disorder. *Nature Genetics*, *37*, 733–738.

Durand, B., Dufour, B., Fraisse, D., Defour, S., Duhem, K., & Le-Barillec, K. (2008). Levels of PCDDs, PCDFs and dioxin-like PCBs in raw cow's milk collected in France in 2006. *Chemosphere*, *70*, 689–693.

Eagleson, K. L., Xie, Z., & Levitt, P. (2017). The pleiotropic MET receptor network: circuit development and the neural-medical interface of autism. *Biological Psychiatry*, *81*, 424–433.

Ebert, D. H., & Greenberg, M. E. (2013). Activity-dependent neuronal signalling and autism spectrum disorder. *Nature*, *493*, 327–337.

Edbauer, D., Neilson, J. R., Foster, K. A., Wang, C. F., Seeburg, D. P., Batterton, M. N., et al. (2010). Regulation of synaptic structure and function by FMRP-associated microRNAs miR-125b and miR-132. *Neuron*, *65*, 373–384.

Eguchi, A., Nishizawa-Jotaki, S., Tanabe, H., Rahmutulla, B., Watanabe, M., Miyaso, H., et al. (2019). An altered DNA methylation status in the human umbilical cord is correlated with maternal exposure to polychlorinated biphenyls. *International Journal of Environmental Research and Public Health*, *16*.

Eisner, A., Gao, Y., Collier, F., Drummond, K., Thomson, S., Burgner, D., et al. (2022). Cord blood immune profile: Associations with higher prenatal plastic chemical levels. *Environmental Pollution*, *120332*.

El-Fishawy, P., & State, M. W. (2010). The genetics of autism: Key issues, recent findings, and clinical implications. *The Psychiatric Clinics of North America*, *33*, 83–105.

Eskenazi, B., Rauch, S. A., Tenerelli, R., Huen, K., Holland, N. T., Lustig, R. H., et al. (2017). In utero and childhood DDT, DDE, PBDE and PCBs exposure and sex hormones in adolescent boys: The CHAMACOS study. *International Journal of Hygiene and Environmental Health*, *220*, 364–372.

Feng, J., Zhou, Y., Campbell, S. L., Le, T., Li, E., Sweatt, J. D., et al. (2010). Dnmt1 and Dnmt3a maintain DNA methylation and regulate synaptic function in adult forebrain neurons. *Nature Neuroscience*, *13*, 423–430.

Foerster, C., Zuniga-Venegas, L., Enriquez, P., Rojas, J., Zamora, C., Munoz, X., et al. (2021). Levels of polychlorinated dibenzo-p-dioxins/furans (PCDD/Fs) and dioxin-like polychlorinated biphenyls (DL-PCBs) in human breast milk in chile: A pilot study. *International Journal of Environmental Research and Public Health*, *18*(9), 4825. https://doi.org/10.3390/ijerph18094825.

Frazier, T. W., Thompson, L., Youngstrom, E. A., Law, P., Hardan, A. Y., Eng, C., et al. (2014). A twin study of heritable and shared environmental contributions to autism. *Journal of Autism and Developmental Disorders*, *44*, 2013–2025.

Fu, P., Guo, X., Cheung, F. M. H., & Yung, K. K. L. (2019). The association between PM2.5 exposure and neurological disorders: A systematic review and meta-analysis. *Sci Total Environ*, *655*, 1240–1248.

Gadient, R. A., Lein, P., Higgins, D., & Patterson, P. H. (1998). Effect of leukemia inhibitory factor (LIF) on the morphology and survival of cultured hippocampal neurons and glial cells. *Brain Research*, *798*, 140–146.

Gama, J., Neves, B., & Pereira, A. (2022). Chronic effects of dietary pesticides on the gut microbiome and neurodevelopment. *Frontiers in Microbiology*, *13*, 931440.

Garay, P. A., & McAllister, A. K. (2010). Novel roles for immune molecules in neural development: implications for neurodevelopmental disorders. *Front Synaptic Neurosci*, *2*, 136.
Geschwind, D. H. (2011). Genetics of autism spectrum disorders. *Trends in Cognitive Sciences*, *15*, 409–416.
Geschwind, D. H., & Levitt, P. (2007). Autism spectrum disorders: developmental disconnection syndromes. *Current Opinion in Neurobiology*, *17*, 103–111.
Glynn, A., Thuvander, A., Aune, M., Johannisson, A., Darnerud, P. O., Ronquist, G., et al. (2008). Immune cell counts and risks of respiratory infections among infants exposed pre- and postnatally to organochlorine compounds: A prospective study. *Environmental Health*, *7*, 62.
Goines, P. E., & Ashwood, P. (2013). Cytokine dysregulation in autism spectrum disorders (ASD): Possible role of the environment. *Neurotoxicology and Teratology*, *36*, 67–81.
Goines, P., & Van de Water, J. (2010). The immune system's role in the biology of autism. *Current Opinion in Neurology*, *23*, 111–117.
Goldey, E. S., & Crofton, K. M. (1998). Thyroxine replacement attenuates hypothyroxinemia, hearing loss, and motor deficits following developmental exposure to Aroclor 1254 in rats. *Toxicological Sciences*, *45*, 94–105.
Goncharov, A., Rej, R., Negoita, S., Schymura, M., Santiago-Rivera, A., Morse, G., et al. (2009). Lower serum testosterone associated with elevated polychlorinated biphenyl concentrations in Native American men. *Environmental Health Perspectives*, *117*, 1454–1460.
Granillo, L., Sethi, S., Keil, K. P., Lin, Y., Ozonoff, S., Iosif, A. M., et al. (2019). Polychlorinated biphenyls influence on autism spectrum disorder risk in the MARBLES cohort. *Environmental Research*, *171*, 177–184.
Grether, J. K., Rosen, N. J., Smith, K. S., & Croen, L. A. (2009). Investigation of shifts in autism reporting in the California Department of Developmental Services. *Journal of Autism and Developmental Disorders*, *39*, 1412–1419.
Grimm, F. A., Hu, D., Kania-Korwel, I., Lehmler, H. J., Ludewig, G., Hornbuckle, K. C., et al. (2015). Metabolism and metabolites of polychlorinated biphenyls. *Critical Reviews in Toxicology*, *45*, 245–272.
Guang, S., Pang, N., Deng, X., Yang, L., He, F., Wu, L., et al. (2018). Synaptopathology involved in autism spectrum disorder. *Frontiers in Cellular Neuroscience*, *12*, 470.
Guedes, J. R., Ferreira, P. A., Costa, J. M., Cardoso, A. L., & Peca, J. (2022). Microglia-dependent remodeling of neuronal circuits. *Journal of Neurochemistry*, *163*, 74–93.
Guo, X., Chandrasekaran, V., Lein, P., Kaplan, P. L., & Higgins, D. (1999). Leukemia inhibitory factor and ciliary neurotrophic factor cause dendritic retraction in cultured rat sympathetic neurons. *The Journal of Neuroscience*, *19*, 2113–2121.
Guo, X., Metzler-Northrup, J., Lein, P., Rueger, D., & Higgins, D. (1997). Leukemia inhibitory factor and ciliary neurotrophic factor regulate dendritic growth in cultures of rat sympathetic neurons. *Brain Research. Developmental Brain Research*, *104*, 101–110.
Hagerman, R., Au, J., & Hagerman, P. (2011). FMR1 premutation and full mutation molecular mechanisms related to autism. *Journal of Neurodevelopmental Disorders*, *3*, 211–224.
Hallmayer, J., Cleveland, S., Torres, A., Phillips, J., Cohen, B., Torigoe, T., et al. (2011). Genetic heritability and shared environmental factors among twin pairs with autism. *Archives of General Psychiatry*, *68*, 1095–1102.
Hannah, T. J., Megson, D., & Sandau, C. D. (2022). A review of the mechanisms of by-product PCB formation in pigments, dyes and paints. *Sci Total Environ*, *852*, 158529.
Hansen, K. F., Karelina, K., Sakamoto, K., Wayman, G. A., Impey, S., & Obrietan, K. (2013). miRNA-132: A dynamic regulator of cognitive capacity. *Brain Structure & Function*, *218*, 817–831.

Hansen, K. F., Sakamoto, K., Aten, S., Price, K. H., Loeser, J., Hesse, A. M., et al. (2016). Targeted deletion of miR-132/-212 impairs memory and alters the hippocampal transcriptome. *Learning & Memory*, *23*, 61–71.

Hara, Y., Ago, Y., Takano, E., Hasebe, S., Nakazawa, T., Hashimoto, H., et al. (2017). Prenatal exposure to valproic acid increases miR-132 levels in the mouse embryonic brain. *Molecular Autism*, *8*, 33.

Harrill, J. A., Robinette, B. L., & Mundy, W. R. (2011). Use of high content image analysis to detect chemical-induced changes in synaptogenesis in vitro. *Toxicology In Vitro: an International Journal Published in Association with BIBRA*, *25*, 368–387.

Heilmann, C., Budtz-Jorgensen, E., Nielsen, F., Heinzow, B., Weihe, P., & Grandjean, P. (2010). Serum concentrations of antibodies against vaccine toxoids in children exposed perinatally to immunotoxicants. *Environmental Health Perspectives*, *118*, 1434–1438.

Henningsson, S., Jonsson, L., Ljunggren, E., Westberg, L., Gillberg, C., Rastam, M., et al. (2009). Possible association between the androgen receptor gene and autism spectrum disorder. *Psychoneuroendocrinology*, *34*, 752–761.

Herbert, M. R. (2010). Contributions of the environment and environmentally vulnerable physiology to autism spectrum disorders. *Current Opinion in Neurology*, *23*, 103–110.

Herkert, N. J., Jahnke, J. C., & Hornbuckle, K. C. (2018). Emissions of tetrachlorobiphenyls (PCBs 47, 51, and 68) from polymer resin on kitchen cabinets as a non-aroclor source to residential air. *Environmental Science & Technology*, *52*, 5154–5160.

Hertz-Picciotto, I., Bergman, A., Fangstrom, B., Rose, M., Krakowiak, P., Pessah, I., et al. (2011). Polybrominated diphenyl ethers in relation to autism and developmental delay: A case-control study. *Environmental Health*, *10*, 1.

Hertz-Picciotto, I., & Delwiche, L. (2009). The rise in autism and the role of age at diagnosis. *Epidemiology*, *20*, 84–90.

Hertz-Picciotto, I., Schmidt, R. J., & Krakowiak, P. (2018). Understanding environmental contributions to autism: Causal concepts and the state of science. *Autism Research*, *11*, 554–586.

Hertz-Picciotto, I., Schmidt, R. J., Walker, C. K., Bennett, D. H., Oliver, M., Shedd-Wise, K. M., et al. (2018). A prospective study of environmental exposures and early biomarkers in autism spectrum disorder: Design, protocols, and preliminary data from the MARBLES Study. *Environmental Health Perspectives*, *126*, 117004.

Hisada, A., Shimodaira, K., Okai, T., Watanabe, K., Takemori, H., Takasuga, T., et al. (2014). Associations between levels of hydroxylated PCBs and PCBs in serum of pregnant women and blood thyroid hormone levels and body size of neonates. *International Journal of Hygiene and Environmental Health*, *217*, 546–553.

Hochstenbach, K., van Leeuwen, D. M., Gmuender, H., Gottschalk, R. W., Stolevik, S. B., Nygaard, U. C., et al. (2012). Toxicogenomic profiles in relation to maternal immunotoxic exposure and immune functionality in newborns. *Toxicological Sciences*, *129*, 315–324.

Holland, E. B., Feng, W., Zheng, J., Dong, Y., Li, X., Lehmler, H. J., et al. (2017). An extended structure-activity relationship of nondioxin-like PCBs evaluates and supports modeling predictions and identifies picomolar potency of PCB 202 towards ryanodine receptors. *Toxicological Sciences*, *155*, 170–181.

Hong, M., Lee, S. M., Park, S., Yoon, S. J., Kim, Y. E., & Oh, I. H. (2020). Prevalence and economic burden of autism spectrum disorder in South Korea using national health insurance data from 2008 to 2015. *Journal of Autism and Developmental Disorders*, *50*, 333–339.

Howard, A. S., Fitzpatrick, R., Pessah, I., Kostyniak, P., & Lein, P. J. (2003). Polychlorinated biphenyls induce caspase-dependent cell death in cultured embryonic rat hippocampal but not cortical neurons via activation of the ryanodine receptor. *Toxicology and Applied Pharmacology*, *190*, 72–86.

Hu, D., & Hornbuckle, K. C. (2010). Inadvertent polychlorinated biphenyls in commercial paint pigments. *Environmental Science & Technology*, *44*, 2822–2827.

Huang, G., Chen, S., Chen, X., Zheng, J., Xu, Z., Doostparast Torshizi, A., et al. (2019). Uncovering the functional link between SHANK3 deletions and deficiency in neurodevelopment using iPSC-derived human neurons. *Frontiers in Neuroanatomy*, *13*, 23.

Hutsler, J. J., & Zhang, H. (2010). Increased dendritic spine densities on cortical projection neurons in autism spectrum disorders. *Brain Research*, *1309*, 83–94.

Irwin, M. R., & Cole, S. W. (2011). Reciprocal regulation of the neural and innate immune systems. *Nature Reviews. Immunology*, *11*, 625–632.

Irwin, S. A., Patel, B., Idupulapati, M., Harris, J. B., Crisostomo, R. A., Larsen, B. P., et al. (2001). Abnormal dendritic spine characteristics in the temporal and visual cortices of patients with fragile-X syndrome: A quantitative examination. *American Journal of Medical Genetics*, *98*, 161–167.

Itoh, S., Baba, T., Yuasa, M., Miyashita, C., Kobayashi, S., Araki, A., et al. (2018). Association of maternal serum concentration of hydroxylated polychlorinated biphenyls with maternal and neonatal thyroid hormones: The Hokkaido birth cohort study. *Environmental Research*, *167*, 583–590.

Itoh, H., Iwasaki, M., Kasuga, Y., Yokoyama, S., Onuma, H., Nishimura, H., et al. (2014). Association between serum organochlorines and global methylation level of leukocyte DNA among Japanese women: A cross-sectional study. *Sci Total Environ*, *490*, 603–609.

Jang, Y. J., Park, H. R., Kim, T. H., Yang, W. J., Lee, J. J., Choi, S. Y., et al. (2012). High dose bisphenol A impairs hippocampal neurogenesis in female mice across generations. *Toxicology*, *296*, 73–82.

Johansson, N., Fredriksson, A., & Eriksson, P. (2008). Neonatal exposure to perfluorooctane sulfonate (PFOS) and perfluorooctanoic acid (PFOA) causes neurobehavioural defects in adult mice. *Neurotoxicology*, *29*, 160–169.

Johnstone, A. F., Gross, G. W., Weiss, D. G., Schroeder, O. H., Gramowski, A., & Shafer, T. J. (2010). Microelectrode arrays: A physiologically based neurotoxicity testing platform for the 21st century. *Neurotoxicology*, *31*, 331–350.

Kalikiri, M. K., Mamidala, M. P., Rao, A. N., & Rajesh, V. (2017). Analysis and functional characterization of sequence variations in ligand binding domain of thyroid hormone receptors in autism spectrum disorder (ASD) patients. *Autism Research*, *10*, 1919–1928.

Kalkbrenner, A. E., Schmidt, R. J., & Penlesky, A. C. (2014). Environmental chemical exposures and autism spectrum disorders: A review of the epidemiological evidence. *Current Problems in Pediatric and Adolescent Health Care*, *44*(10), 277–318. https://doi.org/10.1016/j.cppeds.2014.06.001.

Kalkman, H. O. (2012). A review of the evidence for the canonical Wnt pathway in autism spectrum disorders. *Molecular Autism*, *3*, 10.

Kang, D. W., Adams, J. B., Gregory, A. C., Borody, T., Chittick, L., Fasano, A., et al. (2017). Microbiota Transfer Therapy alters gut ecosystem and improves gastrointestinal and autism symptoms: An open-label study. *Microbiome*, *5*, 10.

Kania-Korwel, I., & Lehmler, H. J. (2016). Toxicokinetics of chiral polychlorinated biphenyls across different species—A review. *Environmental Science and Pollution Research International*, *23*, 2058–2080.

Keil, K. P., & Lein, P. J. (2016). DNA methylation: A mechanism linking environmental chemical exposures to risk of autism spectrum disorders? *Environ Epigenet*, *2*(1), dvv012. https://doi.org/10.1093/eep/dvv012.

Keil, K. P., Miller, G. W., Chen, H., Sethi, S., Schmuck, M. R., Dhakal, K., et al. (2018). PCB 95 promotes dendritic growth in primary rat hippocampal neurons via mTOR-dependent mechanisms. *Archives of Toxicology*, *92*, 3163–3173.

Keil, K. P., Sethi, S., & Lein, P. J. (2018). Sex-dependent effects of 2,2',3,5',6-pentachlorobiphenyl (PCB 95) on dendritic arborization of primary mouse neurons. *Toxicological Sciences*, *168*, 95–109.

Keil, K. P., Sethi, S., Wilson, M. D., Silverman, J. L., Pessah, I. N., & Lein, P. J. (2019). Genetic mutations in Ca(2+) signaling alter dendrite morphology and social approach in juvenile mice. *Genes, Brain, and Behavior, 18*(1), e12526. https://doi.org/10.1111/gbb.12526.

Keil Stietz, K. P., Sethi, S., Klocke, C. R., de Ruyter, T. E., Wilson, M. D., Pessah, I. N., et al. (2021). Sex and genotype modulate the dendritic effects of developmental exposure to a human-relevant polychlorinated biphenyls mixture in the juvenile mouse. *Frontiers in Neuroscience, 15*, 766802.

Kenet, T., Froemke, R. C., Schreiner, C. E., Pessah, I. N., & Merzenich, M. M. (2007). Perinatal exposure to a noncoplanar polychlorinated biphenyl alters tonotopy, receptive fields, and plasticity in rat primary auditory cortex. *Proceedings of the National Academy of Sciences of the United States of America, 104*, 7646–7651.

Kennedy, C. L., Spiegelhoff, A., Lavery, T., Wang, K., Manuel, R. S., Wang, Z., et al. (2022). Developmental polychlorinated biphenyl (PCB) exposure alters voiding physiology in young adult male and female mice. *Am J Clin Exp Urol, 10*, 82–97.

Keown, C. L., Shih, P., Nair, A., Peterson, N., Mulvey, M. E., & Muller, R. A. (2013). Local functional overconnectivity in posterior brain regions is associated with symptom severity in autism spectrum disorders. *Cell Reports, 5*, 567–572.

Khan, S., Michmizos, K., Tommerdahl, M., Ganesan, S., Kitzbichler, M. G., Zetino, M., et al. (2015). Somatosensory cortex functional connectivity abnormalities in autism show opposite trends, depending on direction and spatial scale. *Brain, 138*, 1394–1409.

Kim, I. J., Beck, H. N., Lein, P. J., & Higgins, D. (2002). Interferon gamma induces retrograde dendritic retraction and inhibits synapse formation. *The Journal of Neuroscience, 22*, 4530–4539.

Kim, K., Bennett, D. H., Calafat, A. M., Hertz-Picciotto, I., & Shin, H. M. (2020). Temporal trends and determinants of serum concentrations of per- and polyfluoroalkyl substances among Northern California mothers with a young child, 2009-2016. *Environmental Research, 186*, 109491.

Kim, K. H., Bose, D. D., Ghogha, A., Riehl, J., Zhang, R., Barnhart, C. D., et al. (2011). Para- and ortho-substitutions are key determinants of polybrominated diphenyl ether activity toward ryanodine receptors and neurotoxicity. *Environmental Health Perspectives, 119*, 519–526.

Kim, S., Eom, S., Kim, H. J., Lee, J. J., Choi, G., Choi, S., et al. (2018). Association between maternal exposure to major phthalates, heavy metals, and persistent organic pollutants, and the neurodevelopmental performances of their children at 1 to 2years of age- CHECK cohort study. *Sci Total Environ, 624*, 377–384.

Kim, J. H., Jarvik, G. P., Browning, B. L., Rajagopalan, R., Gordon, A. S., Rieder, M. J., et al. (2013). Exome sequencing reveals novel rare variants in the ryanodine receptor and calcium channel genes in malignant hyperthermia families. *Anesthesiology, 119*, 1054–1065.

Kim, K. Y., Kim, D. S., Lee, S. K., Lee, I. K., Kang, J. H., Chang, Y. S., et al. (2010). Association of low-dose exposure to persistent organic pollutants with global DNA hypomethylation in healthy Koreans. *Environmental Health Perspectives, 118*, 370–374.

Kimura-Kuroda, J., Nagata, I., & Kuroda, Y. (2005). Hydroxylated metabolites of polychlorinated biphenyls inhibit thyroid-hormone-dependent extension of cerebellar Purkinje cell dendrites. *Brain Research. Developmental Brain Research, 154*, 259–263.

Kimura-Kuroda, J., Nagata, I., & Kuroda, Y. (2007). Disrupting effects of hydroxy-polychlorinated biphenyl (PCB) congeners on neuronal development of cerebellar Purkinje cells: A possible causal factor for developmental brain disorders? *Chemosphere, 67*, S412–S420.

King, M., & Bearman, P. (2009). Diagnostic change and the increased prevalence of autism. *International Journal of Epidemiology, 38*, 1224–1234.

Klein, M. E., Lioy, D. T., Ma, L., Impey, S., Mandel, G., & Goodman, R. H. (2007). Homeostatic regulation of MeCP2 expression by a CREB-induced microRNA. *Nature Neuroscience*, *10*, 1513–1514.

Klocke, C., & Lein, P. J. (2020). Evidence implicating non-dioxin-like congeners as the key mediators of polychlorinated biphenyl (PCB) developmental neurotoxicity. *International Journal of Molecular Sciences*, *21*(3), 1013. https://doi.org/10.3390/ijms21031013.

Klocke, C., Sethi, S., & Lein, P. J. (2019). The developmental neurotoxicity of legacy vs. contemporary polychlorinated biphenyls (PCBs): Similarities and differences. *Environmental Science and Pollution Research International.*

Kong, X., Gong, Z., Zhang, L., Sun, X., Ou, Z., Xu, B., et al. (2019). JAK2/STAT3 signaling mediates IL-6-inhibited neurogenesis of neural stem cells through DNA demethylation/methylation. *Brain, Behavior, and Immunity*, *79*, 159–173.

Kong, Y., Zhou, W., & Sun, Z. (2020). Nuclear receptor corepressors in intellectual disability and autism. *Molecular Psychiatry*, *25*, 2220–2236.

Kouidhi, S., & Clerget-Froidevaux, M. S. (2018). Integrating thyroid hormone signaling in hypothalamic control of metabolism: crosstalk between nuclear receptors. *International Journal of Molecular Sciences*, *19*(7), 2017. https://doi.org/10.3390/ijms19072017.

Krakowiak, P., Goines, P. E., Tancredi, D. J., Ashwood, P., Hansen, R. L., Hertz-Picciotto, I., et al. (2017). Neonatal cytokine profiles associated with autism spectrum disorder. *Biological Psychiatry*, *81*, 442–451.

Kramer, S., Hikel, S. M., Adams, K., Hinds, D., & Moon, K. (2012). Current status of the epidemiologic evidence linking polychlorinated biphenyls and non-hodgkin lymphoma, and the role of immune dysregulation. *Environmental Health Perspectives*, *120*, 1067–1075.

Kreitinger, J. M., Beamer, C. A., & Shepherd, D. M. (2016). Environmental immunology: Lessons learned from exposure to a select panel of immunotoxicants. *Journal of Immunology*, *196*, 3217–3225.

Krueger, D. D., & Bear, M. F. (2011). Toward fulfilling the promise of molecular medicine in fragile X syndrome. *Annual Review of Medicine*, *62*, 411–429.

Kummer, K. K., Zeidler, M., Kalpachidou, T., & Kress, M. (2021). Role of IL-6 in the regulation of neuronal development, survival and function. *Cytokine*, *144*, 155582.

Lai, K. P., Chung, Y. T., Li, R., Wan, H. T., & Wong, C. K. (2016). Bisphenol A alters gut microbiome: Comparative metagenomics analysis. *Environmental Pollution*, *218*, 923–930.

Lai, M. C., Lombardo, M. V., & Baron-Cohen, S. (2014). Autism. *Lancet*, *383*, 896–910.

Landrigan, P. J. (2010). What causes autism? Exploring the environmental contribution. *Current Opinion in Pediatrics*, *22*, 219–225.

Landrigan, P. J., Lambertini, L., & Birnbaum, L. S. (2012). A research strategy to discover the environmental causes of autism and neurodevelopmental disabilities. *Environmental Health Perspectives*, *120*, a258–a260.

LaSalle, J. M., & Yasui, D. H. (2009). Evolving role of MeCP2 in Rett syndrome and autism. *Epigenomics*, *1*, 119–130.

Laufer, B. I., Neier, K., Valenzuela, A. E., Yasui, D. H., Schmidt, R. J., Lein, P. J., et al. (2022). Placenta and fetal brain share a neurodevelopmental disorder DNA methylation profile in a mouse model of prenatal PCB exposure. *Cell Reports*, *38*, 110442.

Laumonnier, F., Roger, S., Guerin, P., Molinari, F., M'Rad, R., Cahard, D., et al. (2006). Association of a functional deficit of the BKCa channel, a synaptic regulator of neuronal excitability, with autism and mental retardation. *The American Journal of Psychiatry*, *163*, 1622–1629.

Lebovitz, Y., Ringel-Scaia, V. M., Allen, I. C., & Theus, M. H. (2018). Emerging developments in microbiome and microglia research: Implications for neurodevelopmental disorders. *Frontiers in Immunology*, *9*, 1993.

Leehey, M. A., & Hagerman, P. J. (2012). Fragile X-associated tremor/ataxia syndrome. *Handbook of Clinical Neurology*, *103*, 373–386.

Leibson, C., Weaver, A., Myers, S., Long, K., Ransom, J., Voigt, R., et al. (2020). Objective estimates of direct-medical costs among persons aged 3 to 38 years with and without research-defined autism spectrum disorder ascertained during childhood: A population-based birth-cohort study. *Value in Health*, *23*, 595–605.

Leijs, M. M., Gan, L., De Boever, P., Esser, A., Amann, P. M., Ziegler, P., et al. (2019). Altered gene expression in dioxin-like and non-dioxin-like PCB exposed peripheral blood mononuclear cells. *International Journal of Environmental Research and Public Health*, *16*(12), 2090. https://doi.org/10.3390/ijerph16122090.

Lein, P., Silbergeld, E., Locke, P., & Goldberg, A. M. (2005). In vitro and other alternative approaches to developmental neurotoxicity testing (DNT). *Environmental Toxicology and Pharmacology*, *19*, 735–744.

Lesiak, A., Zhu, M., Chen, H., Appleyard, S. M., Impey, S., Lein, P. J., et al. (2014). The environmental neurotoxicant PCB 95 promotes synaptogenesis via ryanodine receptor-dependent miR132 upregulation. *The Journal of Neuroscience*, *34*, 717–725.

Levitt, P., & Campbell, D. B. (2009). The genetic and neurobiologic compass points toward common signaling dysfunctions in autism spectrum disorders. *The Journal of Clinical Investigation*, *119*, 747–754.

Li, X., Hefti, M. M., Marek, R. F., Hornbuckle, K. C., Wang, K., & Lehmler, H. J. (2022). Assessment of polychlorinated biphenyls and their hydroxylated metabolites in postmortem human brain samples: Age and brain region differences. *Environmental Science & Technology*, *56*, 9515–9526.

Li, C. Y., Lee, S., Cade, S., Kuo, L. J., Schultz, I. R., Bhatt, D. K., et al. (2017). Novel interactions between gut microbiome and host drug-processing genes modify the hepatic metabolism of the environmental chemicals polybrominated diphenyl ethers. *Drug Metabolism and Disposition*, *45*, 1197–1214.

Li, X., Lim, J. J., Wang, K., Prasad, B., Bhatt, D. K., Cui, J. Y., et al. (2022). The disposition of polychlorinated biphenyls (PCBs) differs between germ-free and conventional mice. *Environmental Toxicology and Pharmacology*, *92*, 103854.

Li, T., Tian, D., Lu, M., Wang, B., Li, J., Xu, B., et al. (2022). Gut microbiota dysbiosis induced by polychlorinated biphenyl 126 contributes to increased brain proinflammatory cytokines: Landscapes from the gut-brain axis and fecal microbiota transplantation. *Ecotoxicology and Environmental Safety*, *241*, 113726.

Li, R. J., Xu, J., Fu, C., Zhang, J., Zheng, Y. G., Jia, H., et al. (2016). Regulation of mTORC1 by lysosomal calcium and calmodulin. *eLife*, *5*, e19360. https://doi.org/10.7554/eLife.19360.

Liew, Z., Ritz, B., von Ehrenstein, O. S., Bech, B. H., Nohr, E. A., Fei, C., et al. (2015). Attention deficit/hyperactivity disorder and childhood autism in association with prenatal exposure to perfluoroalkyl substances: A nested case-control study in the Danish National Birth Cohort. *Environmental Health Perspectives*, *123*, 367–373.

Lim, J. J., Li, X., Lehmler, H. J., Wang, D., Gu, H., & Cui, J. Y. (2020). Gut microbiome critically impacts PCB-induced changes in metabolic fingerprints and the hepatic transcriptome in mice. *Toxicological Sciences*, *177*, 168–187.

Lind, L., Penell, J., Luttropp, K., Nordfors, L., Syvanen, A. C., Axelsson, T., et al. (2013). Global DNA hypermethylation is associated with high serum levels of persistent organic pollutants in an elderly population. *Environment International*, *59*, 456–461.

Liu, J., Koscielska, K. A., Cao, Z., Hulsizer, S., Grace, N., Mitchell, G., et al. (2012). Signaling defects in iPSC-derived fragile X premutation neurons. *Human Molecular Genetics*, *21*, 3795–3805.

Liu, S., Li, E., Sun, Z., Fu, D., Duan, G., Jiang, M., et al. (2019). Altered gut microbiota and short chain fatty acids in Chinese children with autism spectrum disorder. *Scientific Reports*, *9*, 287.

Liu, Z., Lu, Y., Zhong, K., Wang, C., & Xu, X. (2022). The associations between endocrine disrupting chemicals and markers of inflammation and immune responses: A systematic review and meta-analysis. *Ecotoxicology and Environmental Safety*, *234*, 113382.

Liu, J., Tan, Y., Song, E., & Song, Y. (2020). A critical review of polychlorinated biphenyls metabolism, metabolites, and their correlation with oxidative stress. *Chemical Research in Toxicology*, *33*, 2022–2042.

Liu, W., Zheng, M., Wang, D., Xing, Y., Zhao, X., Ma, X., et al. (2004). Formation of PCDD/Fs and PCBs in the process of production of 1,4-dichlorobenzene. *Chemosphere*, *57*, 1317–1323.

Lord, C., Elsabbagh, M., Baird, G., & Veenstra-Vanderweele, J. (2018). Autism spectrum disorder. *Lancet*, *392*, 508–520.

Lu, A. T., & Cantor, R. M. (2012). Allowing for sex differences increases power in a GWAS of multiplex Autism families. *Molecular Psychiatry*, *17*, 215–222.

Lutz, A. K., Perez Arevalo, A., Ioannidis, V., Stirmlinger, N., Demestre, M., Delorme, R., et al. (2021). SHANK2 mutations result in dysregulation of the ERK1/2 pathway in human induced pluripotent stem cells-derived neurons and Shank2(-/-) mice. *Frontiers in Molecular Neuroscience*, *14*, 773571.

Lyall, K., Croen, L. A., Sjodin, A., Yoshida, C. K., Zerbo, O., Kharrazi, M., et al. (2017). Polychlorinated biphenyl and organochlorine pesticide concentrations in maternal mid-pregnancy serum samples: Association with autism spectrum disorder and intellectual disability. *Environmental Health Perspectives*, *125*, 474–480.

Lyall, K., Croen, L. A., Weiss, L. A., Kharrazi, M., Traglia, M., Delorenze, G. N., et al. (2017). Prenatal serum concentrations of brominated flame retardants and autism spectrum disorder and intellectual disability in the early markers of autism study: A population-based case-control study in California. *Environmental Health Perspectives*, *125*, 087023.

Lyall, K., Yau, V. M., Hansen, R., Kharrazi, M., Yoshida, C. K., Calafat, A. M., et al. (2018). Prenatal maternal serum concentrations of per- and polyfluoroalkyl substances in association with autism spectrum disorder and intellectual disability. *Environmental Health Perspectives*, *126*, 017001.

Majerczyk, D., Ayad, E. G., Brewton, K. L., Saing, P., & Hart, P. C. (2022). Systemic maternal inflammation promotes ASD via IL-6 and IFN-gamma. *Bioscience Reports*, *42*(11), BSR20220713. https://doi.org/10.1042/BSR20220713.

Maleki, M., Noorimotlagh, Z., Mirzaee, S. A., Jaafarzadeh, N., Martinez, S. S., Rahim, F., et al. (2022). An updated systematic review on the maternal exposure to environmental pesticides and involved mechanisms of autism spectrum disorder (ASD) progression risk in children. *Reviews on Environmental Health*.

Malm, H., Artama, M., Brown, A. S., Gissler, M., Gyllenberg, D., Hinkka-Yli-Salomaki, S., et al. (2012). Infant and childhood neurodevelopmental outcomes following prenatal exposure to selective serotonin reuptake inhibitors: Overview and design of a Finnish Register-Based Study (FinESSI). *BMC Psychiatry*, *12*, 217.

Mandic-Maravic, V., Coric, V., Mitkovic-Voncina, M., Djordjevic, M., Savic-Radojevic, A., Ercegovac, M., et al. (2019). Interaction of glutathione S-transferase polymorphisms and tobacco smoking during pregnancy in susceptibility to autism spectrum disorders. *Scientific Reports*, *9*, 3206.

Marcantoni, A., Calorio, C., Hidisoglu, E., Chiantia, G., & Carbone, E. (2020). Cav1.2 channelopathies causing autism: New hallmarks on Timothy syndrome. *Pflügers Archiv*, *472*, 775–789.

Marek, D., Papin, S., Eellefsen, K., Niederhauser, J., Isidor, N., Ransijn, A., et al. (2012). Carriers of the fragile X mental retardataion 1 (FMR1) premutation allele present with increased levels of cytokine IL-10. *Journal of Neuroinflammation, 9*, 1–8. https://doi.org/10.1186/1742-2094-9-238.
Mari-Bauset, S., Peraita-Costa, I., Donat-Vargas, C., Llopis-Gonzalez, A., Mari-Sanchis, A., Llopis-Morales, J., et al. (2022). Systematic review of prenatal exposure to endocrine disrupting chemicals and autism spectrum disorder in offspring. *Autism, 26*, 6–32.
Masini, E., Loi, E., Vega-Benedetti, A. F., Carta, M., Doneddu, G., Fadda, R., et al. (2020). An overview of the main genetic, epigenetic and environmental factors involved in autism spectrum disorder focusing on synaptic activity. *International Journal of Molecular Sciences, 21*.
Matelski, L., Keil Stietz, K. P., Sethi, S., Taylor, S. T., Van de Water, J., & Lein, P. J. (2020). The influence of sex, genotype, and dose on serum and hippocampal cytokine levels in juvenile mice developmentally exposed to a human-relevant mixture of polychlorinated biphenyls. *Current Research in Toxicology, 1*, 85–103.
Matelski, L., & Van de Water, J. (2016). Risk factors in autism: Thinking outside the brain. *Journal of Autoimmunity, 67*, 1–7.
Matteoli, G., & Boeckxstaens, G. E. (2013). The vagal innervation of the gut and immune homeostasis. *Gut, 62*, 1214–1222.
Mayer, E. A., Padua, D., & Tillisch, K. (2014). Altered brain-gut axis in autism: Comorbidity or causative mechanisms? *Bioessays, 36*, 933–939.
McCarthy, M. M. (2016a). Multifaceted origins of sex differences in the brain. *Philosophical Transactions of the Royal Society of London. Series B, Biological Sciences, 371*, 20150106.
McCarthy, M. M. (2016b). Sex differences in the developing brain as a source of inherent risk. *Dialogues in Clinical Neuroscience, 18*, 361–372.
McCarthy, M. M., Arnold, A. P., Ball, G. F., Blaustein, J. D., & De Vries, G. J. (2012). Sex differences in the brain: The not so inconvenient truth. *The Journal of Neuroscience, 32*, 2241–2247.
McCarthy, M. M., Herold, K., & Stockman, S. L. (2018). Fast, furious and enduring: Sensitive versus critical periods in sexual differentiation of the brain. *Physiology & Behavior, 187*, 13–19.
McCarthy, M. M., Nugent, B. M., & Lenz, K. M. (2017). Neuroimmunology and neuroepigenetics in the establishment of sex differences in the brain. *Nature Reviews. Neuroscience, 18*, 471–484.
McCarthy, M. M., & Wright, C. L. (2017). Convergence of sex differences and the neuroimmune system in autism spectrum disorder. *Biological Psychiatry, 81*, 402–410.
McLellan, J., Kim, D. H. J., Bruce, M., Ramirez-Celis, A., & Van de Water, J. (2022). Maternal immune dysregulation and autism-understanding the role of cytokines, chemokines and autoantibodies. *Frontiers in Psychiatry, 13*, 834910.
Mehri, F., Bashirian, S., Khazaei, S., & Jenabi, E. (2021). Association between pesticide and polychlorinated biphenyl exposure during pregnancy and autism spectrum disorder among children: A meta-analysis. *Clin Exp Pediatr, 64*, 286–292.
Meltzer, A., & Van de Water, J. (2017). The role of the immune system in autism spectrum disorder. *Neuropsychopharmacology, 42*, 284–298.
Melymuk, L., Blumenthal, J., Sanka, O., Shu-Yin, A., Singla, V., Sebkova, K., et al. (2022). Persistent problem: Global challenges to managing PCBs. *Environmental Science & Technology, 56*, 9029–9040.
Miller, M. T., Stromland, K., Ventura, L., Johansson, M., Bandim, J. M., & Gillberg, C. (2005). Autism associated with conditions characterized by developmental errors in early embryogenesis: A mini review. *International Journal of Developmental Neuroscience, 23*, 201–219.

Mitchell, M. M., Woods, R., Chi, L. H., Schmidt, R. J., Pessah, I. N., Kostyniak, P. J., et al. (2012). Levels of select PCB and PBDE congeners in human postmortem brain reveal possible environmental involvement in 15q11-q13 duplication autism spectrum disorder. *Environmental and Molecular Mutagenesis*, *53*, 589–598.

Modafferi, S., Zhong, X., Kleensang, A., Murata, Y., Fagiani, F., Pamies, D., et al. (2021). Gene-environment interactions in developmental neurotoxicity: A case study of synergy between chlorpyrifos and CHD8 knockout in human brain spheres. *Environmental Health Perspectives*, *129*, 77001.

Moosa, A., Shu, H., Sarachana, T., & Hu, V. W. (2018). Are endocrine disrupting compounds environmental risk factors for autism spectrum disorder? *Hormones and Behavior*, *101*, 13–21.

Mordaunt, C. E., Park, B. Y., Bakulski, K. M., Feinberg, J. I., Croen, L. A., Ladd-Acosta, C., et al. (2019). A meta-analysis of two high-risk prospective cohort studies reveals autism-specific transcriptional changes to chromatin, autoimmune, and environmental response genes in umbilical cord blood. *Molecular Autism*, *10*, 36.

Mshaty, A., Haijima, A., Takatsuru, Y., Ninomiya, A., Yajima, H., Kokubo, M., et al. (2020). Neurotoxic effects of lactational exposure to perfluorooctane sulfonate on learning and memory in adult male mouse. *Food and Chemical Toxicology*, *145*, 111710.

Muller, P. A., Schneeberger, M., Matheis, F., Wang, P., Kerner, Z., Ilanges, A., et al. (2020). Microbiota modulate sympathetic neurons via a gut-brain circuit. *Nature*, *583*(7816), 441–446. https://doi.org/10.1038/s41586-020-2474-7. Epub 2020 Jul 8. Erratum in: Nature. 2020 Sep;585(7823):E2. PMID: 32641826. PMCID: PMC7367767.

Nagai, K., Fukuno, S., Suzuki, H., & Konishi, H. (2016). Higher gene expression of CYP1A2, 2B1 and 2D2 in the brain of female compared with male rats. *Pharmazie*, *71*, 334–336.

Nanou, E., & Catterall, W. A. (2018). Calcium channels, synaptic plasticity, and neuropsychiatric disease. *Neuron*, *98*, 466–481.

Napolitano, A., Schiavi, S., La Rosa, P., Rossi-Espagnet, M. C., Petrillo, S., Bottino, F., et al. (2022). Sex differences in autism spectrum disorder: Diagnostic, neurobiological, and behavioral features. *Frontiers in Psychiatry*, *13*, 889636.

Nave, K. A., & Werner, H. B. (2014). Myelination of the nervous system: Mechanisms and functions. *Annual Review of Cell and Developmental Biology*, *30*, 503–533.

Needham, B. D., Funabashi, M., Adame, M. D., Wang, Z., Boktor, J. C., Haney, J., et al. (2022). A gut-derived metabolite alters brain activity and anxiety behaviour in mice. *Nature*, *602*, 647–653.

Ngounou Wetie, A. G., Wormwood, K. L., Charette, L., Ryan, J. P., Woods, A. G., & Darie, C. C. (2015). Comparative two-dimensional polyacrylamide gel electrophoresis of the salivary proteome of children with autism spectrum disorder. *Journal of Cellular and Molecular Medicine*, *19*, 2664–2678.

Nguyen, R. L., Medvedeva, Y. V., Ayyagari, T. E., Schmunk, G., & Gargus, J. J. (2018). Intracellular calcium dysregulation in autism spectrum disorder: An analysis of converging organelle signaling pathways. *Biochimica et Biophysica Acta, Molecular Cell Research*, *1865*, 1718–1732.

Nielsen, T. C., Nassar, N., Shand, A. W., Jones, H. F., Han, V. X., Patel, S., et al. (2022). Association of maternal autoimmune disease and early childhood infections with offspring autism spectrum disorder: A population-based cohort study. *Autism Research*. https://doi.org/10.1002/aur.2824.

Ninomiya, A., Mshaty, A., Haijima, A., Yajima, H., Kokubo, M., Khairinisa, M. A., et al. (2022). The neurotoxic effect of lactational PFOS exposure on cerebellar functional development in male mice. *Food and Chemical Toxicology*, *159*, 112751.

Nomiyama, K., Eguchi, A., Takaguchi, K., Yoo, J., Mizukawa, H., Oshihoi, T., et al. (2019). Targeted metabolome analysis of the dog brain exposed to PCBs suggests inhibition of oxidative phosphorylation by hydroxylated PCBs. *Toxicology and Applied Pharmacology*, *377*, 114620.

Nowack, N., Wittsiepe, J., Kasper-Sonnenberg, M., Wilhelm, M., & Scholmerich, A. (2015). Influence of low-level prenatal exposure to PCDD/Fs and PCBs on empathizing, systemizing and autistic traits: Results from the Duisburg Birth Cohort Study. *PLoS One*, *10*, e0129906.

Oh, J., Bennett, D. H., Calafat, A. M., Tancredi, D., Roa, D. L., Schmidt, R. J., et al. (2021). Prenatal exposure to per- and polyfluoroalkyl substances in association with autism spectrum disorder in the MARBLES study. *Environment International*, *147*, 106328.

Oh, J., Schmidt, R. J., Tancredi, D., Calafat, A. M., Roa, D. L., Hertz-Picciotto, I., et al. (2021). Prenatal exposure to per- and polyfluoroalkyl substances and cognitive development in infancy and toddlerhood. *Environmental Research*, *196*, 110939.

Okano, M., Bell, D. W., Haber, D. A., & Li, E. (1999). DNA methyltransferases Dnmt3a and Dnmt3b are essential for de novo methylation and mammalian development. *Cell*, *99*, 247–257.

Oppenheimer, J. H., & Schwartz, H. L. (1997). Molecular basis of thyroid hormone-dependent brain development. *Endocrine Reviews*, *18*, 462–475.

Oulhote, Y., Steuerwald, U., Debes, F., Weihe, P., & Grandjean, P. (2016). Behavioral difficulties in 7-year old children in relation to developmental exposure to perfluorinated alkyl substances. *Environment International*, *97*, 237–245.

Panesar, H. K., Kennedy, C. L., Keil Stietz, K. P., & Lein, P. J. (2020). Polychlorinated biphenyls (PCBs): Risk factors for autism spectrum disorder? *Toxics*, *8*(3), 70. https://doi.org/10.3390/toxics8030070.

Park, J., Lee, K., Kim, K., & Yi, S. J. (2022). The role of histone modifications: From neurodevelopment to neurodiseases. *Signal Transduction and Targeted Therapy*, 7, 217.

Parner, E. T., Schendel, D. E., & Thorsen, P. (2008). Autism prevalence trends over time in Denmark: Changes in prevalence and age at diagnosis. *Archives of Pediatrics & Adolescent Medicine*, *162*, 1150–1156.

Parra, A. S., & Johnston, C. A. (2022). Emerging roles of RNA-binding proteins in neurodevelopment. *J Dev Biol*, *10*(2), 23. https://doi.org/10.3390/jdb10020023.

Pasca, S. P., Portmann, T., Voineagu, I., Yazawa, M., Shcheglovitov, A., Pasca, A. M., et al. (2011). Using iPSC-derived neurons to uncover cellular phenotypes associated with Timothy syndrome. *Nature Medicine*, *17*, 1657–1662.

Pelch, K. E., Bolden, A. L., & Kwiatkowski, C. F. (2019). Environmental chemicals and autism: A scoping review of the human and animal research. *Environmental Health Perspectives*, *127*, 46001.

Peng, Y., Lu, Z., Li, G., Piechowicz, M., Anderson, M., Uddin, Y., et al. (2016). The autism-associated MET receptor tyrosine kinase engages early neuronal growth mechanism and controls glutamatergic circuits development in the forebrain. *Molecular Psychiatry*, *21*, 925–935.

Penzes, P., Cahill, M. E., Jones, K. A., VanLeeuwen, J. E., & Woolfrey, K. M. (2011). Dendritic spine pathology in neuropsychiatric disorders. *Nature Neuroscience*, *14*, 285–293.

Percy, A. K. (2011). Rett syndrome: exploring the autism link. *Archives of Neurology*, *68*, 985–989.

Persichetti, A. S., Shao, J., Gotts, S. J., & Martin, A. (2022). Maladaptive laterality in cortical networks related to social communication in autism spectrum disorder. *The Journal of Neuroscience*. https://doi.org/10.1523/JNEUROSCI.1229-22.2022. JN-RM-1229-22.

Pessah, I. N., Cherednichenko, G., & Lein, P. J. (2010). Minding the calcium store: Ryanodine receptor activation as a convergent mechanism of PCB toxicity. *Pharmacology & Therapeutics*, *125*, 260–285.

Pessah, I. N., & Lein, P. J. (2008). Evidence for environmental susceptibility in autism: What we need to know about gene x environment interactions. In A. Zimmerman (Ed.), *Autism: Current theories and evidence* (pp. 409–428). Totowa, NJ: Humana Press.

Pessah, I. N., Lein, P. J., Seegal, R. F., & Sagiv, S. K. (2019). Neurotoxicity of polychlorinated biphenyls and related organohalogens. *Acta Neuropathologica*, *138*, 363–387.

Petriello, M. C., Hoffman, J. B., Vsevolozhskaya, O., Morris, A. J., & Hennig, B. (2018). Dioxin-like PCB 126 increases intestinal inflammation and disrupts gut microbiota and metabolic homeostasis. *Environmental Pollution*, *242*, 1022–1032.

Pilsner, J. R., Hu, H., Ettinger, A., Sanchez, B. N., Wright, R. O., Cantonwine, D., et al. (2009). Influence of prenatal lead exposure on genomic methylation of cord blood DNA. *Environmental Health Perspectives*, *117*, 1466–1471.

Polanska, K., Jurewicz, J., & Hanke, W. (2012). Exposure to environmental and lifestyle factors and attention-deficit/hyperactivity disorder in children—A review of epidemiological studies. *International Journal of Occupational Medicine and Environmental Health*, *25*, 330–355.

Pourtavakoli, A., & Ghafouri-Fard, S. (2022). Calcium signaling in neurodevelopment and pathophysiology of autism spectrum disorders. *Molecular Biology Reports*, *49*(11), 10811–10823. https://doi.org/10.1007/s11033-022-07775-6.

Purushotham, S. S., Reddy, N. M. N., D'Souza, M. N., Choudhury, N. R., Ganguly, A., Gopalakrishna, N., et al. (2022). A perspective on molecular signalling dysfunction, its clinical relevance and therapeutics in autism spectrum disorder. *Experimental Brain Research*, *240*, 2525–2567.

Qiu, S., Aldinger, K. A., & Levitt, P. (2012). Modeling of autism genetic variations in mice: focusing on synaptic and microcircuit dysfunctions. *Developmental Neuroscience*, *34*, 88–100.

Ramirez, V., Gonzalez-Palacios, P., Baca, M. A., Gonzalez-Domenech, P. J., Fernandez-Cabezas, M., Alvarez-Cubero, M. J., et al. (2022). Effect of exposure to endocrine disrupting chemicals in obesity and neurodevelopment: The genetic and microbiota link. *Sci Total Environ*, *852*, 158219.

Ransohoff, R. M., & Brown, M. A. (2012). Innate immunity in the central nervous system. *The Journal of Clinical Investigation*, *122*, 1164–1171.

Ristori, M. V., Quagliariello, A., Reddel, S., Ianiro, G., Vicari, S., Gasbarrini, A., et al. (2019). Autism, gastrointestinal symptoms and modulation of gut microbiota by nutritional interventions. *Nutrients*, *11*.

Robin, G., Lopez, J. R., Espinal, G. M., Hulsizer, S., Hagerman, P. J., & Pessah, I. N. (2017). Calcium dysregulation and Cdk5-ATM pathway involved in a mouse model of fragile X-associated tremor/ataxia syndrome. *Human Molecular Genetics*, *26*, 2649–2666.

Robinette, B. L., Harrill, J. A., Mundy, W. R., & Shafer, T. J. (2011). In vitro assessment of developmental neurotoxicity: use of microelectrode arrays to measure functional changes in neuronal network ontogeny. *Frontiers in Neuroengineering*, *4*, 1.

Rock, K. D., & Patisaul, H. B. (2018). Environmental mechanisms of neurodevelopmental toxicity. *Curr Environ Health Rep*, *5*, 145–157.

Rodier, P. M., Ingram, J. L., Tisdale, B., & Croog, V. J. (1997). Linking etiologies in humans and animal models: studies of autism. *Reproductive Toxicology*, *11*, 417–422.

Rodier, P., Miller, R. K., & Brent, R. L. (2011). Does treatment of premature labor with terbutaline increase the risk of autism spectrum disorders? *American Journal of Obstetrics and Gynecology*, *204*, 91–94.

Rosenfeld, C. S. (2017). Gut dysbiosis in animals due to environmental chemical exposures. *Frontiers in Cellular and Infection Microbiology*, 7, 396.

Rosenquist, A. H., Hoyer, B. B., Julvez, J., Sunyer, J., Pedersen, H. S., Lenters, V., et al. (2017). Prenatal and postnatal PCB-153 and p,p'-DDE exposures and behavior scores at 5-9 years of age among children in Greenland and Ukraine. *Environmental Health Perspectives, 125*, 107002.

Rossignol, D. A., Genuis, S. J., & Frye, R. E. (2014). Environmental toxicants and autism spectrum disorders: A systematic review. *Translational Psychiatry, 4*, e360.

Rovet, J. F. (2014). The role of thyroid hormones for brain development and cognitive function. *Endocrine Development, 26*, 26–43.

Roy, H. A., & Green, A. L. (2019). The central autonomic network and regulation of bladder function. *Frontiers in Neuroscience, 13*, 535.

Rubenstein, J. L., & Merzenich, M. M. (2003). Model of autism: Increased ratio of excitation/inhibition in key neural systems. *Genes, Brain, and Behavior, 2*, 255–267.

Rude, K. M., Keogh, C. E., & Gareau, M. G. (2019). The role of the gut microbiome in mediating neurotoxic outcomes to PCB exposure. *Neurotoxicology, 75*, 30–40.

Rude, K. M., Pusceddu, M. M., Keogh, C. E., Sladek, J. A., Rabasa, G., Miller, E. N., et al. (2019). Developmental exposure to polychlorinated biphenyls (PCBs) in the maternal diet causes host-microbe defects in weanling offspring mice. *Environmental Pollution, 253*, 708–721.

Rudie, J. D., Hernandez, L. M., Brown, J. A., Beck-Pancer, D., Colich, N. L., Gorrindo, P., et al. (2012). Autism-associated promoter variant in MET impacts functional and structural brain networks. *Neuron, 75*, 904–915.

Rusiecki, J. A., Baccarelli, A., Bollati, V., Tarantini, L., Moore, L. E., & Bonefeld-Jorgensen, E. C. (2008). Global DNA hypomethylation is associated with high serum-persistent organic pollutants in Greenlandic Inuit. *Environmental Health Perspectives, 116*, 1547–1552.

Russo, A. J., Albert, A., & Bowman, J. (2017). Decreased phosphorylated CREB and AKT in individuals with autism normalized after zinc therapy. *Acad J Ped Neonatol, 5*, 57–60.

Salazar, P., Villaseca, P., Cisternas, P., & Inestrosa, N. C. (2021). Neurodevelopmental impact of the offspring by thyroid hormone system-disrupting environmental chemicals during pregnancy. *Environmental Research, 200*, 111345.

Samaco, R. C., Hogart, A., & LaSalle, J. M. (2005). Epigenetic overlap in autism-spectrum neurodevelopmental disorders: MECP2 deficiency causes reduced expression of UBE3A and GABRB3. *Human Molecular Genetics, 14*, 483–492.

Sanders, S. J., Murtha, M. T., Gupta, A. R., Murdoch, J. D., Raubeson, M. J., Willsey, A. J., et al. (2012). De novo mutations revealed by whole-exome sequencing are strongly associated with autism. *Nature, 485*, 237–241.

Sandin, S., Lichtenstein, P., Kuja-Halkola, R., Larsson, H., Hultman, C. M., & Reichenberg, A. (2014). The familial risk of autism. *JAMA, 311*, 1770–1777.

Sarachana, T., Zhou, R., Chen, G., Manji, H. K., & Hu, V. W. (2010). Investigation of post-transcriptional gene regulatory networks associated with autism spectrum disorders by microRNA expression profiling of lymphoblastoid cell lines. *Genome Medicine, 2*, 23.

Schmunk, G., & Gargus, J. J. (2013). Channelopathy pathogenesis in autism spectrum disorders. *Frontiers in Genetics, 4*, 222.

Schofield, D., Zeppel, M. J. B., Tanton, R., Veerman, J. L., Kelly, S. J., Passey, M. E., et al. (2019). Intellectual disability and autism: socioeconomic impacts of informal caring, projected to 2030. *The British Journal of Psychiatry, 215*(5), 654–660. https://doi.org/10.1192/bjp.2019.204.

Sciacca, M., Marino, L., Vitaliti, G., Falsaperla, R., & Marino, S. (2022). NRXN1 deletion in two twins' genotype and phenotype: A clinical case and literature review. *Children (Basel), 9*.

Scott-Van Zeeland, A. A., Abrahams, B. S., Alvarez-Retuerto, A. I., Sonnenblick, L. I., Rudie, J. D., Ghahremani, D., et al. (2010). Altered functional connectivity in frontal lobe circuits is associated with variation in the autism risk gene CNTNAP2. *Science Translational Medicine*, *2*, 56ra80.

Sechman, A., Batoryna, M., Antos, P. A., & Hrabia, A. (2016). Effects of PCB 126 and PCB 153 on secretion of steroid hormones and mRNA expression of steroidogenic genes (STAR, HSD3B, CYP19A1) and estrogen receptors (ERalpha, ERbeta) in prehierarchical chicken ovarian follicles. *Toxicology Letters*, *264*, 29–37.

Serajee, F. J., Nabi, R., Zhong, H., & Huq, M. (2004). Polymorphisms in xenobiotic metabolism genes and autism. *Journal of Child Neurology*, *19*, 413–417.

Sethi, S., Chen, X., Kass, P. H., & Puschner, B. (2017). Polychlorinated biphenyl and polybrominated diphenyl ether profiles in serum from cattle, sheep, and goats across California. *Chemosphere*, *181*, 63–73.

Sethi, S., Keil, K. P., Chen, H., Hayakawa, K., Li, X., Lin, Y., et al. (2017). Detection of 3,3'-dichlorobiphenyl in human maternal plasma and its effects on axonal and dendritic growth in primary rat neurons. *Toxicological Sciences*.

Sethi, S., Keil, K. P., & Lein, P. J. (2018). 3,3'-Dichlorobiphenyl (PCB 11) promotes dendritic arborization in primary rat cortical neurons via a CREB-dependent mechanism. *Archives of Toxicology*, *92*, 3337–3345.

Sethi, S., Keil Stietz, K. P., Valenzuela, A. E., Klocke, C. R., Silverman, J. L., Puschner, B., et al. (2021). Developmental exposure to a human-relevant polychlorinated biphenyl mixture causes behavioral phenotypes that vary by sex and genotype in juvenile mice expressing human mutations that modulate neuronal calcium. *Frontiers in Neuroscience*, *15*, 766826.

Sethi, S., Morgan, R. K., Feng, W., Lin, Y., Li, X., Luna, C., et al. (2019). Comparative analyses of the 12 most abundant PCB congeners detected in human maternal serum for activity at the thyroid hormone receptor and ryanodine receptor. *Environmental Science & Technology*, *53*, 3948–3958.

Sharon, G., Cruz, N. J., Kang, D. W., Gandal, M. J., Wang, B., Kim, Y. M., et al. (2019). Human gut microbiota from autism spectrum disorder promote behavioral symptoms in mice. *Cell*, *177*, 1600–1618. e1617.

Shelton, J. F., Hertz-Picciotto, I., & Pessah, I. N. (2012). Tipping the balance of autism risk: Potential mechanisms linking pesticides and autism. *Environmental Health Perspectives*, *120*, 944–951.

Shin, H. M., Bennett, D. H., Calafat, A. M., Tancredi, D., & Hertz-Picciotto, I. (2020). Modeled prenatal exposure to per- and polyfluoroalkyl substances in association with child autism spectrum disorder: A case-control study. *Environmental Research*, *186*, 109514.

Sima, M., Rossnerova, A., Simova, Z., & Rossner, P., Jr. (2021). The impact of air pollution exposure on the microRNA machinery and lung cancer development. *J Pers Med*, *11*(1), 60. https://doi.org/10.3390/jpm11010060.

Skafidas, E., Testa, R., Zantomio, D., Chana, G., Everall, I. P., & Pantelis, C. (2014). Predicting the diagnosis of autism spectrum disorder using gene pathway analysis. *Molecular Psychiatry*, *19*, 504–510.

Skogheim, T. S., Weyde, K. V. F., Aase, H., Engel, S. M., Suren, P., Oie, M. G., et al. (2021). Prenatal exposure to per- and polyfluoroalkyl substances (PFAS) and associations with attention-deficit/hyperactivity disorder and autism spectrum disorder in children. *Environmental Research*, *202*, 111692.

Sotgiu, S., Manca, S., Gagliano, A., Minutolo, A., Melis, M. C., Pisuttu, G., et al. (2020). Immune regulation of neurodevelopment at the mother-foetus interface: the case of autism. *Clin Transl Immunology*, *9*, e1211.

Soueid, J., Kourtian, S., Makhoul, N. J., Makoukji, J., Haddad, S., Ghanem, S. S., et al. (2016). RYR2, PTDSS1 and AREG genes are implicated in a Lebanese population-based study of copy number variation in autism. *Scientific Reports*, *6*, 19088.
Splawski, I., Timothy, K. W., Sharpe, L. M., Decher, N., Kumar, P., Bloise, R., et al. (2004). Ca(V)1.2 calcium channel dysfunction causes a multisystem disorder including arrhythmia and autism. *Cell*, *119*, 19–31.
Spratt, P. W. E., Ben-Shalom, R., Keeshen, C. M., Burke, K. J., Jr., Clarkson, R. L., Sanders, S. J., et al. (2019). The autism-associated gene Scn2a contributes to dendritic excitability and synaptic function in the prefrontal cortex. *Neuron*, *103*, 673–685. e675.
Stamou, M., Streifel, K. M., Goines, P. E., & Lein, P. J. (2013). Neuronal connectivity as a convergent target of gene x environment interactions that confer risk for Autism Spectrum Disorders. *Neurotoxicology and Teratology*, *36*, 3–16.
Stolevik, S. B., Nygaard, U. C., Namork, E., Haugen, M., Kvalem, H. E., Meltzer, H. M., et al. (2011). Prenatal exposure to polychlorinated biphenyls and dioxins is associated with increased risk of wheeze and infections in infants. *Food and Chemical Toxicology*, *49*, 1843–1848.
Stolevik, S. B., Nygaard, U. C., Namork, E., Haugen, M., Meltzer, H. M., Alexander, J., et al. (2013). Prenatal exposure to polychlorinated biphenyls and dioxins from the maternal diet may be associated with immunosuppressive effects that persist into early childhood. *Food and Chemical Toxicology*, *51*, 165–172.
Sudhof, T. C. (2008). Neuroligins and neurexins link synaptic function to cognitive disease. *Nature*, *455*, 903–911.
Sullivan, O., & Ciernia, A. V. (2022). Work hard, play hard: how sexually differentiated microglia work to shape social play and reproductive behavior. *Frontiers in Behavioral Neuroscience*, *16*, 989011.
Ta, T. A., & Pessah, I. N. (2007). Ryanodine receptor type 1 (RyR1) possessing malignant hyperthermia mutation R615C exhibits heightened sensitivity to dysregulation by non-coplanar 2,2',3,5',6-pentachlorobiphenyl (PCB 95). *Neurotoxicology*, *28*, 770–779.
Takeuchi, S., Anezaki, K., & Kojima, H. (2017). Effects of unintentional PCBs in pigments and chemical products on transcriptional activity via aryl hydrocarbon and nuclear hormone receptors. *Environmental Pollution*, *227*, 306–313.
Takeuchi, S., Shiraishi, F., Kitamura, S., Kuroki, H., Jin, K., & Kojima, H. (2011). Characterization of steroid hormone receptor activities in 100 hydroxylated polychlorinated biphenyls, including congeners identified in humans. *Toxicology*, *289*, 112–121.
Talebizadeh, Z., Butler, M. G., & Theodoro, M. F. (2008). Feasibility and relevance of examining lymphoblastoid cell lines to study role of microRNAs in autism. *Autism Research*, *1*, 240–250.
Tehrani, R., & Van Aken, B. (2014). Hydroxylated polychlorinated biphenyls in the environment: sources, fate, and toxicities. *Environmental Science and Pollution Research International*, *21*, 6334–6345.
Therien, V. D., Degre-Pelletier, J., Barbeau, E. B., Samson, F., & Soulieres, I. (2022). Differential neural correlates underlying mental rotation processes in two distinct cognitive profiles in autism. *Neuroimage Clin*, *36*, 103221.
Thompson, M. R., & Boekelheide, K. (2013). Multiple environmental chemical exposures to lead, mercury and polychlorinated biphenyls among childbearing-aged women (NHANES 1999-2004): Body burden and risk factors. *Environmental Research*, *121*, 23–30.
Thony, B., Auerbach, G., & Blau, N. (2000). Tetrahydrobiopterin biosynthesis, regeneration and functions. *The Biochemical Journal*, *347*(Pt 1), 1–16.

Todd, P. K., & Mack, K. J. (2001). Phosphorylation, CREB, and mental retardation. *Pediatric Research*, *50*, 672.

Tremblay, M. W., & Jiang, Y. H. (2019). DNA methylation and susceptibility to autism spectrum disorder. *Annual Review of Medicine*, *70*, 151–166.

Uwimana, E., Ruiz, P., Li, X., & Lehmler, H. J. (2019). Human CYP2A6, CYP2B6, AND CYP2E1 atropselectively metabolize polychlorinated biphenyls to hydroxylated metabolites. *Environmental Science & Technology*, *53*, 2114–2123.

Vereyken, E. J., Bajova, H., Chow, S., de Graan, P. N., & Gruol, D. L. (2007). Chronic interleukin-6 alters the level of synaptic proteins in hippocampus in culture and in vivo. *The European Journal of Neuroscience*, *25*, 3605–3616.

Vester, A., & Caudle, W. M. (2016). The synapse as a central target for neurodevelopmental susceptibility to pesticides. *Toxics*, *4*(3), 18. https://doi.org/10.3390/toxics4030018.

Viberg, H., Lee, I., & Eriksson, P. (2013). Adult dose-dependent behavioral and cognitive disturbances after a single neonatal PFHxS dose. *Toxicology*, *304*, 185–191.

Vikman, K. S., Owe-Larsson, B., Brask, J., Kristensson, K. S., & Hill, R. H. (2001). Interferon-gamma-induced changes in synaptic activity and AMPA receptor clustering in hippocampal cultures. *Brain Research*, *896*, 18–29.

von Gontard, A., Baeyens, D., Van Hoecke, E., Warzak, W. J., & Bachmann, C. (2011). Psychological and psychiatric issues in urinary and fecal incontinence. *The Journal of Urology*, *185*, 1432–1436.

Vuong, H. E., Pronovost, G. N., Williams, D. W., Coley, E. J. L., Siegler, E. L., Qiu, A., et al. (2020). The maternal microbiome modulates fetal neurodevelopment in mice. *Nature*, *586*, 281–286.

Vukcevic, M., Zorzato, F., Keck, S., Tsakiris, D. A., Keiser, J., Maizels, R. M., et al. (2013). Gain of function in the immune system caused by a ryanodine receptor 1 mutation. *Journal of Cell Science*, *126*, 3485–3492.

Vuong, A. M., Yolton, K., Dietrich, K. N., Braun, J. M., Lanphear, B. P., & Chen, A. (2018). Exposure to polybrominated diphenyl ethers (PBDEs) and child behavior: Current findings and future directions. *Hormones and Behavior*, *101*, 94–104.

Wahlang, B., Alexander, N. C., 2nd, Li, X., Rouchka, E. C., Kirpich, I. A., & Cave, M. C. (2021). Polychlorinated biphenyls altered gut microbiome in CAR and PXR knockout mice exhibiting toxicant-associated steatohepatitis. *Toxicology Reports*, *8*, 536–547.

Walker, D. M., Goetz, B. M., & Gore, A. C. (2014). Dynamic postnatal developmental and sex-specific neuroendocrine effects of prenatal polychlorinated biphenyls in rats. *Molecular Endocrinology*, *28*, 99–115.

Wan, L., Liu, D., Xiao, W. B., Zhang, B. X., Yan, X. X., Luo, Z. H., et al. (2022). Association of SHANK family with neuropsychiatric disorders: An update on genetic and animal model discoveries. *Cellular and Molecular Neurobiology*, *42*, 1623–1643.

Wayman, G. A., Bose, D. D., Yang, D., Lesiak, A., Bruun, D., Impey, S., et al. (2012). PCB-95 modulates the calcium-dependent signaling pathway responsible for activity-dependent dendritic growth. *Environmental Health Perspectives*, *120*, 1003–1009.

Wayman, G. A., Impey, S., Marks, D., Saneyoshi, T., Grant, W. F., Derkach, V., et al. (2006). Activity-dependent dendritic arborization mediated by CaM-kinase I activation and enhanced CREB-dependent transcription of Wnt-2. *Neuron*, *50*, 897–909.

Wayman, G. A., Yang, D., Bose, D. D., Lesiak, A., Ledoux, V., Bruun, D., et al. (2012). PCB-95 promotes dendritic growth via ryanodine receptor-dependent mechanisms. *Environmental Health Perspectives*, *120*, 997–1002.

Weisglas-Kuperus, N., Patandin, S., Berbers, G. A., Sas, T. C., Mulder, P. G., Sauer, P. J., et al. (2000). Immunologic effects of background exposure to polychlorinated biphenyls and dioxins in Dutch preschool children. *Environmental Health Perspectives*, *108*, 1203–1207.

Weiss, B. (2012). The intersection of neurotoxicology and endocrine disruption. *Neurotoxicology*, *33*, 1410–1419.

Wen, Z., Cheng, T. L., Li, G. Z., Sun, S. B., Yu, S. Y., Zhang, Y., et al. (2017). Identification of autism-related MECP2 mutations by whole-exome sequencing and functional validation. *Molecular Autism*, *8*, 43.

Williams, G. R. (2008). Neurodevelopmental and neurophysiological actions of thyroid hormone. *Journal of Neuroendocrinology*, *20*, 784–794.

Winans, B., Humble, M. C., & Lawrence, B. P. (2011). Environmental toxicants and the developing immune system: a missing link in the global battle against infectious disease? *Reproductive Toxicology*, *31*, 327–336.

Winneke, G., Ranft, U., Wittsiepe, J., Kasper-Sonnenberg, M., Furst, P., Kramer, U., et al. (2014). Behavioral sexual dimorphism in school-age children and early developmental exposure to dioxins and PCBs: A follow-up study of the Duisburg Cohort. *Environmental Health Perspectives*, *122*, 292–298.

Woods, R., Vallero, R. O., Golub, M. S., Suarez, J. K., Ta, T. A., Yasui, D. H., et al. (2012). Long-lived epigenetic interactions between perinatal PBDE exposure and Mecp2308 mutation. *Human Molecular Genetics*, *21*, 2399–2411.

Wu, Q., Zhou, Z. J., & Ohsako, S. (2006). Effect of environmental contaminants on DNA methyltransferase activity of mouse preimplantation embryos. *Wei sheng yan jiu = Journal of Hygiene Research*, *35*, 30–32.

Xiang, Y., Xing, Z., Liu, J., Qin, W., & Huang, X. (2020). Recent advances in the biodegradation of polychlorinated biphenyls. *World Journal of Microbiology and Biotechnology*, *36*, 145.

Xu, L., Guo, X., Li, N., Pan, Q., & Ma, Y. Z. (2019). Effects of quercetin on Aroclor 1254-induced expression of CYP450 and cytokines in pregnant rats. *Journal of Immunotoxicology*, *16*, 140–148.

Xu, N., Li, A. D., Ji, L. L., Ye, Y., Wang, Z. Y., & Tong, L. (2019). miR-132 regulates the expression of synaptic proteins in APP/PS1 transgenic mice through C1q. *European Journal of Histochemistry*, *63*.

Xu, L., Long, J., Shi, C., Zhang, N., Lv, Y., Feng, J., et al. (2018). Effect of leukocyte inhibitory factor on neuron differentiation from human induced pluripotent stem cell-derived neural precursor cells. *International Journal of Molecular Medicine*, *41*, 2037–2049.

Yaghoobi, B., Miller, G. W., Holland, E. B., Li, X., Harvey, D., Li, S., et al. (2022). Ryanodine receptor-active non-dioxin-like polychlorinated biphenyls cause neurobehavioral deficits in larval zebrafish. *Front Toxicol*, *4*, 947795.

Yang, D., Kania-Korwel, I., Ghogha, A., Chen, H., Stamou, M., Bose, D. D., et al. (2014). PCB 136 atropselectively alters morphometric and functional parameters of neuronal connectivity in cultured rat hippocampal neurons via ryanodine receptor-dependent mechanisms. *Toxicological Sciences*, *138*, 379–392.

Yang, D., Kim, K. H., Phimister, A., Bachstetter, A. D., Ward, T. R., Stackman, R. W., et al. (2009). Developmental exposure to polychlorinated biphenyls interferes with experience-dependent dendritic plasticity and ryanodine receptor expression in weanling rats. *Environmental Health Perspectives*, *117*, 426–435.

Yang, D., & Lein, P. J. (2010). Polychlorinated biphenyls increase apoptosis in the developing rat brain. *Curr Neurobiol*, *1*, 70–76.

Yousefi, B., Kokhaei, P., Mehranfar, F., Bahar, A., Abdolshahi, A., Emadi, A., et al. (2022). The role of the host microbiome in autism and neurodegenerative disorders and effect of epigenetic procedures in the brain functions. *Neuroscience and Biobehavioral Reviews*, *132*, 998–1009.

Yu, H., Liu, Y., Shu, X., Ma, L., & Pan, Y. (2020). Assessment of the spatial distribution of organochlorine pesticides (OCPs) and polychlorinated biphenyls (PCBs) in urban soil of China. *Chemosphere*, *243*, 125392.

Yu, X., Rahman, M. M., Wang, Z., Carter, S. A., Schwartz, J., Chen, Z., et al. (2022). Evidence of susceptibility to autism risks associated with early life ambient air pollution: A systematic review. *Environmental Research*, *208*, 112590.
Yu, G., Su, Q., Chen, Y., Wu, L., Wu, S., & Li, H. (2021). Epigenetics in neurodegenerative disorders induced by pesticides. *Genes Environ*, *43*, 55.
Yuen, B., Boncompagni, S., Feng, W., Yang, T., Lopez, J. R., Matthaei, K. I., et al. (2012). Mice expressing T4826I-RYR1 are viable but exhibit sex- and genotype-dependent susceptibility to malignant hyperthermia and muscle damage. *The FASEB Journal*, *26*, 1311–1322.
Zablotsky, B., Black, L. I., & Blumberg, S. J. (2017). Estimated prevalence of children with diagnosed developmental disabilities in the United States, 2014-2016. *NCHS Data Brief*, 1–8.
Zahalka, E. A., Ellis, D. H., Goldey, E. S., Stanton, M. E., & Lau, C. (2001). Perinatal exposure to polychlorinated biphenyls Aroclor 1016 or 1254 did not alter brain catecholamines nor delayed alternation performance in Long-Evans rats. *Brain Research Bulletin*, *55*, 487–500.
Zeidan, J., Fombonne, E., Scorah, J., Ibrahim, A., Durkin, M. S., Saxena, S., et al. (2022). Global prevalence of autism: A systematic review update. *Autism Research*, *15*, 778–790.
Zhang, Y. H., Wang, T., Li, Y. F., Deng, Y. N., & Shen, F. G. (2022). Roles of the Notch signaling pathway and microglia in autism. *Behavioural Brain Research*, *437*, 114131.
Zhang, H., Yolton, K., Webster, G. M., Sjodin, A., Calafat, A. M., Dietrich, K. N., et al. (2017). Prenatal PBDE and PCB exposures and reading, cognition, and externalizing behavior in children. *Environmental Health Perspectives*, *125*, 746–752.
Zheng, F., Kasper, L. H., Bedford, D. C., Lerach, S., Teubner, B. J., & Brindle, P. K. (2016). Mutation of the CH1 domain in the histone acetyltransferase CREBBP results in autism-relevant behaviors in mice. *PLoS One*, *11*, e0146366.
Zhu, Y., Mordaunt, C. E., Yasui, D. H., Marathe, R., Coulson, R. L., Dunaway, K. W., et al. (2019). Placental DNA methylation levels at CYP2E1 and IRS2 are associated with child outcome in a prospective autism study. *Human Molecular Genetics*, *28*, 2659–2674.
Zimmermann, M., Zimmermann-Kogadeeva, M., Wegmann, R., & Goodman, A. L. (2019). Mapping human microbiome drug metabolism by gut bacteria and their genes. *Nature*, *570*, 462–467.
Zoeller, R. T. (2007). Environmental chemicals impacting the thyroid: targets and consequences. *Thyroid*, *17*, 811–817.
Zoghbi, H. Y., & Bear, M. F. (2012). Synaptic dysfunction in neurodevelopmental disorders associated with autism and intellectual disabilities. *Cold Spring Harbor Perspectives in Biology*, *4*(3), a009886. https://doi.org/10.1101/cshperspect.a009886.

Kolibri 360 GmbH
Ferdinand-Jühlke-Straße 7
99095 Erfurt, Deutschland
produktsicherheit@kolibri360.de